I0484123

FEDERAL EXECUTIVE TEAM

Director, Climate Change Science Program ....William J. Brennan

Director, Climate Change Science Program Office ....Peter A. Schultz

Lead Agency Principal Representative to CCSP;
Deputy Under Secretary of Commerce for Oceans and Atmosphere,
National Oceanic and Atmospheric Administration ....Mary M. Glackin

Product Lead, Earth System Research Laboratory,
National Oceanic and Atmospheric Administration ................................A.R. Ravishankara

Synthesis and Assessment Product Advisory
Group Chair; Associate Director, EPA National
Center for Environmental Assessment..Michael W. Slimak

Synthesis and Assessment Product Coordinator,
Climate Change Science Program Office ....Fabien J.G. Laurier

Special Advisor, National Oceanic
and Atmospheric Administration ....Chad A. McNutt

EDITORIAL AND PRODUCTION TEAM

Co-Chair...A.R. Ravishankara, NOAA
Co-Chair...Michael J. Kurylo, NASA
Scientific and Technical Editor ....Christine A. Ennis, NOAA/CIRES
Scientific Editor ....Jessica Blunden, STG, Inc.
Scientific Editor ....Anne M. Waple, STG, Inc.
Scientific Editor ....Christian Zamarra, STG, Inc.
Technical Advisor ....David J. Dokken, USGCRP
Graphic Design Lead ....Sara W. Veasey, NOAA
Graphic Design Co-Lead ....Deborah B. Riddle, NOAA
Designer..Brandon Farrar, STG, Inc.
Designer..Glenn M. Hyatt, NOAA
Designer..Deborah Misch, STG, Inc.
Copy Editor...Anne Markel, STG, Inc.
Copy Editor...Lesley Morgan, STG, Inc.
Copy Editor...Susan Osborne, STG, Inc.
Copy Editor...Susanne Skok, STG, Inc.
Copy Editor...Mara Sprain, STG, Inc.
Copy Editor...Brooke Stewart, STG, Inc.
Graphics Support ....Debra A. Dailey-Fisher, NOAA
Technical Support..Jesse Enloe, STG, Inc.

This Synthesis and Assessment Product, described in the U.S. Climate Change Science Program (CCSP) Strategic Plan, was prepared in accordance with Section 515 of the Treasury and General Government Appropriations Act for Fiscal Year 2001 (Public Law 106-554) and the information quality act guidelines issued by the Department of Commerce and NOAA pursuant to Section 515 <http://www.noaanews.noaa.gov/stories/iq.htm>. The CCSP Interagency Committee relies on Department of Commerce and NOAA certifications regarding compliance with Section 515 and Department guidelines as the basis for determining that this product conforms with Section 515. For purposes of compliance with Section 515, this CCSP Synthesis and Assessment Product is an "interpreted product" as that term is used in NOAA guidelines and is classified as "highly influential." This document does not express any regulatory policies of the United States or any of its agencies, or provide recommendations for regulatory action.

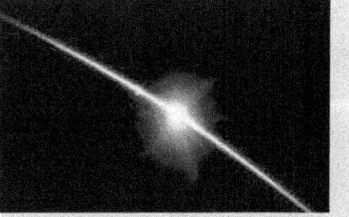

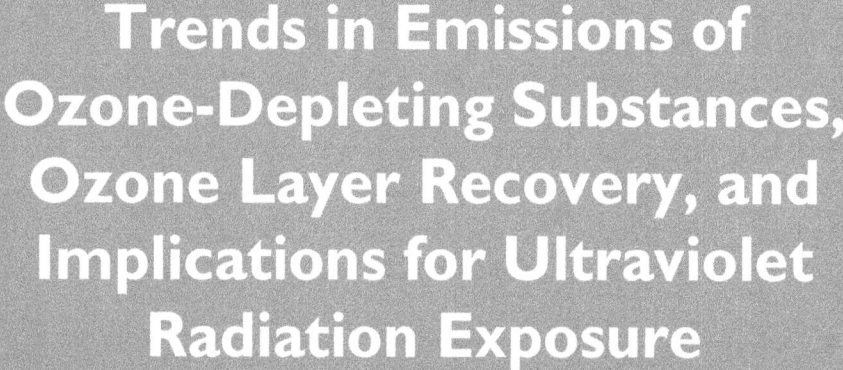

Trends in Emissions of Ozone-Depleting Substances, Ozone Layer Recovery, and Implications for Ultraviolet Radiation Exposure

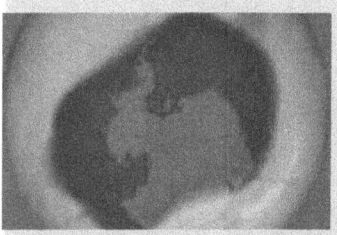

Synthesis and Assessment Product 2.4
Report by the U.S. Climate Change Science Program
and the Subcommittee on Global Change Research

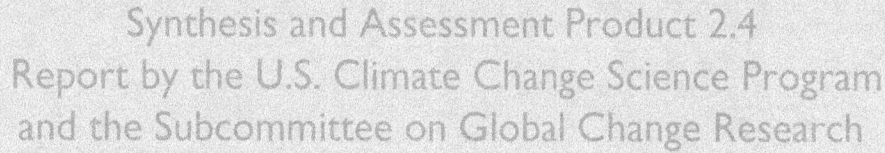

EDITED BY:
A.R. Ravishankara, Michael J. Kurylo, and Christine A. Ennis

Transmittal letter to Congress will go here.

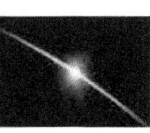

Contents:

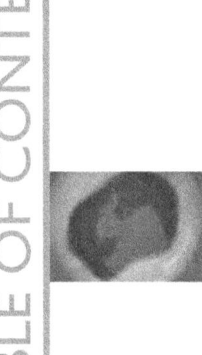
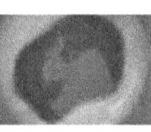

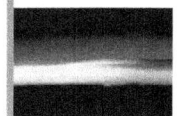

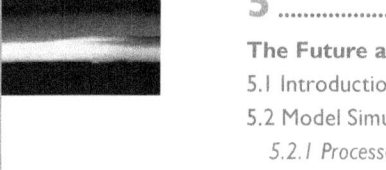

5 133

The Future and Recovery

6 155

Implications for the United States

Twenty Questions and Answers About the Ozone Layer: 2006 Update

Preface　　　　**Authors:** A.R. Ravishankara, NOAA; Michael J. Kurylo, NASA

Executive Summary　**Convening Lead Authors:** A.R. Ravishankara, NOAA; Michael J. Kurylo, NASA
　　　　　　　　Lead Authors: Richard Bevilacqua, Naval Research Laboratory; Jeff Cohen, U.S. EPA;
　　　　　　　　John S. Daniel, NOAA; Anne R. Douglass, NASA; David W. Fahey, NOAA; Jay R.
　　　　　　　　Herman, NASA; Terry Keating, U.S. EPA; Malcolm Ko, NASA; Stephen A. Montzka,
　　　　　　　　NOAA; Paul A. Newman, NASA; V. Ramaswamy, NOAA; Anne-Marie Schmoltner,
　　　　　　　　NSF; Richard Stolarski, NASA; Kenneth Vick, USDA

Chapter 1　　　**Convening Lead Authors:** A.R. Ravishankara, NOAA; Michael J. Kurylo, NASA;
　　　　　　　　Anne-Marie Schmoltner, NSF

Chapter 2　　　**Convening Lead Author:** Stephen A. Montzka
　　　　　　　　Lead Authors: John S. Daniel, NOAA; Jeff Cohen, U.S. EPA; Kenneth Vick, USDA

Chapter 3　　　**Convening Lead Authors:** Paul A. Newman, NASA; Jay R. Herman, NASA
　　　　　　　　Lead Authors: Richard Bevilacqua, Naval Research Laboratory; Richard Stolarski,
　　　　　　　　NASA; Terry Keating, U.S. EPA

Chapter 4　　　**Convening Lead Author:** David W. Fahey, NOAA
　　　　　　　　Lead Authors: Anne R. Douglass, NASA; V. Ramaswamy, NOAA; Anne-Marie
　　　　　　　　Schmoltner, NSF

Chapter 5　　　**Convening Lead Author:** Malcolm Ko, NASA
　　　　　　　　Lead Authors: John S. Daniel, NOAA; Jay R. Herman, NASA; Paul A. Newman,
　　　　　　　　ASA; V. Ramaswamy, NOAA

Chapter 6　　　**Convening Lead Authors:** A.R. Ravishankara, NOAA; Michael J. Kurylo, NASA
　　　　　　　　Lead Authors: John S. Daniel, NOAA; David W. Fahey, NOAA; Jay R. Herman, NASA;
　　　　　　　　Stephen A. Montzka, NOAA; Malcolm Ko, NASA; Paul A. Newman, NASA; Richard
　　　　　　　　Stolarski, NASA

ACKNOWLEDGEMENT

CCSP Synthesis and Assessment Product 2.4 (SAP 2.4) was developed with the benefit of a scientifically rigorous, first draft peer review conducted by a committee appointed by the National Research Council (NRC). Prior to their delivery to the SAP 2.4 Author Team, the NRC review comments, in turn, were reviewed in draft form by a second group of highly qualified experts to ensure that the review met NRC standards. The resultant NRC Review Report was instrumental in shaping the final version of SAP 2.4, and in improving its completeness, sharpening its focus, communicating its conclusions and recommendations, and improving its general readability.

We thank the members of the NRC Review Committee: M. Joan Alexander (Chair), NorthWest Research Associates, Boulder, Colorado; Derek Cunnold, Georgia Institute of Technology, Atlanta; Terry Deshler, University of Wyoming, Laramie; Steven Lloyd, The Johns Hopkins Applied Physics Laboratory, Laurel, Maryland; Mack McFarland, DuPont Fluoroproducts, Wilmington, Delaware; Michelle Santee, Jet Propulsion Laboratory, Pasadena, California; Theodore G. Shepherd, University of Toronto, Ontario, Canada; Margaret Tolbert, University of Colorado, Boulder; and Donald Wuebbles, University of Illinois, Urbana; and also the NRC Staff members who coordinated the process: Chris Elfring, Director, Board on Atmospheric Sciences and Climate; Leah Probst, Study Director; and Katherine Weller, Senior Program Assistant.

We also thank the individuals who reviewed the NRC Report in its draft form: James G. Anderson, Harvard University, Cambridge, Massachusetts; Greg Bodeker, National Institute of Water and Atmospheric Research, Lauder, New Zealand; Mary Anne Carroll, University of Michigan, Ann Arbor; Veronika Eyring, Institut für Physik der Atmosphäre, Wessling, Germany; Vitali Fioletov, Environment Canada, Downsview, Ontario, Canada; and Ross J. Salawitch, University of Maryland, College Park; and also Marvin Geller, The State University of New York, Stony Brook, the overseer of the NRC review.

We also thank the NOAA Research Council for coordinating a review conducted in preparation for the final clearance of this report. This review provided valuable comments from the following internal NOAA reviewers:

Craig Long (Climate Prediction Center)
Robert Portmann (Earth System Research Laboratory)
Susan Solomon (Earth System Research Laboratory)

The review process for SAP 2.4 also included a public review of the Second Draft, and we thank the individuals who participated in this cycle. The Author Team carefully considered all comments submitted, and a substantial number resulted in further improvements and clarity of SAP 2.4.

We are grateful to Daniel L. Albritton (NOAA) and Philip L. DeCola (NASA), who led the initial discussions that resulted in the inclusion of this topic in the suite of Synthesis and Assessment Products of the U.S. Climate Change Science Program. We also thank the following individuals for their insightful comments during the initial information-gathering stage: Alkiviadis Bais, Aristotle University of Thessaloniki; Donald Blake, University of California at Irvine; Greg Bodeker, National Institute of Water and Atmospheric Research, New Zealand; Martin Dameris, Institut für Physik der Atmosphäre, Germany; Veronika Eyring, Institut für Physik der Atmosphäre, Germany; Vitali Fioletov, Environment Canada; Claire Granier, NOAA/Cooperative Institute for Research in Environmental Sciences; Neil Harris, University of Cambridge, United Kingdom; Randy Kawa, NASA; Kathy Lantz, NOAA/Cooperative Institute for Research in Environmental Sciences; Rolf Müller, Forschungszentrum Jülich, Germany; Ronald Prinn, Massachusetts Institute of Technology; Ross Salawitch, University of Maryland; Michelle Santee, Jet Propulsion Laboratory; Guus Velders, Netherlands Environmental Assessment Agency; and Ray Weiss, Scripps Institution of Oceanography.

We also greatly appreciate the assistance of Krisa Arzayus (NOAA) and Fabien Laurier (CCSP) for expert guidance and coordination throughout the process of preparing SAP 2.4, and Chad McNutt (NOAA) and Rebecca Feldman (NOAA) for assistance with the approval processes.

Finally, it should be noted that the respective review bodies were not asked to endorse the final version of SAP 2.4, as this was the responsibility of the National Science and Technology Council.

Depletion of the stratospheric ozone layer by human-produced ozone-depleting substances has been recognized as a global environmental issue for more than three decades, and the international effort to address the issue via the United Nations Montreal Protocol marked its 20-year anniversary in 2007. Scientific understanding underpinned the Protocol at its inception and ever since. As scientific knowledge advanced and evolved, the Protocol evolved through amendment and adjustment. Policy-relevant science has documented the rise, and now the beginning decline, of the atmospheric abundances of many ozone-depleting substances in response to actions taken by the nations of the world. Projections are for a return of ozone-depleting chemicals (compounds containing chlorine and bromine) to their "pre-ozone-depletion" (pre-1980) levels by the middle of this century for the midlatitudes; the polar regions are expected to follow suit within 20 years after that. Since the 1980s, global ozone sustained a depletion of about 5 percent in the midlatitudes of both the Northern Hemisphere and Southern Hemisphere, where most of the Earth's population resides; it is now showing signs of turning the corner towards increasing ozone. The large seasonal depletions in the polar regions are likely to continue over the next decade but are expected to subside over the next few decades. Ozone-depleting substances should have a negligible effect on ozone in all regions beyond 2070, assuming continued compliance with the Montreal Protocol.

Large increases in surface ultraviolet (UVB; 280-315 nm) radiation and the associated impacts on human health and ecosystems would likely have occurred if atmospheric abundances of ozone-depleting substances had continued to grow. Scientific findings regarding the role of ozone-depleting chemicals, projected ozone losses, and the potential UV impacts galvanized international decision making in the 1980s. As a result of the worldwide adherence to the 1987 Montreal Protocol and its Amendments and Adjustments, the large impacts were avoided, and future trends in UVB and UVA (315-400 nm) at the surface are expected to be more influenced by factors other than stratospheric ozone depletion (such as changes in clouds, atmospheric fine particles, and air quality in the lower atmosphere).

Emissions of ozone-depleting substances by the United States have been significant throughout the history of the ozone depletion issue. At the same time, the United States has played a leading role in advancing the scientific understanding, leading the international decision making, and leading industry's actions to reduce usage of ozone-depleting substances. Continued future declines in emissions of ozone-depleting substances from the United States, along with those from other nations, will play a key role in ensuring the ozone layer's recovery.

Projections of a changing climate have added a new dimension to the issue of the stratospheric ozone layer and its recovery, and scientific knowledge is emerging on the interconnections between these two global issues. Climate change is expected to alter the timing of the recovery of the ozone layer. Ozone-depleting chemicals and ozone depletion are known to influence climate change. The curtailment of the ozone-depleting substances not only helped the ozone layer but also very likely lessened the forcing of climate (i.e., how it alters climate).

Climate change and ozone layer depletion are coupled; this has led to new scientific and decision-making challenges. The recovery of the ozone layer will occur in an atmosphere that is different from where we started roughly three decades back. Our scientific understanding of the connections between climate change and ozone layer depletion is at an early but rapidly advancing stage. That topic will remain a focus for the scientific community's efforts over the next few decades.

RECOMMENDED CITATIONS

For the Report as a whole:
CCSP, 2008: *Trends in Emissions of Ozone-Depleting Substances, Ozone Layer Recovery, and Implications for Ultraviolet Radiation Exposure.* A Report by the U.S. Climate Change Science Program and the Subcommittee on Global Change Research. [Ravishankara, A.R., M.J. Kurylo, and C.A. Ennis (eds.)]. Department of Commerce, NOAA's National Climatic Data Center, Asheville, NC, 240 pp.

For the Preface:
Ravishankara, A.R. and M.J. Kurylo, 2008: Preface. In: *Trends in Emissions of Ozone-Depleting Substances, Ozone Layer Recovery, and Implications for Ultraviolet Radiation Exposure.* A Report by the U.S. Climate Change Science Program and the Subcommittee on Global Change Research. [Ravishankara, A.R., M.J. Kurylo, and C.A. Ennis (eds.)]. Department of Commerce, NOAA's National Climatic Data Center, Asheville, NC, pp.XI–XIII.

For the Executive Summary:
Ravishankara, A.R., M.J. Kurylo, R. Bevilacqua, J. Cohen, J.S. Daniel, A.R. Douglass, D.W. Fahey, J.R. Herman, T. Keating, M. Ko, S.A. Montzka, P.A. Newman, V. Ramaswamy, A-M. Schmoltner, R. Stolarski, and K. Vick, 2008: Executive Summary. In: *Trends in Emissions of Ozone-Depleting Substances, Ozone Layer Recovery, and Implications for Ultraviolet Radiation Exposure.* A Report by the U.S. Climate Change Science Program and the Subcommittee on Global Change Research. [Ravishankara, A.R., M.J. Kurylo, and C.A. Ennis (eds.)]. Department of Commerce, NOAA's National Climatic Data Center, Asheville, NC, pp. 15–22.

For Chapter 1:
Ravishankara, A.R., M.J. Kurylo, and A-M. Schmoltner, 2008: Introduction. In: *Trends in Emissions of Ozone-Depleting Substances, Ozone Layer Recovery, and Implications for Ultraviolet Radiation Exposure.* A Report by the U.S. Climate Change Science Program and the Subcommittee on Global Change Research. [Ravishankara, A.R., M.J. Kurylo, and C.A. Ennis (eds.)]. Department of Commerce, NOAA's National Climatic Data Center, Asheville, NC, pp. 23–28.

For Chapter 2:
Montzka, S.A., J.S. Daniel, J. Cohen, and K. Vick, 2008: Current Trends, Mixing Ratios, and Emissions of Ozone-Depleting Substances and Their Substitutes. In: *Trends in Emissions of Ozone-Depleting Substances, Ozone Layer Recovery, and Implications for Ultraviolet Radiation Exposure.* A Report by the U.S. Climate Change Science Program and the Subcommittee on Global Change Research. [Ravishankara, A.R., M.J. Kurylo, and C.A. Ennis (eds.)]. Department of Commerce, NOAA's National Climatic Data Center, Asheville, NC, pp. 29–78.

For Chapter 3:
Newman, P.A., J.R. Herman, R. Bevilacqua, R. Stolarski, and T. Keating, 2008: Ozone and UV Observations. In: *Trends in Emissions of Ozone-Depleting Substances, Ozone Layer Recovery, and Implications for Ultraviolet Radiation Exposure.* A Report by the U.S. Climate Change Science Program and the Subcommittee on Global Change Research. [Ravishankara, A.R., M.J. Kurylo, and C.A. Ennis (eds.)]. Department of Commerce, NOAA's National Climatic Data Center, Asheville, NC, pp. 79–110.

For Chapter 4:
Fahey, D.W., A.R. Douglass, V. Ramaswamy, and A-M. Schmoltner, 2008: How Do Climate Change and Stratospheric Ozone Loss Interact? In: *Trends in Emissions of Ozone-Depleting Substances, Ozone Layer Recovery, and Implications for Ultraviolet Radiation Exposure.* A Report by the U.S. Climate Change Science Program and the Subcommittee on Global Change Research. [Ravishankara, A.R., M.J. Kurylo, and C.A. Ennis (eds.)]. Department of Commerce, NOAA's National Climatic Data Center, Asheville, NC, pp.111–132.

RECOMMENDED CITATIONS

For Chapter 5:

Ko, M., J.S. Daniel, J.R. Herman, P.A. Newman, and V. Ramaswamy, 2008: The Future and Recovery. In: *Trends in Emissions of Ozone-Depleting Substances, Ozone Layer Recovery, and Implications for Ultraviolet Radiation Exposure*. A Report by the U.S. Climate Change Science Program and the Subcommittee on Global Change Research. [Ravishankara, A.R., M.J. Kurylo, and C.A. Ennis (eds.)]. Department of Commerce, NOAA's National Climatic Data Center, Asheville, NC, pp. 133–154.

For Chapter 6:

Ravishankara, A.R., M.J. Kurylo, J.S. Daniel, D.W. Fahey, J.R. Herman, S.A. Montzka, M. Ko, P.A. Newman, and R. Stolarski, 2008: Implications for the United States. In: *Trends in Emissions of Ozone-Depleting Substances, Ozone Layer Recovery, and Implications for Ultraviolet Radiation Exposure*. A Report by the U.S. Climate Change Science Program and the Subcommittee on Global Change Research. [Ravishankara, A.R., M.J. Kurylo, and C.A. Ennis (eds.)]. Department of Commerce, NOAA's National Climatic Data Center, Asheville, NC, pp. 155–166.

For Appendix A:

Fahey, D.W. (Lead Author), *Twenty Questions and Answers About the Ozone Layer: 2006 Update*, 50 pp., World Meteorological Organization, Geneva, Switzerland, 2007. [Reprinted from *Scientific Assessment of Ozone Depletion: 2006*, Global Ozone Research and Monitoring Project—Report no. 50, 572 pp., World Meteorological Organization, Geneva, Switzerland, 2007.]

PREFACE

Report Motivation and Guidance for Using this Synthesis/Assessment Product

Authors: A.R. Ravishankara, NOAA; Michael J. Kurylo, NASA

A primary objective of the U.S. Climate Change Science Program (CCSP) is to provide the best possible scientific information to support public discussion, as well as government and private sector decision making, on key climate-related issues. To help meet this objective, the CCSP has identified an initial set of 21 Synthesis and Assessment Products (SAPs) that address its highest priority research, observation, and decision support needs.

This report, CCSP SAP 2.4, addresses Goal 2 of the CCSP Strategic Plan: Improve quantification of the forces bringing about changes in the Earth's climate and related systems. The Atmospheric Composition chapter of the CCSP Strategic Plan describes a vision to produce a Synthesis and Assessment Product (SAP) on "Trends in emissions of ozone-depleting substances, ozone layer recovery, and implications for ultraviolet radiation (UV) exposure–SAP 2.4." The report provides a synthesis and integration of the current knowledge of the stratospheric ozone layer, ozone-depleting substances, and ultraviolet radiation reaching the Earth's surface.

P.1 CONTEXT FOR THIS SYNTHESIS AND ASSESSMENT PRODUCT

SAP 2.4 contributes to the ongoing and iterative international process of producing and refining climate-related assessments and decision support tools. SAP 2.4 integrates findings from the World Meteorological Organization (WMO) / United Nations Environment Programme (UNEP) 2006 assessment on the ozone layer (*Scientific Assessment of Ozone Depletion: 2006*) and the 2005 Special Report of the Intergovernmental Panel on Climate Change (IPCC) and the Technology and Economic Assessment Panel (TEAP) on *Safeguarding the Ozone Layer and the Global Climate System – Issues Related to Hydrofluorocarbons and Perfluorocarbons*. Both of these assessments have been extensively reviewed prior to their publication. SAP 2.4 discusses these assess-

ments from both the global perspective and in the specific context of the United States of America; this SAP 2.4 gives the United States-specific perspective of a global issue for decision-makers in the United States. The SAP discusses ozone changes over North America, the contributions of the United States to ozone-depleting substances, and the UV changes due to the ozone layer changes over the North American continent. This SAP takes advantage of these thoroughly vetted scientific assessments to prepare a product that can be used to inform domestic and international decision makers in government and industry, scientists, and the public. This SAP was planned and initiated in August 2005, before the release of the Fourth Assessment Report (AR4) of the Intergovernmental Panel on Climate Change (*Climate Change 2007: The Physical Science Basis*). Therefore, this report does not rely on the IPCC AR4; however, some key pertinent issues from the IPCC report are used in a few instances where updated information was essential. They are noted as such in the chapters.

P.2 AUDIENCE AND INTENDED USE

The audience for SAP 2.4 includes scientists, decision makers in the public sector (federal, state, and local governments), the private sector (chemical industry, transportation, and agriculture; and climate policy and health-related interest groups), the international community, and the general public. This broad audience is indicative of the diversity of stakeholder groups interested in knowledge of the stratospheric ozone layer, ozone-depleting substances, and ultraviolet radiation, and of how such knowledge might be used to inform decisions. The primary users of SAP 2.4 are intended to include, but are not limited to, officials involved in formulating climate and environmental policy, individuals responsible for managing emissions of ozone-depleting substances, and scientists involved in assessing and/or advancing the frontier of knowledge. The plan for this SAP was presented at the CCSP workshop, "U.S. Climate Change Science Program, Climate Sciences in Support of Decision Making," held in Arlington, Virginia, during 14-16 November 2005.

SAP 2.4 is intended to be used:

- as a state-of-the-art assessment of our knowledge of the stratospheric ozone layer, ozone-depleting substances, and ultraviolet radiation at the surface;
- to provide the scientific basis for decision support to guide management and policy decisions that affect the ozone layer and emissions of ozone-depleting substances;
- as a means of informing policymakers and the public concerning the general state of our knowledge of the stratospheric ozone layer and emissions of ozone-depleting substances with respect to the contributions of and impacts on the United States; and
- to provide scientific information on the ozone layer to inform important stakeholder groups. Examples of these groups include: the chemical industry that produces ozone-depleting substances and substitutes for ozone-depleting substances; agencies in the United States and sectors of the United States economy that request exemptions from emissions of substances banned by the Montreal Protocol and its Amendments; and the climate-science community.

Senior managers and the general public may use the Executive Summary of SAP 2.4 to improve their overall understanding of what is known and unknown about the effects of United States emissions on the stratospheric ozone layer and ultraviolet radiation at the surface. It will also provide an estimate of the impacts of the ozone layer changes on the country.

P.3 TOPICS AND CONTENT

The focus of this Report follows the Prospectus guidelines developed by the Climate Change Science Program and posted on its website at (http://www.climatescience.gov). SAP 2.4 addresses key issues related to the stratospheric ozone layer, including its changes in the past and expected levels in the future. Also, it takes account of the current abundances and emissions of ozone-depleting substances. Further, it synthesizes the best available information on the past and future levels of ultraviolet radiation at the Earth's surface. Lastly, it explores the interactions between climate change and stratospheric ozone changes. The discussion of these topics is carried out within the context of both the globe and the United States to distill a regional assessment from the global assessments. More specifically, SAP 2.4:

- Quantifies current information on sources, sinks, and abundances of ozone-depleting substances and associated uncertainties.
- Discusses levels of ozone in various regions of the stratosphere, including the polar regions. It pays special attention to the Antarctic ozone hole and to ozone above the continental United States.

- Provides information on the past, current, and future levels of ultraviolet radiation, both generally and for the continental United States.
- Provides an assessment of the impact of climate and compositional changes on the future of the ozone layer, and provides some qualitative discussion of the impacts of the ozone layer on climate.
- Describes how these findings relate especially to the United States.
- Identifies the gaps in understanding where research is critical for future assessments of the ozone layer.

The questions addressed by this report include:

- What is the current state of the stratospheric ozone layer?
- What are the recorded changes in the emissions and concentrations of ozone-depleting substances?
- What do the observations indicate about the abundances and trends of stratospheric ozone?
- What is the trend in the occurrence, depth, duration, and extent of the Antarctic ozone hole?
- What is the state of ozone depletion in the Arctic region?
- When can one expect recovery of the global ozone layer and of the Antarctic ozone hole?
- What are the influences of climate change on the recovery of the ozone layer?
- How has surface ultraviolet radiation changed in the past and what is expected for the future?
- What are the findings specific to the United States on the topics of ozone-depleting substances, stratospheric ozone depletion, surface ultraviolet radiation changes, and expectations for the future ozone layer?
- What are the various possible emission scenarios that can be considered for any further policy actions on emissions of ozone-depleting gases?

P.4 OUTLINE OF THE REPORT

The above questions provide the basis for information presented in the six chapters of SAP 2.4. The chapters are written in a style consistent with major authoritative international scientific assessments (*e.g.*, IPCC assessments, and the reports of the Global Ozone Research and Monitoring Project of WMO). However, additional explanatory material is included both within the Chapters and as an Appendix to aid the diverse readership of this SAP. The Executive Summary, which presents the key findings from the main body of the Report, as well as Chapters 1 and 6, are intended to be useful especially for those involved with policy-related ozone layer issues. Chapter 1 is intended as a background "primer" for those less familiar with the topic of stratospheric ozone depletion. Chapters 2 through 5 provide the detailed material that supports the findings of the Executive

Summary. Though they are written at a more technical level, they incorporate material to aid their accessibility to the broad readership of this SAP. The chapters of SAP 2.4 are:

- Chapter 1: Introduction
- Chapter 2: Current Trends, Mixing Ratios, and Emissions of Ozone-Depleting Substances and Their Substitutes
- Chapter 3: Ozone and UV Observations
- Chapter 4: How Do Climate Change and Stratospheric Ozone Loss Interact?
- Chapter 5: The Future and Recovery
- Chapter 6: Implications for the United States

For those interested readers who are not specialists on the ozone-layer issue, an Appendix gives additional scientific background on the topics of this SAP. A glossary and a list of acronyms are included at the end of the report.

P.5 THE SYNTHESIS AND ASSESSMENT PRODUCT TEAM

The authors for this SAP were chosen based on their expertise and participation in the international assessments from which this product derives a great deal of information. The SAP 2.4 Author Team and their roles are:

- Dr. A. R. Ravishankara, NOAA, Overall Lead
- Dr. Michael J. Kurylo, NASA, Overall Lead
- Dr. Richard Bevilacqua, NRL/DoD, Scientific Content
- Dr. Jeff Cohen, USEPA, Scientific Content
- Dr. John Daniel, NOAA, Scientific Content
- Dr. Anne Douglass, NASA, Scientific Content
- Dr. David Fahey, NOAA, Scientific Content
- Dr. Jay Herman, NASA, Scientific Content
- Dr. Terry Keating, USEPA, Scientific Content
- Dr. Malcolm Ko, NASA, Scientific Content
- Dr. Stephen Montzka, NOAA, Scientific Content
- Dr. Paul Newman, NASA, Scientific Content
- Dr. V. Ramaswamy, NOAA, Scientific Content
- Dr. Anne-Marie Schmoltner, NSF, Scientific Content
- Dr. Richard Stolarski, NASA, Scientific Content
- Dr. Kenneth Vick, USDA, Scientific Content

Those who served as Convening Lead Authors (CLAs) and Lead Authors (LAs) are shown at the beginning of each chapter. An Editorial Staff managed the assembly, formatting, and preparation of the Report.

The U.S. Climate Change Science Program
Preface

EXECUTIVE SUMMARY

Convening Lead Authors: A.R. Ravishankara, NOAA; Michael J. Kurylo, NASA

Lead Authors: Richard Bevilacqua, NRL; Jeff Cohen, U.S. EPA; John S. Daniel, NOAA; Anne R. Douglass, NASA; David W. Fahey, NOAA; Jay R. Herman, NASA; Terry Keating, U.S. EPA; Malcolm Ko, NASA; Stephen A. Montzka, NOAA; Paul A. Newman, NASA; V. Ramaswamy, NOAA; Anne-Marie Schmoltner, NSF; Richard Stolarski, NASA; Kenneth Vick, USDA

Synopsis

Depletion of the stratospheric ozone layer by human-produced ozone-depleting substances has been recognized as a global environmental issue for more than three decades, and the international effort to address the issue via the United Nations Montreal Protocol marked its 20-year anniversary in 2007. Scientific understanding underpinned the Protocol at its inception and ever since. As scientific knowledge advanced and evolved, the Protocol evolved through amendment and adjustment. Policy-relevant science has documented the rise, and now the beginning decline, of the atmospheric abundances of many ozone-depleting substances in response to actions taken by the nations of the world. Projections are for a return of ozone-depleting chemicals (compounds containing chlorine and bromine) to their "pre-ozone-depletion" (pre-1980) levels by the middle of this century for the midlatitudes; the polar regions are expected to follow suit within 20 years after that. Since the 1980s, global ozone sustained a depletion of about 5 percent in the midlatitudes of both the Northern Hemisphere and Southern Hemisphere, where most of the Earth's population resides; it is now showing signs of turning the corner towards increasing ozone. The large seasonal depletions in the polar regions are likely to continue over the next decade but are expected to subside over the next few decades. Ozone-depleting substances should have a negligible effect on ozone in all regions beyond 2070, assuming continued compliance with the Montreal Protocol.

Large increases in surface ultraviolet (UVB; 280-315 nm) radiation and the associated impacts on human health and ecosystems would likely have occurred if atmospheric abundances of ozone-depleting substances had continued to grow. Scientific findings regarding the role of ozone-depleting chemicals, projected ozone losses, and the potential UV impacts galvanized international decision making in the 1980s. As a result of the worldwide adherence to the 1987 Montreal Protocol and its Amendments and Adjustments, the large impacts were avoided, and future trends in UVB and UVA (315-400 nm) at the surface are expected to be more influenced by factors other than stratospheric ozone depletion (such as changes in clouds, atmospheric fine particles, and air quality in the lower atmosphere).

Emissions of ozone-depleting substances by the United States have been significant throughout the history of the ozone depletion issue. At the same time, the United States has played a leading role in advancing the scientific understanding, leading the international decision making, and leading industry's actions to reduce usage of ozone-depleting substances. Continued future declines in emissions of ozone-depleting substances from the United States, along with those from other nations, will play a key role in ensuring the ozone layer's recovery.

Projections of a changing climate have added a new dimension to the issue of the stratospheric ozone layer and its recovery, and scientific knowledge is emerging on the interconnections between these two global issues. Climate change is expected to alter the timing of the recovery of the ozone layer. Ozone-depleting chemicals and ozone depletion are known to influence climate change. The curtailment of the ozone-depleting substances not only helped the ozone layer but also very likely lessened the forcing of climate (i.e., how it alters climate).

Climate change and ozone layer depletion are coupled; this has led to new scientific and decision-making challenges. The recovery of the ozone layer will occur in an atmosphere that is different from where we started roughly three decades back. Our scientific understanding of the connections between climate change and ozone layer depletion is at an early but rapidly advancing stage. That topic will remain a focus for the scientific community's efforts over the next few decades.

ES.I WHAT IS OZONE LAYER DEPLETION AND WHY IS IT A CONCERN?

The depletion of the ozone layer can lead to enhancements of ultraviolet (UV) radiation that reaches Earth's surface, with consequences for human health, the Earth's ecosystems, and physical materials.

The stratospheric ozone layer lies in a region of the atmosphere approximately 15 to 45 kilometers (roughly 9 to 28 miles) above Earth's surface. The ozone layer acts as a protective shield, preventing most of the Sun's harmful ultraviolet (UV) radiation from reaching the surface. The depletion of the ozone layer can therefore lead to enhancements of the UV radiation that reaches Earth's surface, with consequences for human health, the Earth's ecosystems, and physical materials. The ozone layer and its changes can also alter the atmosphere's temperature structure and weather/climate-related circulation patterns.

Effects of the Montreal Protocol and Its Amendments

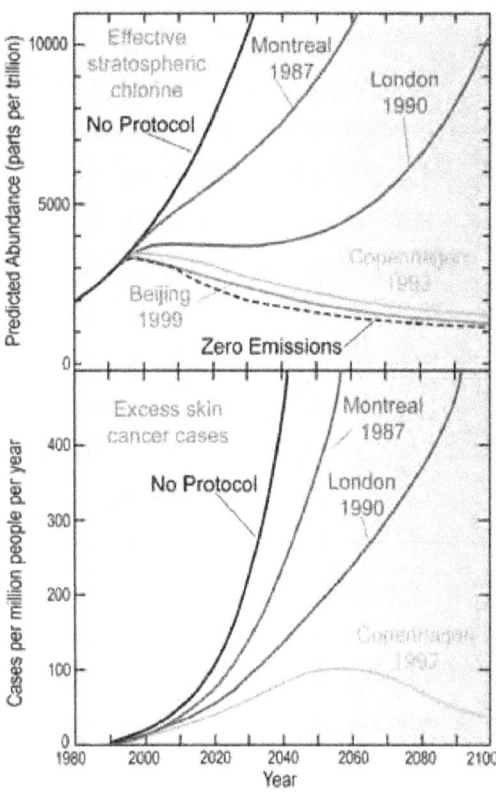

Figure ES.I Effect of the Montreal Protocol. The top panel gives a measure of the projected future abundance of ozone-depleting substances in the stratosphere, without and with the Protocol and its various Amendments. The bottom panel shows similar projections for how excess skin cancer cases might have increased (adapted from Appendix A of this Report).

Research in the 1970s and early 1980s showed that the ozone-depleting substances (ODSs), mainly chlorofluorocarbons (CFCs) and certain compounds containing bromine, would deplete stratospheric ozone. The discovery of the springtime Antarctic ozone hole in 1985 showed that ozone depletion was real and occurring at that time, and was not just a prediction for the future.

Faced with the scientific consensus that ozone depletion was real and due to human-produced ozone-depleting substances, nations throughout the world agreed to the Montreal Protocol and its subsequent Amendments and Adjustments. The United States is a signatory to this protocol. The Protocol and its Amendments were successfully implemented starting in the late 1980s. Thus, this Protocol was one of the first international agreements to address a global environmental problem. The Montreal Protocol has had clear benefits in reducing ozone-depleting substances, placing the ozone layer on a path to recovery, and protecting human health (Figure ES.1).

Ozone layer depletion, like climate change, is a global issue with regional impacts. The depletion of the ozone layer is caused by the collective emissions of human-produced ozone-depleting substances at Earth's surface from various regions and countries. These ozone-depleting substances persist long enough in the atmosphere to be quite well mixed in the lower atmosphere and then be transported to the stratosphere, where their interaction with the harsh UV radiation releases chlorine and bromine. Thus, they pose a global threat, regardless of where on Earth's surface they are emitted. Emissions of ozone-depleting substances arise from their use as coolants, fire-extinguishing chemicals, electronics cleaning agents, and in foam blowing and other applications. The contributions to the global atmospheric burden of these ozone-depleting substances vary by regions and countries. There are large variations in the extent and timing of ozone depletion in various regions, and the impacts are also different. Consequently, the impacts of ozone layer depletion can be different in different regions of the world.

The findings from this Synthesis and Assessment Product are summarized in three parts. Section ES.2 of this Executive Summary lists the findings to inform the public in general nontechnical terms, and Section ES.3 summarizes findings for those involved in potential policy formulation. The Executive Summary findings are backed up by a more technical set of findings, primarily for scientists and secondarily for those who want to delve more into the details. These technical findings are listed near the beginning of Chapters 2 through 5, and in Chapter 6 on Policy Implications for the United States. Appendix A of this Synthesis and Assessment Product provides extensive background material on the science regarding the ozone layer, ozone-depleting substances, surface ultraviolet radiation, and connections to climate change.

ES.2 KEY FINDINGS ABOUT THE OZONE LAYER, SURFACE UV, OZONE-DEPLETING SUBSTANCES, AND CONNECTIONS TO CLIMATE CHANGE

ES.2.1 The Ozone Layer, Ozone-Depleting Substances, and Climate Change: What Are the Connections?

Ozone layer changes caused by ozone-depleting substances are intertwined with the issue of climate change, even though the two issues have been distinct in most policy formulations.

Over the course of the past 20 years, the close connections between stratospheric ozone depletion and climate change issues have become clearer (Figure ES.2).

- Ozone-depleting substances and many of the chemicals being used to replace them are potent greenhouse gases that influence the Earth's climate by trapping terrestrial infrared (heat) radiation that would otherwise escape to space.
- Ozone is itself a greenhouse gas. The stratospheric ozone layer heats the stratosphere and, indirectly, the lower atmosphere (troposphere). Thus, stratospheric ozone is a key component that affects climate. Depletion of the ozone layer has a cooling effect on climate,

though large uncertainties exist regarding this effect, which is a combination of multiple contributing factors.

- The recovery of the ozone layer is influenced not only by the decreases in ozone-depleting substances required by the Montreal Protocol, but also by changes to climate and Earth's atmospheric composition.

Ozone-depleting substances are continuing to make a significant contribution to global climate change, but in the future ODSs are expected to make a smaller and smaller contribution. The direct ODS contribution to global climate change between 1750 (pre-industrial times) and 2005, as measured by a quantity called radiative forcing that is a metric for the ability to force climate change, is approximately 20% of that from carbon dioxide (CO_2), the largest human-caused contributor to global radiative forcing (Figure ES.3). The combined radiative forcing from ODSs and substitutes including hydrofluorocarbons (HFCs) is still increasing, but at a much slower rate than in the 1980s. The total contribution of human-produced ODSs and substitutes in 2005 was about 15% of the contribution from the major greenhouse gases (CO_2, methane [CH_4], nitrous oxide [N_2O]). The ODS contribution is expected to decline in coming decades as ODS emissions decline and CO_2 emissions continue to rise.

The recovery of the ozone layer is influenced not only by the decreases in ozone-depleting substances required by the Montreal Protocol, but also by changes to climate and Earth's atmospheric composition.

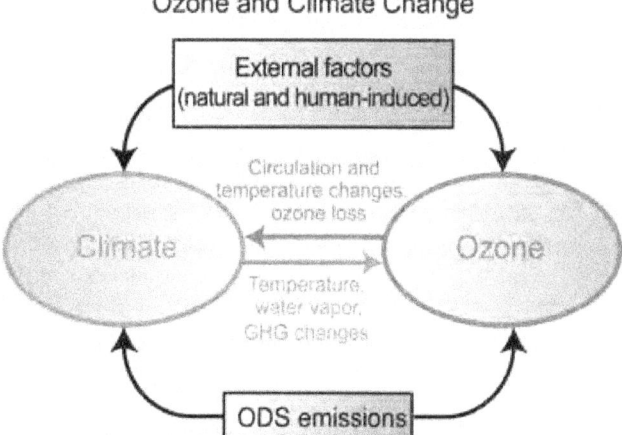

Figure ES.2 Simplified schematic of some of the processes that interconnect the issues of ozone layer depletion and climate change (adapted from Chapter 4 of this Report).

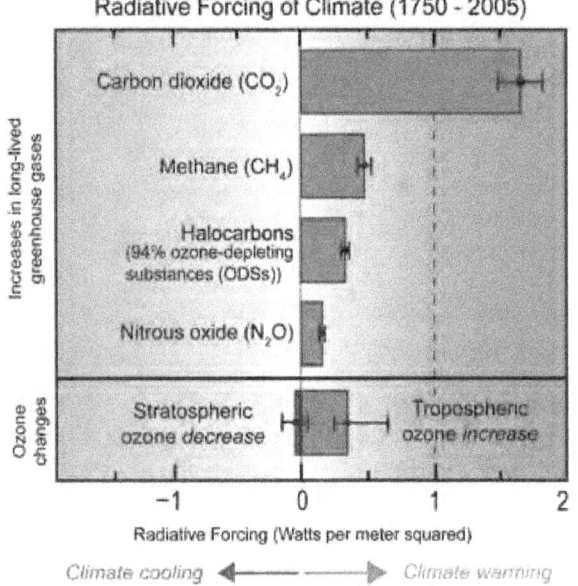

Figure ES.3 Radiative forcing values for the principal contributions to climate change from atmospheric gas changes since preindustrial times, including halogen-containing gases such as ODSs, and the cooling caused by depletion of stratospheric ozone. These climate influences are expressed as radiative forcings, a metric for the ability to force climate change (adapted from IPCC, 2007).

Total global production and consumption of ozone-depleting substances (ODSs) have declined substantially since the late 1980s in response to the Montreal Protocol. Hence, the total amount of ODSs in the atmosphere is now decreasing both in the troposphere and stratosphere.

Depletion of stratospheric ozone since about 1980 is estimated to have caused a slight *negative* (cooling) radiative forcing of climate (approximately −0.05 Watts per meter squared [W per m²] with a range of −0.15 to +0.05 W per m²) (Figure ES.3). While this forcing is likely to be a cooling term (*i.e.*, in the opposite direction to climate forcing by the ODSs that caused the depletion) it has large uncertainties. Globally averaged, it may even represent a warming within the error bars, or it could offset a large portion (up to 44%) of the ODS warming, while the current best estimate is an offset of approximately 15%. This estimate is based on observed ozone changes and assumes that they are due entirely to ODSs. Recent research has shown that ozone cooling and ODS warming often occur in different places and times, making it less appropriate to consider the two terms as offsetting one another than previously thought.

Climate change will lead to either increases or decreases in ozone abundances depending on the location in the atmosphere and the magnitude of climate change. While the surface temperature has increased, observed stratospheric temperature decreased starting in the 1960s and it is expected to continue to decrease. The global average trend is attributed mainly to ozone depletion, increased CO₂, and changes in water vapor. Dynamical changes are also likely to be important for local temperature changes, but are not significant for global mean stratospheric temperature trends. Stratospheric temperatures influence ozone amounts through chemical and transport processes. Stratospheric water vapor influences stratospheric ozone through chemistry, formation of polar stratospheric clouds, and changes in temperature.

ES.2.2 Ozone-Depleting Substances: Past, Present, and Future

The Montreal Protocol has been effective in reducing the use of ozone-depleting substances. Assuming continued compliance with the Protocol, the atmospheric abundance of ODSs is expected to decline back to its pre-1980 level by the middle of this century.

Total global production and consumption of ODSs have declined substantially since the late 1980s in response to the Montreal Protocol. By 2005, the annual aggregated production and consumption magnitudes of the ODSs, after accounting for their differences in ozone depletion capabilities, had declined 95 percent from peak amounts produced and consumed in the late 1980s.

In response to these global production and consumption changes, global ODS emissions have declined. Hence, the total amount of ODSs in the atmosphere, as measured by their combined ability to deplete the ozone layer, is now decreasing both in the troposphere and stratosphere.

In this Report, future halocarbon emissions are derived using a new bottom-up approach for estimating emissions from the sizes of the banks (ODSs produced but not yet released). The new method gives future CFC emissions that are higher than previously estimated in WMO (2003). There are still some uncertainties in the future abundances of ODSs.

The effective sum of chlorine and bromine in the stratosphere, with bromine weighted by its larger per-atom efficiency in depleting ozone, is estimated to recover to the 1980 value between 2040 and 2050 in the midlatitudes (Figure ES.4) and between 2060 and 2070 in the polar regions.

ES.2.3 Ozone in the Stratosphere: Past, Present, and Future

Total global ozone, as well as seasonal springtime ozone in both southern and northern polar regions, exhibited declines since the early 1980s, but recent observations show that ozone depletion is not worsening and in some atmospheric regions is showing signs that recovery has started. Ozone in the future is projected to recover as the atmospheric amounts of ODSs decline over the next few decades (with recovery above midlatitudes and the Arctic preceding Antarctic recovery). With continued adherence to the Montreal Protocol, ozone-depleting substances identified in the Protocol should have a negligible effect on ozone in all regions beyond 2070.

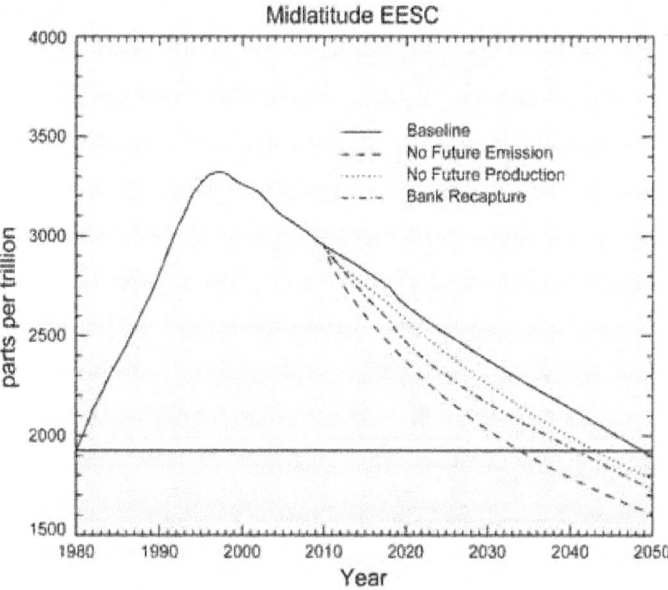

Figure ES.4 Estimates (presented in parts per trillion, [ppt]) of the effective sum of ozone-depleting chlorine and bromine in the stratosphere (called Equivalent Effective Stratospheric Chlorine, [EESC]), a metric that accounts for the differences in ozone depletion capabilities of chlorine and bromine. Estimates in the past are based upon observations, and estimates in the future are based upon a baseline scenario and three comparative test cases. The horizontal line represents the 1980 ("pre-ozone-depletion") level of EESC (adapted from WMO, 2007).

Total global ozone declined by roughly 5 percent since the early 1980s but has remained relatively constant over the last four years (2002 to 2006). Northern midlatitude ozone reached a minimum in 1993, and has increased somewhat since then. The 1993 minimum largely resulted from the increase of particles in the stratosphere caused by the eruption of Mt. Pinatubo. Southern midlatitude ozone decreased until the late 1990s, and has been constant since. There are no significant total ozone trends over the tropics.

Ozone depletion in the upper stratosphere, where the influence of chlorine is easiest to detect, has slowed and has closely followed the trends in the sum of total chlorine. Although bromine plays a lesser role than chlorine in controlling ozone in the upper stratosphere, it too shows signs of leveling off in the stratosphere (see Section 2.4.2).

Antarctic ozone depletion can be measured in different ways, such as the total amount of ozone lost (called mass deficit), the minimum values of ozone observed, and the geographical area of the ozone hole. Over the last decade (1995 to 2006), the Antarctic ozone depletion by all these measures has not worsened. The ozone hole area and ozone mass deficit were observed to be below average in some recent winter years while higher minimum column amounts have also been recorded. This variability results from the strong influence of meteorological variability on ozone amounts, and not from any changes in the amounts of chlorine and bromine available for ozone depletion. Declines in the amounts of chlorine and bromine available for ozone depletion are likely quite small in this region.

Arctic spring total ozone values over the last decade were lower than values observed in the 1980s. In addition, spring Arctic ozone is highly variable depending on meteorological conditions. For current halogen levels, human-caused chemical loss and variability in ozone transport are about equally important for year-to-year Arctic ozone variability. Colder-than-average vortex conditions result in larger halogen-driven chemical ozone losses.

If volcanic eruptions that inject material into the stratosphere were to occur in the coming

Recent observations show that ozone depletion is not worsening and in some atmospheric regions is showing signs that recovery has started.

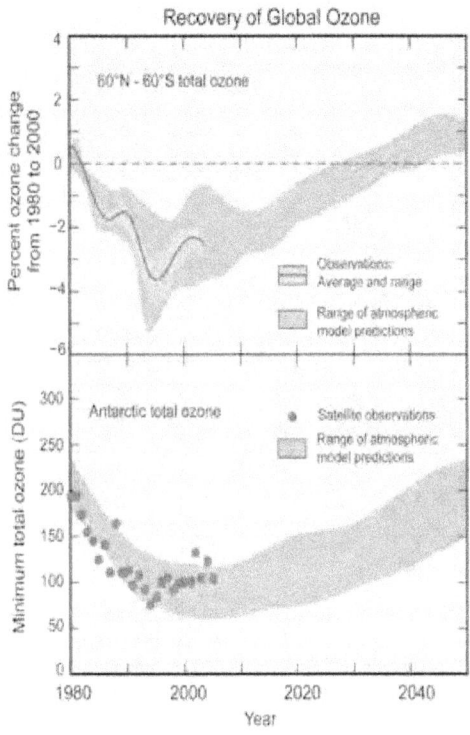

Figure ES.5 Global ozone recovery predictions (derived from Fahey, 2007).

decades, they are expected to cause major temperature and circulation changes in the stratosphere as have occurred after past eruptions. The changes are caused by the large increases in fine particles formed from sulfur dioxide injected into the stratosphere following such eruptions. The increases result in a transient shift in stratospheric ozone levels and climate because natural processes gradually remove the additional sulfate particles after the eruption.

Assuming an absence of volcanic injections into the stratosphere, and based on the projected changes in ozone-depleting substances and changes in the major climate-relevant trace gases, modeling calculations predict the following for the future of the ozone layer (Figure ES.5).

- The ozone content between 60°N and 60°S, between now and 2020, will increase in response to decreases in halogen loading.
- Global ozone is expected to return to its 1980 value up to 15 years earlier than the halogen recovery date because of stratospheric cooling and changes in

circulation associated with greenhouse gas emissions.
- Global ozone abundances (from 60°N to 60°S) are expected to be 2 percent above the 1980 values by 2100 for the assumed scenario for greenhouse gases noted in this report. Values at midlatitudes could be as much as 5 percent higher.

The minimum ozone value for Antarctic ozone is projected to start increasing after 2010 in several model calculations, while another measure of ozone depletion (the ozone mass deficit, the total amount lost in a season) begins decreasing around 2005 in most models.

- Model simulations show that the ozone amount in the Antarctic will reach the 1980 values 10 to 20 years earlier than the 2060 to 2070 time frame of when the ODSs reach their 1980 levels in polar regions.
- Ozone in the Arctic region is expected to increase as ODSs decline in the atmosphere. Because of large interannual variability, the simulated results do not show a smooth monotonic recovery of Arctic ozone. The dates of the minimum ozone from different models occur between 1997 and 2015.
- Most climate chemistry models show Arctic ozone values by 2050 larger than the 1980 values, with the recovery date between 2020 and 2040.

The above projections are based on currently available models. As our scientific understanding and modeling capabilities continue to evolve, our best predictions of the timing and extent of ozone layer recovery will also evolve.

ES.2.4 Surface Ultraviolet Radiation: Past, Present, and Future
The Montreal Protocol and its Amendments have prevented large increases in global surface UVB radiation. As the stratospheric ozone layer recovers over the next few decades, factors such as changes in clouds, atmospheric fine particles, and air quality in the lower atmosphere will be the dominant factors influencing future UV changes.

Surface UVB changes resulting from ozone depletion over Antarctica in early austral spring have been very large. Changes in the surface

UVB due to ozone depletion in most other locations of the world have not been clearly discernable, because the effects have been much smaller compared with changes due to other factors. For example, trends in UV exposure changes at ground level in the midlatitude United States attributable to ozone changes are difficult to discern from ground-based observations, since the observations are also dependent on changes in clouds and pollution from suspended fine particles in the air. What is clear is that in the absence of the Montreal Protocol, ozone depletion would have caused increases in surface UV by 2010 over most of the world, to such an extent that other factors (*e.g.*, clouds, atmospheric fine particles, air quality) would have been of relatively minor importance.

Possible future UV trends at the surface are likely to be influenced more by changes in clouds, atmospheric fine particles, and lower atmosphere air quality than by ozone layer depletion.

ES.3 IMPLICATIONS FOR THE UNITED STATES: IMPACTS, ACCOUNTABILITY, AND POTENTIAL MANAGEMENT OPTIONS

It is not possible to make a simple connection between emissions of ozone-depleting substances from the United States and the depletion of ozone above the country. This is because ODSs persist long enough in the atmosphere to be quite well mixed in the global lower atmosphere, before transport to the stratosphere occurs. Thus, ODSs pose a global threat, regardless of where on Earth's surface they are emitted. However, the depletion of stratospheric ozone over the various regions of the United States, and the contribution of emissions from the United States to the global burden of ozone-depleting substances, can be quantified.

Impacts: Changes in Ozone and Surface Ultraviolet Radiation over the United States

Ozone depletion above the continental United States (*i.e.*, the midlatitudes) has essentially followed the depletion occurring over the

northern midlatitude regions: a decrease to a minimum around the mid-1990s and a slight increase since that time. The minimum total column ozone amounts over the continental United States, reached in 1993, were about 5 to 8 percent below the amounts present prior to 1980. The ozone increase since 1993 has diminished the ozone deficit to about 2 to 5 percent below the pre-1980 amounts. These midlatitude ozone changes are estimated to contain a significant contribution from the ozone depletion that occurs in the Arctic during springtime.

Ozone over Northern high latitudes, such as over northern Alaska, is most influenced by Arctic springtime total ozone values, which in recent years have been lower than those observed in the 1980s. The springtime ozone depletions are highly variable from year to year.

Calculations based on satellite observations of column ozone and surface reflectivity suggest that the averaged erythemal irradiance (which is a weighted combination of UVA and UVB based on skin sensitivity) over the United States had increased roughly by about 7 percent at the time when the ozone minimum was reached in 1993 and is now about 4 percent higher than in 1979. Direct surface-based observations do not show significant trends in UV levels over the United States over the past three decades because effects of clouds and atmospheric fine particles have likely masked the increase in UV due to ozone depletion over this region.

Accountability: United States Contributions to Ozone-Depleting Substances

The contributions of the United States to the emission of ODSs to date have been significant. For example, in terms of dispersive uses of ODSs regulated and restricted by the Montreal Protocol, emissions from the United States accounted for between 15 and 39 percent of the overall atmospheric abundance of ODSs measured between 1994 and 2004. The United States has also contributed significantly to emission reductions of ODSs, thereby helping efforts to achieve the expected recovery of the ozone layer and prevent large surface UV changes.

Emissions from the United States accounted for between 15 and 39% of the overall atmospheric abundance of ODSs measured between 1994 and 2004. The United States has also contributed significantly to emission reductions of ODSs, thereby helping efforts to achieve the expected recovery of the ozone layer.

Future Options

United States emissions of ODSs in the future, like those from other developed nations, will be determined largely by the size of ODS "banks," *i.e.*, those ODSs that are already produced but not yet released to the atmosphere. While global ODS banks are estimated to have been 2960 ODP-kilotons (Kt) in 2005, ODS banks in the United States then were 830 ODP-Kt. Of this U.S. bank, approximately 210 ODP-Kt has been classified as accessible by the U.S. Environmental Protection Agency. The expected future declining emissions of ODSs from the United States and throughout the globe will also aid in reducing the climate forcing from these substances. While global banks amounted to between 5 and 24 Gt CO_2-equivalents, the accessible bank of hydrochlorofluorocarbons (HCFCs) in the United States, for example, amounted to between 0.9 and 1.1 Gt CO_2-equivalents.

While the Montreal Protocol has had a large beneficial effect on current and projected ozone depletion, options remain for the United States, and other countries as well, to reduce ozone depletion arising from ozone-depleting substances over the coming decades. The greatest reduction possible would be obtained from the hypothetical cessation of all future emissions of ozone-depleting substances (including emissions from banks and future production). If such a cessation had been implemented globally in 2007, the anticipated return of the ozone-depleting substances to their 1980 level would be advanced by about 15 years.

Methyl bromide is a potent ODS that has significant quarantine and pre-shipment (QPS) uses that are not restricted by the Montreal Protocol, and Critical Use Exemptions (CUEs) that are currently large compared to restricted uses. The importance of human-emitted methyl bromide to future ozone depletion will depend on the future magnitude of emissions from these unrestricted uses and from CUEs. Reducing these emissions would benefit the ozone layer.

The World Avoided

Without the Montreal Protocol regulations, the levels of ODSs around 2010 likely would have been more than 50 percent larger than currently predicted (Figure ES.1). The abundances in the remaining twenty-first century would have depended on the specific actions taken by humankind. The increases in ODSs would have caused a corresponding substantially greater global ozone depletion. The Antarctic ozone hole would have persisted longer and may have been even larger than what has been observed to date.

Introduction

Convening Lead Authors: A.R. Ravishankara, NOAA; Michael J. Kurylo, NASA; Anne-Marie Schmoltner, NSF

Ozone (O_3) is the triatomic form of oxygen. It is a key atmospheric trace gas that is present everywhere in the atmosphere and is most abundant in the stratosphere. The abundance of ozone in the stratosphere is largest in the region that is roughly between 15 and 35 kilometers (km) height above the Earth's surface, which is referred to as the stratospheric ozone layer. This stratospheric ozone layer (Box 1.1) plays many important roles in the Earth system:

- It protects the lower part of the atmosphere (the troposphere) and the Earth's surface from damaging, or "harsh" ultraviolet[1] (UV) radiation from the sun;
- It influences the chemical composition of the lower atmosphere by altering the amount and type (wavelength distribution) of solar radiation passing through it;
- It changes the temperature structure of the stratosphere and thus influences atmospheric transport and mixing; and
- It contributes ozone to the upper troposphere, where ozone is an important greenhouse gas.

Because of many of the above contributions, ozone in the stratosphere and its changes also play a significant role in the Earth's climate system; changes in the ozone layer are influenced by climate change and also contribute to climate change. Appendix A of this Product contains background information and answers to some of the most frequently asked questions about the stratospheric ozone layer (Fahey, 2007).

The focus of this Product is on key issues related to: (1) the stratospheric ozone layer, including its changes in the past, its current abundances, and expected levels in the future; (2) emissions of ozone-depleting substances (ODSs) and their influences on the ozone layer and climate; and (3) the changes in the ground-level UV radiation associated with stratospheric ozone changes.

The potential for human-produced chemicals, such as chlorofluorocarbons (CFCs), to deplete the stratospheric ozone layer has received a great deal of attention since the early 1970s.

---

[1] 'Harsh' UV radiation indicates the higher energy portion of the UV spectrum.

The nations of the world agreed to protect the ozone layer through the 1985 Vienna Convention. That same year the ozone hole in Antarctica was discovered.

The chemical processes that lead to the formation of ozone, as well as those that remove or destroy it, are distinctly different in the stratosphere from those in the troposphere (Box 1.2). The ever-present balance in the stratosphere between production, removal, and transport determines the abundance of ozone in any given part of the ozone layer. The majority of the removal processes in the stratosphere involve catalytic cycles in which ozone-destroying chemicals are re-formed after destroying ozone. This catalytic capability is a key reason why very small amounts of ozone-destroying chemicals introduced into the atmosphere can vastly influence the ozone layer (Box 1.2).

The potential for human-produced chemicals, such as chlorofluorocarbons (CFCs), to deplete the stratospheric ozone layer has received a great deal of attention since the early 1970s. The depletion by chlorine released from CFCs in the stratosphere was expected to be catalytic in nature, meaning that small amounts of CFCs could destroy vast amounts of ozone. The ozone depletion was predicted to lead to changes in UV radiation at the Earth's surface, with potentially major environmental consequences. The anticipated effects of increased UV radiation included: increased incidence of skin cancer and cataracts in humans; detrimental effects on ecosystems including the aquatic system; and harmful effects on materials, such as rubber and plastics. These potential effects were debated and the nations of the world agreed to protect the ozone layer through the 1985 Vienna Convention. Then the ozone hole in Antarctica was discovered in 1985. Investigation of the causes of this annually recurring

BOX 1.1: The Stratospheric Ozone Layer and Its Role in the Atmosphere

About 90% of the atmospheric ozone resides in the stratosphere, in a region between roughly 15 and 35 km above the Earth's surface, as indicated by the red line in Box Figure 1.1. This region is referred to as the stratospheric ozone layer. The remainder of the atmospheric ozone resides in the troposphere, the lower layer of the atmosphere. Stratospheric ozone is formed and destroyed by chemical reactions, as shown in Box 1.2. Of particular note are the need for higher-energy UV radiation for the formation of ozone and the catalytic nature of the ozone removal processes. The ozone layer in turn shields the lower part of the atmosphere and the surface from damaging UV radiation because ozone itself absorbs UV radiation. Depletion of the ozone layer allows more UVB radiation (wavelength 280 to 315 nanometers) to reach the Earth's surface. This radiation is harmful to humans and many other biological systems and causes damage to materials. The ozone in the lower atmosphere, the troposphere, is formed by methods different from those in the stratosphere, as

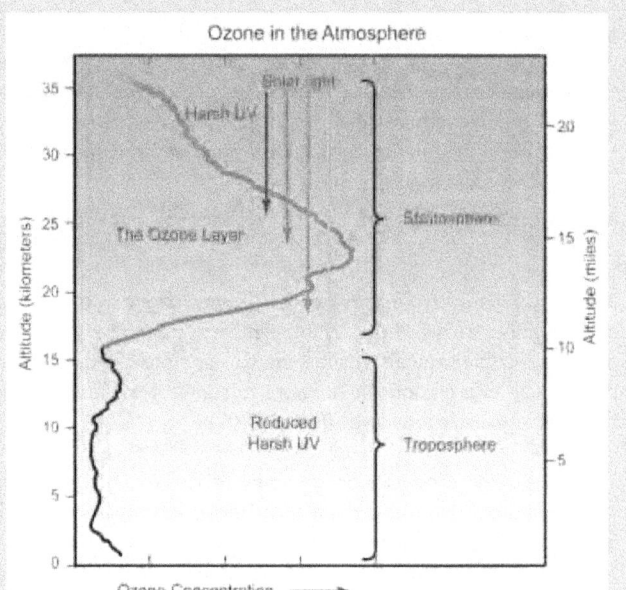

Box Figure 1.1 This figure shows the distribution of ozone in the atmosphere (adapted from Fahey, 2007; see Appendix A of this Report).

shown in Box 1.2. Further, the contribution of this lower atmospheric ozone to the total in the atmosphere is small, of the order of a few percent in the Southern Hemisphere to about 10% in the Northern Hemisphere. The ozone in the lower atmosphere is harmful because, in direct contact, ozone is toxic to biological systems and can deteriorate many materials. It can cause respiratory and other health problems for humans. In addition, ozone and its changes in both the stratosphere and the lower atmosphere are important greenhouse gases and thus their changes influence climate. See Appendix A of this Synthesis and Assessment Product for further background information about ozone.

polar springtime ozone depletion indicated that CFCs and other ozone-depleting chemicals were involved in additional catalytic ozone destruction pathways unique to the extremely cold polar stratosphere. It was also discovered that small particles containing water and nitric and/or sulfuric acid that are found in polar stratospheric clouds (PSCs) play a crucial role in these processes by converting chemically less reactive halogen-containing chemicals into more reactive chemicals, which are more effective in ozone depletion, and involved some catalytic cycles unique to this region.

The Montreal Protocol, a sequel to the Vienna Convention, was agreed to in 1987 in the setting of the scientific knowledge at that time. First, the agreements of the Protocol were to reduce CFC emissions by replacing CFCs with less harmful substances, if possible, that could be used in existing devices for most applications. A few applications utilized not-in-kind (alternative) processes that did not require ozone-depleting chemicals. Many of the replacement chemicals, such as hydrochlorofluorocarbons (HCFCs), still contained chlorine but overall were less harmful to the stratospheric ozone layer than CFCs. Eventually, even the chlorine-containing substitutes were to be replaced by non-chlorine or bromine containing replacements such as hydrofluorocarbons (HFCs).

Over the past three decades of ozone-layer research, it has become clearer that ODSs, as well as many of the CFC-substitutes introduced to comply with the Montreal Protocol, are also potent greenhouse gases. Ozone depletion and climate change are distinct issues but are inextricably linked because ozone itself is a greenhouse gas and many of the ozone-depleting gases

are potent greenhouse gases. To add to the complexity, changes in the major greenhouse gases such as carbon dioxide (CO_2), methane (CH_4), and nitrous oxide (N_2O) also influence ozone depletion. Increases in CO_2 lead to a cooling of the stratosphere, which increases ozone in the upper stratosphere in non-polar regions, but decreases ozone in the polar lower stratosphere. The influence of CH_4 and N_2O on the stratospheric ozone layer is dominated by their chemical interactions. Figure 1.1 captures this influence in a schematic form. An assessment of the climate effects of ODSs must consider both of their roles: as chemicals that deplete ozone, and as greenhouse gases that alter climate.

Because many ozone-depleting substances have lifetimes of many years in the atmosphere, the depletion of stratospheric ozone is a global problem. The observed ozone depletion above a given region will not be directly related to the emissions from that region.

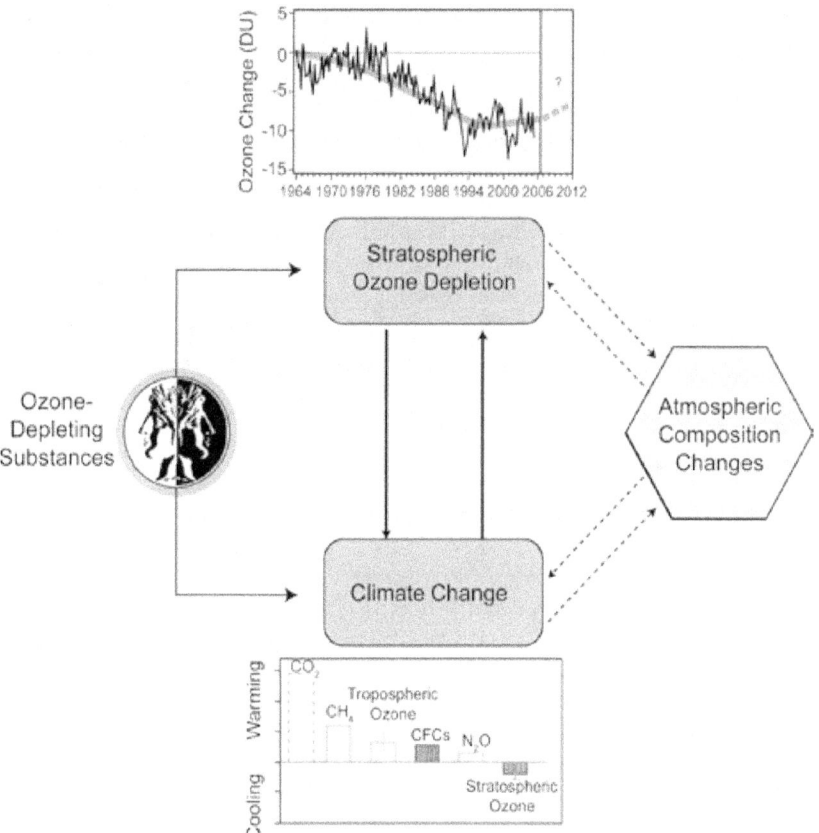

Figure 1.1 The two faces of ozone-depleting substances: their roles as depleting agents of stratospheric ozone, and as greenhouse gases that influence climate. The two roles are further interconnected because ozone itself is a greenhouse gas and because climate change can lead to changes in the ozone layer. The various connections between these two phenomena are shown. A plot of the changes in the observed global ozone illustrates the stratospheric ozone depletion issue. The radiative forcing due to various greenhouse gases, including ODSs, depicts the greenhouse gas issue and stratospheric ozone changes.

Since 1987, there have been many amendments and adjustments to the Montreal Protocol to accelerate efforts to curtail the emissions of ODSs. These actions have come about in response to our evolving knowledge of the ozone layer and its changes, and have led to a reduction in the emissions and, subsequently, in the atmospheric abundances of most ozone-depleting substances. Thus, the projected extremely high atmospheric abundances of ODSs and the associated larger-scale stratospheric ozone depletions were prevented from occurring. However, many key questions remain:

- Are the emission controls working as anticipated, *i.e.*, are the atmospheric abundances of ODSs declining as expected?
- Is the ozone layer recovering due to decreases in emissions of ODSs as predicted?
- Are the changes in UV occurring as expected with changes in ozone?
- What are the influences of other Earth system changes, *e.g.*, climate and atmospheric composition, on the ozone layer and its recovery from the ODS-induced depletion?
- What are the influences of ODSs, and their substitutes, on other aspects of the Earth system, especially climate?

The extent of the ozone layer depletion for a given emission differs depending on the location (*e.g.*, latitude) and time (*e.g.*, season). Because many ODSs have lifetimes of many years in the atmosphere, the depletion of stratospheric ozone is a global problem, and emissions of ODSs anywhere on the globe contribute to the ozone layer depletion. Therefore, the observed ozone depletion in a given region will not be directly related to the emissions from that region. Yet, it is appropriate to ask: what is the contribution of one nation, or region, to the depletion of the global ozone layer? And, how do the ODSs influence stratospheric ozone, and hence UV, in a specific region or over a specific nation? Of course, it may not be feasible to answer these questions completely at the present time, given our current (and evolving) state of knowledge.

The issues of this Synthesis and Assessment Product are addressed within the context of the United States in order to distill a regional assessment from current global assessments.

This Synthesis and Assessment Product (SAP) of the Climate Change Science Program (CCSP), SAP 2.4, addresses key issues related to the stratospheric ozone layer, including its changes in the past and its expected evolution in the future. Also, it takes account of the current abundances and emissions of ozone-depleting substances. Further, it synthesizes the best available information on the past and future levels of UV radiation at the Earth's surface. Lastly, it explores the interactions between climate change and stratospheric ozone changes as well as the ODS changes, and briefly recounts the influence of stratospheric ozone changes on climate change. All of these topics are carried out within the context of the United States in order to distill a regional assessment from current global assessments. More specifically, this document:

- Summarizes current quantitative information on sources (*i.e.*, emissions), sinks (*i.e.*, the removal pathways and their speed), and abundances of ozone-depleting substances as well as associated uncertainties; describes how the combined influence of chlorinated and brominated ODSs in the stratosphere can be quantified, and how all these are likely to change in the future.
- Discusses levels of ozone in various regions of the stratosphere, including the polar regions, paying special attention to the Antarctic ozone hole.
- Provides information on the past, current, and anticipated future levels of ultraviolet radiation.
- Provides an assessment of the impact of changes in both climate and atmospheric composition on the future of the ozone layer.
- Provides a brief assessment of the contribution of ozone-depleting substances on forcing of climate because these chemicals are also greenhouse gases.
- Describes how these findings relate to human activities, with a particular emphasis on the United States. Special emphasis has been placed on quantifying the contributions of the United States to the global

BOX 1.2: A Simplified Representation of the Production and Removal of Ozone in the Atmosphere - the Processes that Determine the Abundance of Ozone

Oxygen molecules (O_2) are broken apart by the harsh UV radiation in the stratosphere to produce atomic oxygen, which reacts further with oxygen molecules to make ozone (O_3). The ozone in the stratosphere is removed predominantly via catalytic chemical reactions that regenerate the catalysts. The catalysts include atoms and radicals produced in the stratosphere from the breakdown of various chemicals emitted at the Earth's surface. They include naturally occurring chemicals such as nitrogen oxides and hydrogen oxides, as well as human-emitted chemicals containing chlorine and bromine atoms, such as chlorofluorocarbons (CFCs) and bromine-containing halons that are used as fire extinguishers. These human-emitted species, referred to as ozone-

depleting substances (ODSs), are of concern for the depletion of the ozone layer. The destruction pathway marked "Pathway 1" in Box Figure 1.2a is predominant outside of the springtime polar regions, while the pathway marked "Pathway 2" is dominant in the springtime polar ozone depletion including the Antarctic ozone hole. Because of the nature of these chemical processes, as discussed above, a very small amount of the catalyst (for example, chlorine atoms from CFCs) can destroy a large amount of stratospheric ozone. In addition to these chemical processes, transport of ozone (redistribution) is key to determining the abundance of ozone in a given location.

In contrast to the stratosphere, in the troposphere ozone is made using near UV and visible radiation (*i.e.*, longer wavelength) because the higher energy, harsh UV (shorter wavelength) is screened out by the stratospheric ozone layer. This tropospheric ozone production process requires nitrogen oxides, mostly from combustion, and volatile organic compounds. Unlike stratospheric ozone, tropospheric ozone is removed not only by chemical reactions but also by other processes including contact with the surface. The transport of ozone from the stratosphere to the troposphere is important as an ozone source in certain regions.

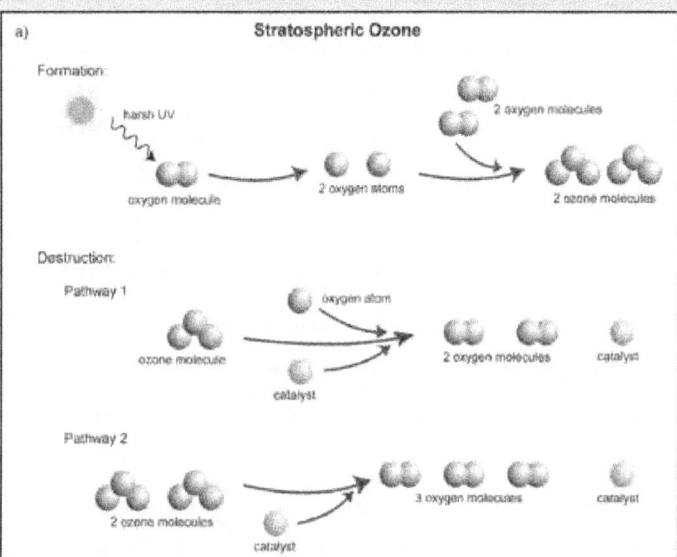

Box Figure 1.2a Highly simplified schematic representation of the chemical processes that lead to the production and removal of stratospheric ozone. The catalysts include both natural and human-emitted species, including chlorine and bromine from ozone-depleting substances. See Chapter 3 and Appendix A for further details.

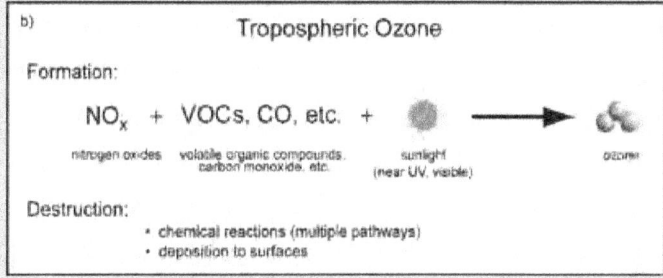

Box Figure 1.2b Schematic representations of the chemical processes that lead to the production and removal of ozone in the troposphere. The VOCs include methane. Other species such as hydrogen also can act as ingredients for ozone production in place of VOCs and CO.

amounts of ODSs. Further, given the influence that ODSs and substitute chemicals have on climate, this Product attempts to calculate the contributions to the relief of climate change via reductions in the emissions of ODSs and switching over to more climate-friendly and ozone-friendly CFC substitutes.

The primary sources of information for this report are the World Meteorological Organization (WMO) / United Nations Environment Programme (UNEP) 2006 assessment on the ozone layer *Scientific Assessment of Ozone Depletion: 2006* (WMO, 2007), and the 2005 Special Report of the Intergovernmental Panel on Climate Change (IPCC) on *Safeguarding the Ozone Layer and the Global Climate System – Issues Related to Hydrofluorocarbons and Perfluorocarbons* (IPCC/TEAP, 2005) and references therein. In addition, this report bases some findings on a few peer-reviewed publications of direct import to this issue that have become available since the finalization of the two international assessments. The report was initiated before the release of the IPCC Fourth Assessment Report (AR4) (IPCC, 2007). Therefore, this report does not rely on the IPCC AR4; however, some key pertinent issues from the IPCC report are used in a few instances where updated information was essential. They are noted as such in those chapters.

CHAPTER

2

Current Trends, Mixing Ratios, and Emissions of Ozone-Depleting Substances and Their Substitutes

Convening Lead Author: Stephen A. Montzka, NOAA

Lead Authors: John S. Daniel, NOAA; Jeff Cohen, U.S. EPA; Kenneth Vick, USDA

KEY FINDINGS

Measures of production, consumption, emission, and atmospheric abundances of ozone-depleting substances (ODSs) and their substitute chemicals provide a coherent picture of how the Montreal Protocol has brought about substantial changes in the chemical composition of the atmosphere. All measures point to a shift away from ozone-depleting substances and toward increases in substitute chemicals. This shift will continue to reduce stratospheric ozone depletion and has had notable climate benefits. These different measures, some of which are independent, are discussed separately here:

From data reported by industry for the globe and for the United States[1]:

- Owing to the Montreal Protocol, by 2005, the summed, global annual production and consumption of ozone-depleting substances for uses regulated by this Protocol had decreased 95% from peak amounts reported during the late 1980s[2]. Summed U.S. production and consumption of these substances for these regulated uses declined by 97-98% over this same period[2].

- Use of substitutes for the more potent ozone-depleting gases has increased over time, but these chemicals are much less efficient at depleting stratospheric ozone than the chemicals they replace[3].

- Declines in overall U.S. consumption of ozone-depleting substances and substitute chemicals through 2005 for uses regulated by the Montreal Protocol have been more rapid than total global declines. When ozone-depletion influences are considered[4], the fractional contribution

---

[1] Production and consumption amounts for uses of ODSs regulated by the Montreal Protocol were obtained from United Nations Environment Programme (UNEP) compilations of data reported to them (UNEP, 2007) and, for magnitudes of unregulated and unrestricted uses, from UNEP Technical Option committee reports (UNEP/MBTOC, 2007; UNEP/CTOC, 2007); global production data for HFCs was taken from IPCC-TEAP (2005).

[2] Consumption is defined here and in the Montreal Protocol as amounts produced plus imports minus exports of a substance or group of substances. Production is defined as amounts produced minus the sum of amounts destroyed or used in feedstock (non-dispersive) applications. Consumption should equal production on a global scale averaged over time. In this Key Finding, magnitudes of production and consumption have been multiplied by weighting factors that are Ozone Depletion Potentials (see footnote #4).

[3] The more potent and abundant ozone-depleting gases referred to here include chlorofluorocarbons, or CFCs, halons, methyl chloroform, and carbon tetrachloride. Chemicals considered to be substitutes include the hydrochlorofluorocarbons, or HCFCs, and the hydrofluorocarbons, or HFCs.

[4] Weighting factors are applied to consumption, production, emission and banks of ODSs throughout this document that approximate the ozone depletion influences and the direct or indirect climate effects of these chemicals so as to allow consideration of them on an equivalent basis and as sums. These weighting factors account for the wide range of influences different chemicals have on ozone and climate. In the case of ozone, the weighting factors are

of the United States to annual global consumption of ozone-depleting substances (ODSs) in data reported to UNEP[5] for all regulated, dispersive uses decreased by more than half, from a mean of 24 (±2)% in 1986-1994 to 10 (±2)%, on average, during 2001-2005. This decline is noted despite an increase in U.S. consumption of methyl bromide (CH_3Br) relative to global consumption in recent years. When direct and indirect climate effects of these chemicals are considered[4], the contribution of the United States to total global consumption of ODSs for regulated, dispersive uses also decreased from the late 1980s to 2005, though the precise magnitude of this decline is sensitive to our understanding of the indirect climate forcing from ODSs related to stratospheric ozone changes.

- Declines in U.S. consumption for uses regulated by the Montreal Protocol have been slightly faster than phase-out schedules for all developed countries in the adjusted and amended Montreal Protocol for most ODSs. Consumption for methyl bromide was notably larger than this scheduled allotment in 2005 and 2006 (by 4.3 and 4.1 ODP-kilotons [Kt], respectively) because of Critical Use Exemptions (CUEs)[6].

- Global consumption of methyl bromide for all fumigation-related uses declined by a factor of two from 1997 to 2005 despite substantial consumption in applications not regulated or restricted by the Montreal Protocol. Nearly half (43%) of the global, industrially-derived emissions of CH_3Br during 2005 arose from QPS[6] consumption not restricted or limited by the Montreal Protocol.

- U.S. consumption of CH_3Br for all fumigation uses declined 40% from 1997 to 2005 despite enhanced Critical Use Exemptions[6] and QPS[6] consumption since 2001. Enhanced Critical Use Exemptions caused the annual U.S. contribution to global CH_3Br consumption for regulated and restricted uses in data reported to UNEP to increase from 23 (±4)% during 2000-2003 to 36 (±1)% during 2004-2005. In the United States during 2001-2006, the additional consumption of methyl bromide for fumigation not restricted or limited by the Protocol (QPS use) was, on average, 57 (±20)% of the amounts used and reported to UNEP[5] for regulated applications and had increased by 13% per year, on average, during 2001-2005.

- The mix of ozone-depleting chemicals produced throughout the globe has changed over time in response to the Montreal Protocol. In 2005, global production weighted by Ozone Depletion Potentials (ODPs) for relevance to ozone depletion was dominated by chlorofluorocarbons, or CFCs, (50%), hydrochlorofluorocarbons or HCFCs (33%), and CH_3Br (11%); in the United States, ODP-weighted consumption was dominated by HCFCs (54%) and CH_3Br (34%). When weighted by overall climate influences [net Global Warming Potentials (GWPs)], global production in 2005 was accounted for primarily by HCFCs and hydrofluorocarbons (HFCs) in similar amounts, a somewhat lesser contribution from CFCs, and very small or negative contributions from halons, CH_3Br, and other chemicals. In the United States, direct and net GWP-weighted consumption was dominated by HFCs and HCFCs with only small contributions from CFCs and other ODSs. Current estimates of global HFC production have large uncertainties owing to restrictions on reporting production magnitudes when less than three manufacturers produce a given chemical.

- Future emission rates from banks[7] will play a substantial role in determining future mixing ratios

---

Ozone Depletion Potentials (ODPs) with units of ODP-Tons or ODP-Kt; 1 ODP-Kt=1 billion grams multiplied by the ODP of a given chemical. In the case of climate, the weighting factors are 100-year direct or net Global Warming Potentials (GWPs), where net GWPs include the radiative influence of stratospheric ozone depletion. Units for quantities weighted by 100-year GWPs are expressed equivalently by, for example, GWP-Tons or gigatons (Gt) CO_2-equivalents. Additional descriptions of these weighting factors appear in Box 2.2 and the main chapter text, and tables of the weighting factors used here appear in Appendix 2A.

[5] The United Nations Environment Programme (UNEP) compiles and publishes global and national statistics on production and consumption of ODSs based upon data reported to them in order to monitor compliance with the adjusted and amended Montreal Protocol (UNEP, 2007).

[6] QPS refers to quarantine and pre-shipment use of an ODS, specifically CH_3Br. Though reporting requirements exist for this dispersive use, it is not restricted or scheduled for phase-out in the Montreal Protocol and these use magnitudes are not included in data published by UNEP to assess compliance with the Protocol. CUEs refer to Critical Use Exemptions for consumption of an ODS above existing Montreal Protocol allotments; they are approved only on a case-by-case basis and are included in amounts reported to and published by UNEP.

[7] Banks represent the amount of a chemical that has been produced but not yet emitted or chemically altered. They exist either in reserve storage or in current applications. Bank magnitudes are derived from the U.S. EPA's Vintaging Model analysis of sales and use of ODSs and substitutes in the U.S. (U.S. EPA, 2007). Owing to a lack of available data at this time, U.S. bank estimates presented here do not include stockpiles of halons.

for some ODSs. Banks in the U.S. and throughout the globe in 2005 are estimated to have been 7 to 16 times larger than emissions during this year, when weighted by their potential influence on climate or ozone depletion. CFCs accounted for the largest fraction of 2005 banks in the United States and throughout the globe. The U.S. EPA has classified approximately one-quarter of U.S. banks in 2005 as being accessible (210 ODP-Kt[8], roughly two-thirds, *i.e.*, approximately 140 ODP-Kt, accounted for by halons; and 1.9 (0.9-2.2) gigatons (Gt) carbon dioxide (CO_2)-equivalents, of which HCFCs account for approximately 1.0 (0.9-1.1) Gt CO_2-equivalents)[8]. Additional halon is likely present in stockpiles, but these amounts are not included in these estimates of U.S. banks owing to a lack of available data at this time.

- Emission histories derived from global ODS production and consumption data and assumed release functions have large uncertainties but suggest strong declines in global emissions of most ODSs other than HCFCs.

From national data quantifying applications that use ozone-depleting substances and substitutes the U.S. EPA has derived U.S. emission histories starting in 1985. Though these emissions estimates are recognized to have substantial uncertainties, they suggest that:

- Total emissions of ODSs and substitutes from the United States have declined substantially since the late 1980s. By 2005, U.S. emissions are estimated by the U.S. EPA to have declined by 81%, when emissions are weighted with factors relevant to ozone depletion. When weighted with factors relevant to climate, annual U.S. emissions of ODSs and substitutes including HFCs declined 74% (between 63 and 76% when indirect climate influences associated with ozone depletion are also included) over this same period.

- The United States accounted for a substantial fraction of global atmospheric mixing ratios of individual ODSs and HFCs measured in 2005, though precise quantification of these contributions is difficult owing to incomplete emission histories for most ODSs. The results suggest that U.S. emissions accounted for between 10 and 50% of the global atmospheric abundances of most ODSs and substitute chemicals measured in 2005, 17-42% of the tropospheric chlorine, 17-35% of the tropospheric bromine, and 15-36% of the tropospheric Equivalent Effective Chlorine (EECl)[9] arising from these chemicals in that year.

- Changes in atmospheric chlorine and bromine inferred from U.S. emissions estimates of chemicals regulated by the Montreal Protocol have less uncertainty than absolute amounts. The data suggest that atmospheric chlorine from U.S. emissions has declined steadily since 1995, but atmospheric bromine from U.S. emissions in 2005 was similar to 1998 levels primarily as a result of recent increases in exempted critical uses[6] and for QPS[6] uses of CH_3Br.

- Atmospheric changes derived from U.S. emissions of chlorinated and brominated ODSs indicate a decline in total reactive halogen (EECl)[9] arising from U.S. emissions through 2005, but a substantially slower rate of decline since 2003. The slower overall decline in 2004-2005 was because of the increases in U.S. emissions of brominated gases during these years (primarily CH_3Br).

- The direct climate influence (as direct radiative forcing)[10] arising from the atmospheric abundances of ozone-depleting substances and substitute chemicals attributable to U.S.

---

[8] Accessible banks are amounts of ODSs in use in fire extinguishers, refrigeration, and air conditioning sectors (not foams). While the accessible bank magnitude given was derived with direct GWP weighting, the ranges of bank magnitudes given in parentheses were derived with net GWPs, *i.e.*, where consideration of the indirect, ozone depletion influences of ODSs are included (see Box 2.2 for more details).

[9] Equivalent Effective Chlorine, or EECl, is an index to approximately quantify the overall effect of ODSs on stratospheric ozone. It is calculated from surface measurements of ozone-depleting substances and accounts for the ODSs having different numbers of chlorine and bromine atoms, for the enhanced efficiency by which bromine atoms destroy ozone relative to chlorine, and the different rates at which ozone-depleting substances decompose in the stratosphere and liberate chemical forms of chlorine and bromine that can participate directly in stratospheric ozone-depleting reactions. Equivalent Effective Stratospheric Chlorine (EESC) is a related index, except that time lags associated with transporting air from the troposphere to the stratosphere are considered. These indices are described additionally in the text and in Box 2.7.

[10] Direct radiative forcing is an estimate of the direct climate influence of a chemical and is expressed as energy per area (Watts per m[2]). It is calculated with knowledge of how a chemical absorbs infrared light in certain wavelength regions (its radiative efficiency) and is directly proportional to its atmospheric abundance for the less abundant greenhouse gases. Direct forcings do not include indirect radiative effects associated with feedbacks, such as those related to ozone depletion. Net forcings discussed in this chapter include the indirect forcings related to stratospheric ozone depletion (see Box 2.2 for further details).

emissions is estimated as having been between 0.067 and 0.16 W per m^2 in 2005. This U.S. contribution amounted to between 19 and 49% of the total global direct climate influence of these chemicals of 0.34 W per m^2. When indirect climate influences of ODSs related to stratospheric ozone depletion are considered, the U.S.-attributed forcing is between 0.04 and 0.18 W per m^2.

Direct observations of the atmosphere provide an independent assessment of the Montreal Protocol's success in reducing atmospheric abundances of ODSs and ozone-depleting chlorine and bromine. These observations show that:

- The global atmospheric abundances of all ODSs are responding to changes in global production and consumption magnitudes. Atmospheric mixing ratios of the most abundant CFCs, the most abundant chlorinated solvents, and CH_3Br are now decreasing. Increases are still observed for halon-1301, HCFCs, and HFCs. Methyl bromide mixing ratios have declined each year since global production was first reduced (1999), despite increases in Critical Use Exemptions recently, continued use in QPS[6] applications, and substantial natural sources over which humans do not exert direct control.

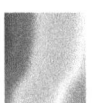

- Global emissions magnitudes derived from global atmospheric data exhibit substantial declines since the 1980s, and provide independent confirmation of the large changes in global production and consumption as shown by UNEP[5] in data reported to them. By 2005, global emissions had declined 77-82% compared to peak years, considering either the climate or ozone-depletion influences of ODSs and substitute chemicals.

- Tropospheric chlorine contained in all regulated ODSs and substitute chemicals has decreased since the early 1990s. Furthermore, measures of stratospheric chlorine show changes consistent with those observed in the troposphere. Stratospheric measurements also confirm that approximately 80% of stratospheric chlorine, which catalyzes ozone destruction, is from ODSs regulated by the Montreal Protocol. The remaining 20% is accounted for primarily by methyl chloride (CH_3Cl), though a small contribution (~2%) is from very short-lived chemicals.

- Tropospheric bromine from ODSs regulated by the Montreal Protocol has declined slowly since 1998. This decline has been dominated by tropospheric changes observed for CH_3Br. Measures of stratospheric bromine show changes consistent with those observed in the troposphere, though a decline in stratospheric bromine is not yet discernable. These stratospheric measurements indicate that approximately 50% of stratospheric bromine is from industrially produced halons and CH_3Br. The remainder is from naturally produced CH_3Br and from very short-lived chemicals produced primarily naturally.

- Observed changes in global atmospheric levels of ODSs containing chlorine and bromine demonstrate a substantial decline in the ozone-depleting halogen content of the atmosphere. The decrease since 1994 in the tropospheric halogen burden (EECl)[9] accounted for by the long-lived ODSs considered here has been 20% of what would be needed to return EECl values to those in 1980 (*i.e.*, before substantial ozone depletion was observed). The decline in the shorter-lived gases methyl chloroform (CH_3CCl_3) and CH_3Br have contributed most to the observed decline. Decreases in stratospheric, ozone-depleting halogen (as Equivalent Effective Stratospheric Chlorine [EESC]) have been smaller because of the time delay associated with mixing tropospheric air into the stratosphere.

- The combined direct radiative forcing from ODSs and substitutes including HFCs is still increasing, but at a slower rate than in the 1980s. This trend arises primarily from slow declines in atmospheric abundances of CFCs and continued increases in abundances of HCFCs and HFCs. The total direct contribution of ODSs and substitutes was 0.34 W per m^2 in 2005 (it is 0.18-0.38 W per m^2 if the indirect ozone depletion forcing is included), compared to a contribution from CO_2, methane (CH_4), and nitrous oxide (N_2O) of 2.3 W per m^2.

INTRODUCTION

In an effort to heal the stratospheric ozone layer, schedules for the global phase-out of manmade ozone-depleting substances (ODSs) were set by the 1987 Montreal Protocol on Substances that Deplete the Ozone Layer and its Amendments and Adjustments. This chapter reviews the changes that have resulted from this international Protocol by assessing reported levels of ODS production and consumption, by deriving emissions with techniques independent of production and consumption estimates, by reporting on how these changes have influenced the atmospheric abundance of ODSs and chemicals used as substitutes, and by assessing

BOX 2.1: Key Issues

To facilitate a rapid phase-out of ODSs, the Montreal Protocol allowed the use of hydrochlorofluorocarbons (HCFCs) as interim substitutes for chlorofluorocarbons (CFCs). Temporary use of HCFCs was allowed because, even though HCFCs contain chlorine and are ODSs, they are much less efficient at causing stratospheric ozone depletion than the ODSs they replaced, and, therefore, have been considered as in-kind replacements to transition to a non-CFC world. Elimination of ODSs (including HCFCs) in nearly all applications is anticipated as the phase-out schedules run their course. Most uses of ODSs have been replaced with the non-ozone-depleting, non-chlorine-, and non-bromine-containing hydrofluorocarbons (HFCs) and other so-called "not-in-kind" alternatives (e.g., non-solvent-based cleaning processes, and hydrocarbon-based refrigerants). These changes have had a measurable influence on the global atmospheric abundance of these gases, with the result that the overall abundance of chlorine and bromine reaching the stratosphere has declined in recent years.

Therefore, the key issues, in the form of questions, that are related to ozone-depleting substances in the atmosphere and that are covered in this chapter, include:
- What is our best information on global production, consumption, and emissions of ozone-depleting substances, primarily CFCs and HCFCs, and HFCs, that are chlorine- and bromine-free, non-ozone-depleting, and longer-term replacements for CFCs and HCFCs? What are the associated uncertainties in these quantities?
- How can the combined influence of chlorinated and brominated ODSs in the stratosphere be quantified, and how is it likely to change in the future?
- What fraction of the produced ODSs is still sequestered and could be potentially released at a later date? (i.e., what are the extents of the so-called "banks"?)
- What do the observations of ODS atmospheric abundances show about the levels of total atmospheric chlorine, bromine, and equivalent chlorine from these long-lived gases? In other words, are the atmospheric abundances actually responding as anticipated to restrictions set forth in the Montreal Protocol?

ODSs and halogenated chemicals used as substitutes have a second important property; they are efficient greenhouse gases (GHGs). As a result, they increase atmospheric heating and can influence climate. By requiring substantial reductions in global emissions of ODSs, the Montreal Protocol has led to societal benefits related to both stratospheric ozone depletion and climate change. The magnitude of this additional climate benefit has been diminished slightly, however, by small offsetting influences such as increased HFC emissions, and possibly by resulting stratospheric ozone increases, which may have a small warming influence. Therefore, it is important to know:
- What are the contributions of the various ODSs, and their substitutes, to climate forcing, in the past, now, and in the future?

Stratospheric ozone depletion is a global environmental issue. Yet, ODS emissions arise from various countries and regions. Also, the impact of ozone depletion is felt to different extents by different regions. Therefore, it is necessary to ask:
- What are the contributions of the United States to production and emissions of ODSs and substitute chemicals in the past?

This chapter attempts to address many of these issues to the extent possible for those that fall within the purview of this document.

how these atmospheric abundance changes have altered the influence of ODSs on stratospheric ozone depletion as well as their influence on climate. Furthermore, because this is a national assessment, this chapter provides estimates of these quantities for the United States and how they have changed over time.

This chapter is organized into six sections. In the first (Section 2.1), changes in reported production and consumption magnitudes of ODSs and substitute chemicals are discussed. These quantities provide important evidence elucidating how the Montreal Protocol has influenced human activities. The Protocol was written to control production and consumption of ODSs. Accordingly, countries report these quantities annually to the United Nations Environment Programme (UNEP) so that compliance with the Protocol can be assessed. The data are derived fundamentally from industry's records of production and international trade and provide the foundation for understanding how emissions of ODSs and substitute chemicals could change as a result of the Montreal Protocol. Limitations of the UNEP data are considered here through comparisons to AFEAS compilations (the Alternative Fluorocarbons Environmental Acceptability Study) (Section 2.1.2) and by considering the magnitudes of production and consumption for uses not regulated or restricted by the Protocol and, therefore, not included in the UNEP compilations (Section 2.1.3). Because the data compiled by UNEP are published on a country-by-country basis, a parallel analysis of U.S. consumption and production of ODSs and substitute chemicals is presented (Sections 2.1.4 and 2.1.5).

In the second section (Section 2.2), emissions magnitudes and changes are assessed because they provide a direct understanding of how policy decisions are altering human influences on the atmosphere. Global emissions are inferred from measured changes in the chemical composition of the remote atmosphere (the "top-down" method of estimation). Emissions derived in this way provide an important independent check on global production and consumption data reported to UNEP. Top-down estimates are also compared to "bottom up" global emission magnitudes estimates, which are derived from

sales data for different applications and time-dependent ODS leak rates from these different applications (AFEAS, 2007; UNEP/TEAP, 2006). As was the case for production and consumption, compound-dependent weighting factors related to stratospheric ozone depletion (Section 2.2.1) and climate (Section 2.2.2) are applied to emissions estimates to add relevance (see Box 2.2). Banks, *i.e.*, amounts of halocarbons that were produced but that have not been emitted to the atmosphere, account for a large fraction of present-day emissions for some halocarbons and are explored in Section 2.2.3. Banks are a particularly important topic because releases from banks account for much of the current emission of some ODSs, yet these releases are not restricted or addressed in the Protocol (Box 2.5). While U.S. regulations address recycling and venting of refrigerant, this represents a relatively small fraction of U.S. ODS banks and most, if not all, of this material will ultimately be vented to the atmosphere unless collected and destroyed. The contribution of emissions from other, non-restricted influences is discussed subsequently (Section 2.2.5).

Annual U.S. emissions of ODSs and substitute chemicals are estimated by U.S. EPA (2007) using a model analysis of sales and use within the United States (Section 2.2.5). U.S. emissions estimates are different from "top-down" global emissions estimates because they rely on the accuracy of industry-related production and sales data or assessments of market demand for ODSs and substitute chemicals. Comprehensive, independent assessments of U.S. emissions from atmospheric observations are not currently possible, though some useful conclusions are drawn from studies conducted to date (Section 2.2.6).

Atmospheric abundances of ODSs and substitute chemicals are discussed in Section 2.3. While emissions estimates provide a useful metric of how changes in human behavior are affecting the atmosphere, the influence of ODSs and substitute chemicals on stratospheric ozone and climate are dependent upon their atmospheric abundance, not rates of emission. The sensitivity of the atmosphere to emission magnitudes is determined by a chemical's persistence, which is quantified as an atmospheric lifetime.

In an effort to heal the stratospheric ozone layer, schedules for the global phase-out of manmade ozone-depleting substances (ODSs) were set by the 1987 Montreal Protocol on Substances that Deplete the Ozone Layer and its Amendments and Adjustments.

Consistency between observed abundances of ODSs and substitute chemicals and calculated or expected abundances requires accurate estimates of both emissions and lifetimes (Section 2.3.1.1)

Halocarbon abundances in the remote atmosphere attributable to U.S. emissions are also derived for past years (Section 2.3.2). The U.S. contributions to global abundances are derived from histories of emissions since 1985 from the U.S. EPA (2007), and, for earlier years, a range of contributions of United States to global halocarbon emissions (Box 2.6).

Subsequently, the overall influences that the wide ranges of changes observed for individual gases are having or will have on ozone depletion (Section 2.4) and on climate forcing (Section 2.5) are discussed. Quantities such as total chlorine, total bromine, and Equivalent Effective Chlorine (EECl and EESC, see Box 2.7) are calculated to assess the changing influences on stratospheric ozone (Section 2.4). Radiative efficiencies are applied to observed atmospheric changes to assess the direct influence these forcings have on climate (Section 2.5). Indirect climate influences related to stratospheric ozone depletion arising from the use of ODSs are also considered.

Finally, though they are included throughout the document, findings related specifically to the United States are reviewed in Section 2.6. Results related to, for example, atmospheric abundances of ODSs calculated from consideration of U.S. emissions are summarized, as are the relative contributions of U.S. emissions to the measured global atmospheric abundances of ODSs and substitute chemicals (Box 2.6). Additional topics with enhanced relevance to U.S. policy are highlighted throughout the text in additional boxes. These include a discussion of methyl bromide (CH_3Br, Box 2.3) and HCFCs (Box 2.4).

Throughout this chapter different weighting factors are applied to quantities such as production, consumption and emission of ODSs. These weighting factors are useful for considering overall changes because different chemicals influence ozone and climate to different extents (see Box 2.2 for more

detail). With regard to stratospheric ozone, the weighting factors are Ozone Depletion Potentials (ODPs). With regard to climate, the weighting factors are 100-year Global Warming Potentials (GWPs). Two main influences are considered in the GWP calculations, the direct effect of a halocarbon on the radiative balance of the atmosphere, and the indirect influence arising from stratospheric ozone changes caused by a halocarbon. Here, the direct influence is accounted for by 100-year direct GWPs and the indirect influence is included by combining direct and indirect GWPs into net GWPs (see Box 2.2 for more detail). Different weighting factors are applied to atmospheric abundances of individual halocarbons to assess their influence on ozone or on the direct radiative forcing of the atmosphere. See Sections 2.4 and 2.5 for further discussion of these factors.

2.1 PRODUCTION AND CONSUMPTION OF OZONE-DEPLETING CHEMICALS AND THEIR SUBSTITUTES DERIVED FROM INDUSTRY ESTIMATES

2.1.1 Production and Consumption: Global Trends

Historical global data on production and consumption or sales of ozone-depleting chemicals are available through databases compiled from different countries by UNEP and from different companies by AFEAS (2007). The data provided by these organizations show how dramatically dispersive, regulated uses of ozone-depleting chemicals have changed over the past 20 years in response to the adjusted and amended 1987 Montreal Protocol on Substances that Deplete the Ozone Layer and to changing market conditions. Data are compiled on an annual basis by UNEP to assess compliance with the Montreal Protocol. The UNEP data provide more complete global coverage in recent years than AFEAS compilations but are not disaggregated by chemical in all instances; only production or consumption data aggregated by compound class are publicly available for CFCs, HCFCs, and halons. Other limitations include the UNEP data only being available for years since 1986, and not all countries have reported production or consumption figures to UNEP for all years. Despite these limitations, the UNEP compilation provides

Different chemicals influence ozone and climate to different extents. For this reason, weighting factors called Ozone Depletion Potentials (ODPs) and Global Warming Potentials (GWPs) are useful for considering overall changes.

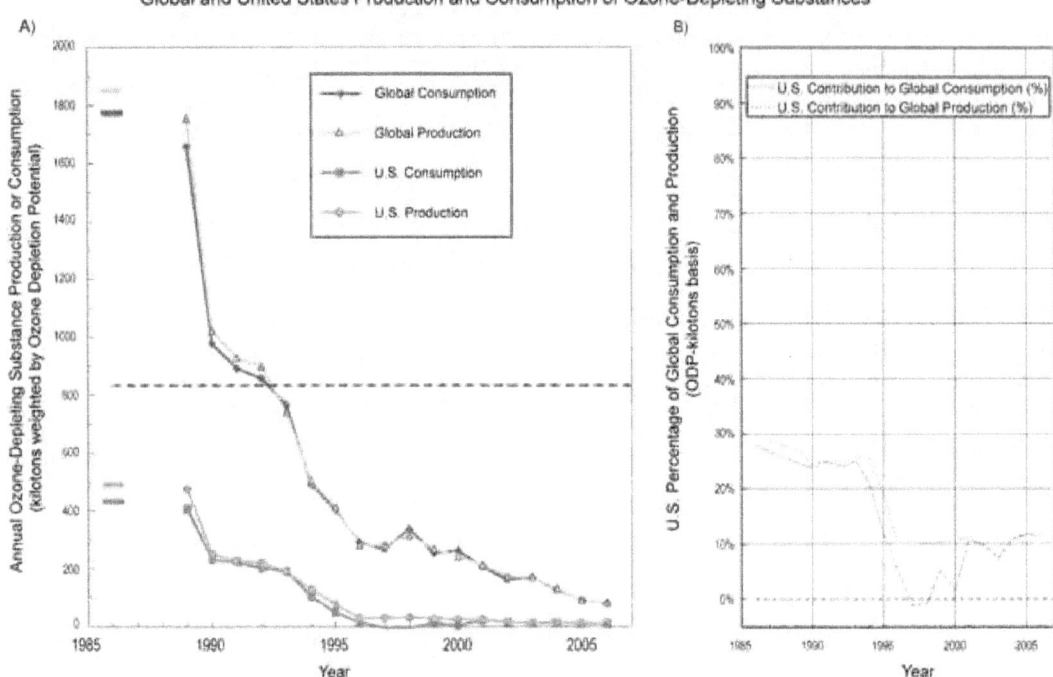

Figure 2.1 Panel A: Annual global production and consumption of all regulated ODSs and substitutes (dark and light blue solid lines) compared to similar quantities for the United States (dark and light red solid lines), as derived from data reported to UNEP (UNEP, 2007). Baseline production and consumption quantities are shown as separate bars with corresponding colors in 1986. All of these data are weighted by compound-dependent ODPs. Panel B: Percentage contributions of U.S. consumption and production to global totals. Negative consumption indicates exports being larger than the sum of imports plus production in a given year (see Section 2.1.4.1).

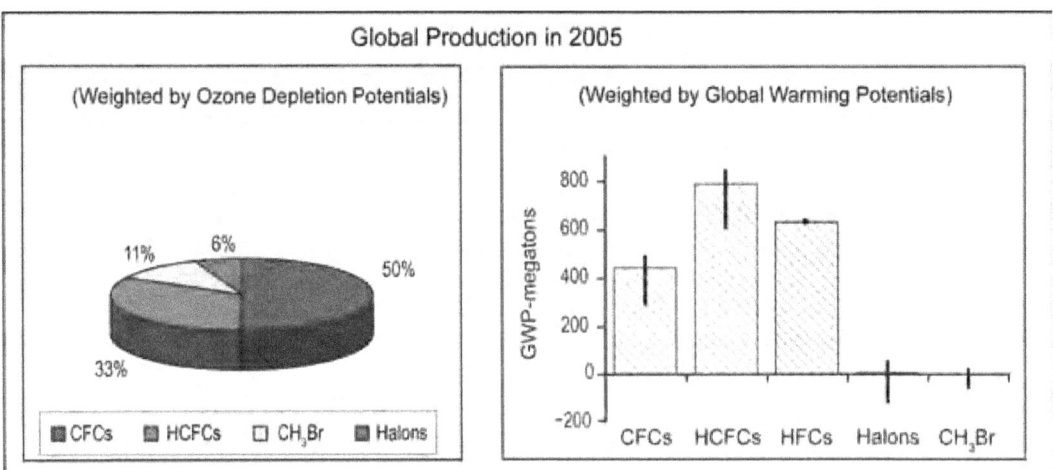

Figure 2.2 The contribution of different compound classes or compounds to total global, ODP-weighted production of ODS and substitute chemicals reported to UNEP for regulated uses for 2005 (left panel), and the global, GWP-weighted production of these chemicals in the same year (right panel) (UNEP, 2007; Personal communication, the UN Ozone Secretariat, 2007; AFEAS, 2007). The hatched bar heights in the right-hand panel were derived with direct GWPs; the given uncertainties represent weighting by net GWPs (see Box 2.2). HFC production includes only the portion of global HFC data reported by AFEAS (2007) for HFC-134a, HFC-125, and HFC-143a. For HFC-23, production was inferred from atmospheric data (Clerbaux and Cunnold et al., 2007). Relative contributions of less than 1% are not included in these charts; note that the global CCl_4 ODP-weighted production was −7% during 2005, though its contribution was not included in this Figure. See Section 2.1.4.1 for additional discussion regarding negative consumption and production values.

Table 2.1 Declines in reported production or consumption and derived emission of ODSs and substitute chemicals (including HFCs) relative to magnitudes in the late 1980s.

Region	Production or Consumption Decline, 1989-2005 (%)[a]	Emission Decline Through 2005 (%)[e]
Weighted by Ozone Depletion Potentials		
Globe	95[b]	82
United States	97-98[c]	81
Weighted by 100-year Global Warming Potentials[d]		
Globe	81 (61-83)[b]	77 (77-84)
United States	87 (81-88)[c]	74 (71-75)

[a] Considers production and consumption of ODSs for only dispersive uses restricted by the Montreal Protocol as shown by UNEP in data reported to them, plus HFC production and consumption or sales data without consideration of use.
[b] Derived from the UNEP (2007) compilation of reported ODS production and consumption; AFEAS (2007) production data for HFC-134a, HFC-125, and HFC-143a; and HFC-23 production inferred from atmospheric data (Clerbaux and Cunnold et al., 2007).
[c] Derived from consumption of ODSs for restricted uses reported to UNEP (UNEP, 2007) but delineated by compound by the U.S. EPA; and the U.S. EPA (2007) vintaging model estimates for HFCs.
[d] Declines indicated are calculated with direct GWP weighting but the ranges given in parentheses indicate the decline calculated when the indirect influence (and uncertainty) related to stratospheric ozone depletion is included in net GWP weighting factors (see Box 2.2).
[e] Derived on a global scale from atmospheric data of ODSs and substitute chemicals and so includes all uses, regulated and not; derived on a U.S. scale from the U.S. EPA (2007) vintaging model estimates of emissions of ODSs and substitute chemicals. HFC global emissions in 2005 were interpolated from 2002 global estimates and the 2015 "business as usual" scenario in IPCC/TEAP (2005).

critical data for assessing global and national changes in production and consumption for regulated uses of ODSs including CFCs, halons, carbon tetrachloride (CCl_4), methyl chloroform (CH_3CCl_3), HCFCs, and CH_3Br, particularly in recent years (UNEP, 2007). The data through 2005 indicate that annual global production and consumption of ODSs and substitutes for ODSs has declined by $1.6\text{-}1.7 \times 10^6$ ODP-Tons since the Montreal Protocol was ratified (Figure 2.1). This corresponds to a 95% decline in both the ODP-weighted production and consumption of these chemicals across the globe by 2005 (Table 2.1) (see Box 2.2). The average total global production and consumption in 2004-2005 was approximately 1.1×10^5 ODP-Tons per year.

In the data reported to UNEP aggregated by compound class, all classes showed declines in total global production and consumption during 2000-2005, though the relative decline was smallest for HCFCs (12 to 16%) and data for CCl_4 are quite variable year-to-year. Production and consumption of CFCs still dominates the ODP-weighted global totals. During 2005 ODP-weighted annual production (consumption) of CFCs accounted for 50% (48%), HCFCs 33% (34%), CH_3Br 11% (14%), and halons 6% (5%)

(CH_3CCl_3 and CCl_4 accounted for less than 1%) (Figure 2.2). Despite small declines in total production of HCFCs since 2000, the relative contribution of HCFCs increased substantially over this period so that by 2005 they accounted for 33% of total ODP-weighted production. Preliminary data suggest that global, ODP-weighted consumption of HCFCs equaled ODP-weighted consumption of CFCs in 2006.

Global production of ODSs and substitutes (unweighted data from UNEP, 2007; IPCC/TEAP, 2005) can be weighted by direct or net GWPs to estimate the potential influence that production could have on climate forcing (see Box 2.2) (Figure 2.3). When weighting by 100-year, direct GWPs is considered, the annual production of ODSs and substitutes declined by 8040 GWP-Mt from 1989 to 2005, which corresponds to a decline of 81% (Table 2.1) (Figure 2.3). A slightly smaller decline of 61 to 83% is calculated for global production through 2005 when net GWPs are used as weighting factors to account for the indirect influence of stratospheric ozone changes arising from the changing mix of ODSs (see Box 2.2). Increases in global HFC production have slowed the overall decline somewhat; production of HFCs

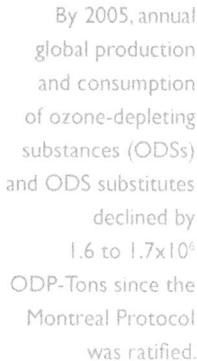

By 2005, annual global production and consumption of ozone-depleting substances (ODSs) and ODS substitutes declined by 1.6 to 1.7×10^6 ODP-Tons since the Montreal Protocol was ratified.

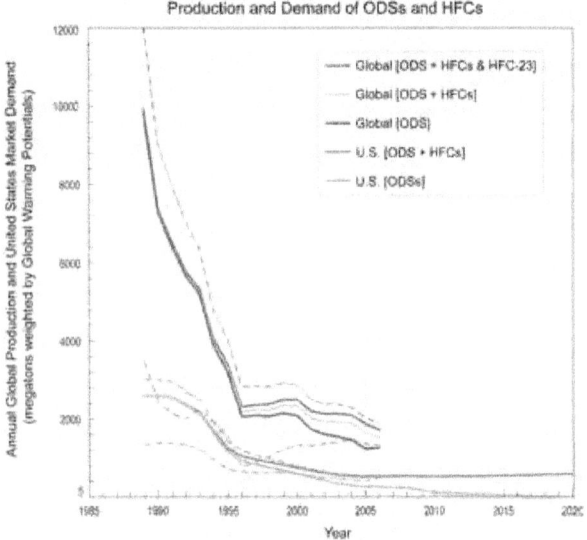

Figure 2.3 Global production of ODSs and HFCs compared to U.S. consumption estimates. Solid lines represent weighting by 100-year, direct GWPs. Dashed lines represent a range of total production or demand calculated with weighting by a range of net GWPs that include the indirect influence of stratospheric ozone on climate. A range of results, rather than a central value, is presented for net GWP weighting because of our incomplete understanding of how stratospheric ozone depletion has influenced climate (see Box 2.2). Results for subsets of different compound classes are also shown. Global ODS production derived from UNEP (2007) compilations (blue line) are compared to U.S. consumption data for ODSs (light red line). The additional influence of HFCs is shown on global (light blue and green lines) and U.S. scales (red line). The light blue line is derived by adding AFEAS global production data of HFC-134a, HFC-125, and HFC-143a (AFEAS, 2007) to the contribution of all ODSs. The green line includes additional inadvertent HFC-23 production derived from measured atmospheric trends (IPCC/TEAP, 2005; Clerbaux and Cunnold et al., 2007). The contribution of HFCs to U.S. GWP-weighted production has been estimated from the U.S. EPA vintaging model (U.S. EPA, 2007). The contributions from other HFCs listed in Table 2.2 are not included here due to a lack of production information on global scales.

By 2004, annual global sales or consumption of CFCs and HCFCs declined by approximately 93% (1×10^6 ODP-Tons) since the late 1980s and since the Montreal Protocol went into force.

in 2005 is estimated here at approximately 630 GWP-Mt (production of HFC-134a, HFC-125, and HFC-143a from AFEAS [2007], and inadvertent production magnitudes of HFC-23 derived from atmospheric measurements [IPCC/TEAP, 2005]) (Figure 2.3).

Based upon these production figures and direct GWPs, CFCs accounted for 24%, HCFCs 42%, and HFCs 34% of the global, CO_2-equivalent production of all ODSs and their substitutes in 2005. The indirect influence associated with stratospheric ozone depletion alters these figures somewhat, though HCFCs and HFCs still account for the largest fraction of both direct and net GWP-weighted global production in 2005 (Figure 2.2). The contribution of HFCs

considered here is an underestimate because global production data on HFCs other than -134a, -125, -143a, and -23 are not currently available, though these four gases alone accounted for 95% of the global total, GWP-weighted demand for HFCs in 2002 (Campbell and Shende et al., 2005).

2.1.2 Production and Consumption: Comparing UNEP and AFEAS Compilations

AFEAS has compiled production and sales data for individual compounds for many years (AFEAS, 2007). Though the data compiled by AFEAS and UNEP are not independent, they do allow for some cross checking and an assessment of consistency in the global totals reported for CFCs and HCFCs (Figure 2.4). The AFEAS compilation only includes data for some ODSs (CFCs, HCFCs, and HFCs) and only for a subset of companies around the globe that are producing ODSs and their substitutes. While this compilation accounted for most of global production and sales of CFCs and HCFCs in the 1980s and early 1990s, it has accounted for a smaller fraction since.

On an ODP-weighted basis, both the AFEAS and UNEP compilations show that by 2004 annual global sales or consumption of CFCs and HCFCs (weighted by chemical-specific ODPs) declined by approximately 1×10^6 ODP-Tons, or by 93%, since the late 1980s and since the ratification of the Montreal Protocol on Substances that Deplete the Ozone Layer by many countries (Figure 2.4). The totals from these two compilations during 1986-2004 are slightly different; annual AFEAS sales figures are 0.1 ($\pm$0.03)$\times 10^6$ ODP-Tons lower than consumption reported to UNEP, on average. This difference may represent errors in accounting or reporting of data, but is most likely the result of consumption outside the companies reporting to AFEAS, such as by countries operating under Article 5 of the Montreal Protocol (so-called "developing" countries). Since 1995, the annual UNEP – AFEAS difference has been 80 $\pm$ 10%, on average, of the consumption reported by these Article 5 countries.

BOX 2.2: Weighting Factors

Accurately assessing the overall effect of changes in production, consumption, and emission of individual gases requires consideration of weighting factors that account for compound-dependent influences on ozone and climate (Clerbaux and Cunnold et al., 2007; Daniel and Velders et al., 2007). When considering the influence of ODS production, consumption, and emission on ozone depletion, Ozone Depletion Potentials (ODPs) are used. Units of these quantities are ODP-tons. When considering the direct or net influence of ODS on climate, direct or net 100-year Global Warming Potentials (GWPs) are used. Units on these quantities are GWP-tons or CO_2-equivalent tons. Ozone Depletion Potentials, direct and indirect Global Warming Potentials, and other compound-specific parameters used in this report are tabulated in Appendix 2A.

ODPs

Ozone Depletion Potentials represent the amount of global ozone destroyed by a particular ODS per unit mass compared to the amount destroyed by a reference gas (usually CFC-11) per unit mass. Ozone Depletion Potentials provide a simple way to compare ODSs with respect to their ability to deplete stratospheric ozone and have proved useful to scientists and policymakers since their initial development (Wuebbles, 1983). Ozone Depletion Potentials take into account the number of chlorine and bromine halogen atoms in a chemical, how rapidly these halogen atoms become released in the stratosphere, how reactive the halogen atoms are for ozone destruction (Cl vs. Br, for example), and how persistent the chemical is throughout the entire atmosphere (its lifetime). Steady-state ODPs are most commonly used and are applicable for longer time periods since they represent the steady-state ozone responses to ODS perturbations. Time-dependent ODPs also have been proposed (Solomon and Albritton, 1992) for use when a particular time horizon or the time-dependence of relative ozone destruction is of interest. Chapter 2 will use steady-state semi-empirical ODPs in this chapter (Solomon et al., 1992), taken from Chapter 8 of the 2006 World Meteorological Organization (WMO) ozone assessment (Daniel and Velders et al., 2007).

GWPs

Global Warming Potentials are analogous indices for comparing the integrated radiative impact of greenhouse gases (IPCC, 2007). They represent the cumulative radiative forcing of a unit mass of a gas relative to the same quantity for a unit mass of a reference gas (generally CO_2) over some time horizon (generally 100 years). Hence, the GWP provides an approximate measure of the relative integrated climate forcing of a GHG. While it is acknowledged to be an imperfect index, it is generally true that emission of a well-mixed gas that is characterized by a larger GWP than another well-mixed gas will lead to a greater climate response.

There are two components to GWPs that we will consider in this chapter. The first is from the direct effect of halocarbons. The addition of an ODS to the atmosphere initially leads to a reduction in the outgoing longwave radiation at the tropopause, causing a globally averaged warming. This results from the strong infrared absorption by the ODS, particularly in the transparent atmospheric window region (8-12 μm). The amount of the net radiative imbalance (down flux minus up flux) at the tropopause per unit mixing ratio increase in an ODS is called the "radiative efficiency" (e.g., units of W per m^2 per parts per billion by mole, ppb). The second component to GWPs considered in this chapter is from the destruction of ozone caused by ODSs. This ozone destruction leads to an additional radiative forcing that can be considered in discussions concerning the overall climate impact of ODSs. It is referred to as an "indirect" forcing because it is caused by the change in ozone (owing to ODSs) and not by the change in ODSs directly.

Our general approach in this chapter is to use direct GWPs as the primary weighting factors when considering climate-relevant magnitudes of production, consumption, or emission of ODSs. Uncertainties quoted on these direct GWP-weighted quantities, however, represent a range of net GWP weightings that include the rather uncertain indirect ODS forcing arising from stratospheric ozone depletion. The indirect and net GWPs are derived here from the indirect GWPs in WMO (2007) by considering the recent revision of the radiative forcing attributed to stratospheric ozone depletion from −0.15±0.10 (IPCC, 2001) to −0.05 ±0.1 W per m^2 (IPCC, 2007) (see Table 2A.3).

BOX 2.2: **Weighting Factors** *cont'd*

Net GWPs

Combining direct and indirect GWPs into net GWPs potentially leads to additional errors and has intentionally been avoided in IPCC/TEAP (2005) and WMO (2007). It is known that the direct and indirect processes will cause different spatial forcing, and it is possible that the climate response to these forcings will differ both in the global mean magnitude and in the spatial pattern (Joshi et al., 2003). Therefore, it is inaccurate to think of direct and indirect GWPs or forcing as being additive. For example, if the direct and indirect GWPs were to exactly cancel each other, it is still expected that there would be a climate response. Nevertheless, in this chapter we have opted to use net GWPs in many situations. Our purpose in doing so is not to present a precise net GWP quantity, but to provide an approximate idea of how ozone destruction may affect some of the conclusions drawn from considering the direct effect alone. Doing so likely provides a more complete and accurate picture of overall climate forcing from ODSs than would be obtained from considering direct GWPs alone. Specifically, throughout the chapter, we provide analyses and conclusions based first on the more accurate direct GWPs and forcing. We then also consider net GWPs and forcing calculated assuming an indirect forcing for a total ODS-induced ozone depletion of -0.15 W per m^2 and $+0.05$ W per m^2. These values are chosen to coincide with uncertainties on the -0.05 ± 0.10 W per m^2 IPCC (2007) ozone forcing estimate ($\pm$ one-standard-deviation uncertainty). We neglect direct GWP uncertainties and other indirect GWP uncertainties, as we are not aware of a complete error analysis of these processes in the current literature.

GWP Uncertainties

The uncertainty in direct GWPs is stated to be about $\pm35\%$ ($\pm$ two-standard-deviation uncertainty), due primarily to uncertainties in the radiative efficiencies and lifetimes of the halocarbons and to uncertainties in the carbon cycle (see, e.g., IPCC, 2001; IPCC/TEAP, 2005; WMO, 2007; IPCC, 2007). Uncertainties in the carbon cycle are thought to lead to an uncertainty of about $\pm15\%$ to the denominator of the GWP, or the CO_2 absolute GWP (IPCC, 2007). This error contributes to the uncertainty in the absolute GWP value, but affects each GWP in the same way. When one is more interested in comparing halocarbon direct GWPs than in the values themselves, an effective uncertainty level of something less than 35% can be assumed.

Uncertainties in the indirect GWPs have not been as well quantified. Two of the most important issues likely involve the absolute amount of forcing caused by the ozone depletion caused by ODSs, and the relationship between ODS-induced ozone forcing with Equivalent Effective Stratospheric Chlorine (EESC). IPCC (2007) has recently reduced the magnitude of the estimated forcing due to stratospheric ozone changes from -0.15 ± 0.10 (IPCC, 2001) to -0.05 ± 0.10 W per m^2. However, this forcing is due to the total ozone change, not only the ozone change due to ODSs. Because a better estimate does not currently exist, we will adopt this latest estimate and ignore the potential contribution of non-ODS processes to the value. It is not possible for us to estimate the error induced by this assumption at the current time. In the procedure for calculating indirect GWPs given in Daniel et al. (1995), there are assumed EESC "thresholds" that lead to discontinuities in the forcing/EESC relationship. Because of these forcing/EESC changes, indirect GWPs exhibit a dependence on the time of emission; this time dependence has likely been relatively small when compared to other uncertainties for emissions between 1970 and 2010 and will not be considered in this chapter. Nevertheless, this oversimplification of the forcing/EESC relationship leads to likely additional errors in the indirect GWPs that have not been quantified. Because of all the previously discussed uncertainties and errors, in this chapter, the indirect and net GWPs are used to provide a very general idea of how ozone depletion might affect conclusions obtained from a direct GWP weighting.

Indirect Forcing

The indirect forcing owing to ozone depletion and attributable to individual compounds, compound classes, or from the aggregate of U.S. emissions was estimated by scaling the global indirect forcing arising from global ozone depletion through 2005 (-0.05 ± 0.10 W per m^2) to the fraction of global Equivalent Effective Chlorine (EECl) attributable to the compound, a compound class, or the U.S. in 2005 (see Figure 2.15). This procedure suggests that the net forcing from ozone depletion attributable to U.S. ODS emissions was between 0.04 and 0.18 W per m^2 in 2005 (Figure 2.17).

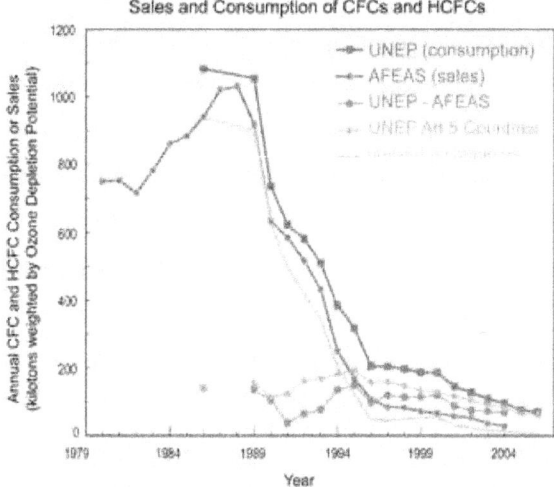

Figure 2.4 Comparison of annual AFEAS sales (green line) and annual UNEP consumption (blue line) totals for the aggregate of CFCs and HCFCs, weighted by Ozone Depletion Potential. Also shown is the annual difference (UNEP consumption minus AFEAS sales), and the magnitude of global consumption in countries operating under Article 5 of the Montreal Protocol, and non-Article 5 countries. Data were compiled by UNEP (2007) and AFEAS (2007).

The differences between the totals compiled by AFEAS and UNEP seem reasonable given the known differences in these databases. Finally, the consistency apparent in global total production and consumption data reported to UNEP suggests that the accounting of export and import activities has been reasonably accurate over time on a global scale (Figure 2.1).

2.1.3 Production and Consumption of ODSs and Substitutes Not Reported by AFEAS or in UNEP Compilations

In UNEP compilations, only production and consumption of ODSs for dispersive uses regulated by the Montreal Protocol are included. There is substantial additional production of ODSs for use as reagents in chemical manufacture of other substances (known as feedstock use) and for treatments to prevent the introduction or spread of pests and diseases during import/export of goods (known as quarantine and pre-shipment (QPS) processes) that are neither restricted by the Montreal Protocol nor included in the production and consumption data compiled by UNEP. Global production for feedstock uses was estimated at 3.2×10^5 ODP-Ton in 2002 (UNEP/CTOC, 2007), or about 1.9 times the total production of ODSs reported for dispersive uses in that year (UNEP, 2007). Emissions from this production are estimated to be 0.5% of amounts produced for feedstock use, but this estimate does not include any additional emissions that may occur during use. ODSs produced substantially as feedstocks include CFC-113, CCl_4, CH_3CCl_3, HCFC-22, HCFC-142b, CH_3Br, and halon-1301.

During the years of highest consumption (1986-1990), the AFEAS compilation accounted for the majority of global consumption of CFCs and HCFCs. During the last decade, however, the data reported to UNEP suggest that 59 (± 8)% of global annual consumption was not included in the AFEAS compilation.

The accuracy of these data hinges on the reliability of sales and import-export magnitudes reported to AFEAS and UNEP by individual companies and nations. This is difficult to assess quantitatively with independent methods, though estimates of global emissions inferred from atmospheric observations provide an independent but qualitative confirmation that large decreases in production and consumption of ODSs have indeed occurred since the late 1980s. The smaller declines noted for emissions as compared to consumption or production (Table 2.1) likely arise in part because emissions of ODSs lag production by months to decades depending upon the specific application. The accuracy of production, consumption, sales, and emission data on a national basis is more difficult to assess by independent methods, though regional estimates of emissions and emission changes are an area of active research (Section 2.2.6).

In addition to feedstock applications, methyl bromide is sold for QPS applications that are not restricted or limited by the Montreal Protocol. In 2005, global production for QPS uses of 0.8×10^4 ODP-tons was similar in magnitude to the non-QPS production reported to UNEP of 1.1×10^4 ODP-tons (UNEP/

MBTOC, 2007). Based on data for CH_3Br use in QPS applications during 1999-2005, including this non-restricted production would increase UNEP-reported, global ODP-weighted production for all ODS by 2 to 9%, and it would influence the estimate of the total decline in ODP-weighted production since the late 1980s given in Table 2.1 only minimally (a decline of 94.3% when QPS is included, compared to 94.7%–rounded to 95% in this Table–when not included). Global production of CH_3Br for QPS is expected to increase in 2006-2007 (UNEP/MBTOC, 2007).

Production magnitudes for three HFCs are currently reported by AFEAS. These data are thought to account for a large fraction of total global HFC production. In 2003, estimates of HFC-134a global production capacity (Campbell and Shende *et al.*, 2005) exceeded AFEAS production data (AFEAS, 2007) for this compound by only 10%. Similar data for other HFCs are not currently available on a global or national basis primarily because of the relatively few number of production facilities. Most of HFC-23 in the atmosphere today arises from overfluorination during the production of HCFC-22 rather than direct production. As a result, production of HFC-23 can be estimated globally based upon emissions inferred from atmospheric measurement records (Clerbaux and Cunnold *et al.*, 2007) though this would be an underestimate if any HFC-23 produced during HCFC-22 manufacture were captured and destroyed. On national scales, HFC-23 production has been estimated from HCFC-22 production magnitudes (U.S. EPA, 2007).

2.1.4 Production and Consumption: United States Trends for ODSs and Substitutes

2.1.4.1 UNITED STATES PRODUCTION AND CONSUMPTION WEIGHTED BY OZONE-DEPLETION POTENTIAL (ODP)

Production and consumption magnitudes of ODSs for regulated, dispersive uses in the United States are reported to UNEP as part of requirements associated with being a signatory to the Montreal Protocol (UNEP, 2007). The data indicate large declines in U.S. production and consumption of most chemicals as a result of the adjusted and amended Montreal Protocol. The total decline in annual U.S. production or

consumption of ozone-depleting substances for dispersive uses restricted by the Protocol since the late 1980s through 2005 was $0.4-0.5\times10^6$ ODP-Tons (Figure 2.1). This represents a 97-98% decline in both U.S. production and consumption of ODSs over this period (Table 2.1). The total U.S. ODP-weighted consumption and production of ODSs reported to UNEP for 2004-2005 averaged $1.2-1.3\times10^4$ ODP-Tons per year.

An analysis of data reported to UNEP reveals that the contribution of the United States to total global ODS production and consumption for regulated, dispersive uses decreased from a mean of 25 (±2)% in 1986-1994 to 10 (±2)%, on average, during 2001-2005 (Figure 2.1). In the interim years (1996-2000) large differences between reported U.S. production and consumption are apparent owing to negative consumption of carbon tetrachloride. Negative consumption is reported when exports outweigh the sum of production plus imports, or when destruction of stockpiles or feedstock use outweighs production in any given year.

Though the mean contribution of the United States to global, ODP-weighted production and consumption of ODSs for regulated, dispersive uses has been 10 (±2)% since 2001, the contribution of different compound classes to this amount varies. Over this period the United States accounted for less than 3% of global annual consumption of CFCs, CH_3CCl_3, and halons, between 20 and 39% of HCFC annual consumption, and between 17 and 37% of CH_3Br annual consumption (UNEP, 2007).

The large range observed for some compounds and compound classes since 2001 reflects changes in U.S. contributions over this period. In data reported to UNEP (2007) during 2003-2005, the United States accounted for 22±2% of global HCFC consumption during these years (19±1% of production). This is notably lower than the U.S. contribution during the previous decade. During 1992-2002 the United States accounted for 38±3% of global HCFC consumption (40±4% of production). HCFCs accounted for over half of total U.S. consumption in 2005 weighted by ODP, the remaining consumption was CH_3Br (34%) and

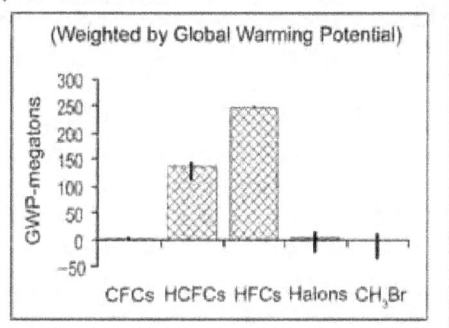

Figure 2.5 Contributions of different compound classes or individual compounds to total United States, ODP-weighted consumption or sales of ODSs and substitutes reported to UNEP for uses restricted by the Montreal Protocol or estimated by the U.S. EPA (left panel), and total United States, GWP-weighted consumption or sales of these chemicals in the same year (right panel) (UNEP, 2007; U.S. EPA, 2007). The hatched bar heights in the right-hand panel were derived with direct GWPs; the given uncertainties represent weighting by net GWPs (see Box.2.2). Relative contributions of less than 1% are not included in these charts; note that the U.S. CCl$_4$ ODP-weighted consumption was −16% during 2005, though this contribution was not included in the total or shown in the pie chart.

The decline in
CO$_2$-equivalent
consumption of
ozone-depleting
substances and
substitute chemicals
decreased slightly
faster in the United
States than across
the globe from
1989 through 2005.

CFCs (12%); other compounds contributed less than 1% (Figure 2.5).

Consumption of CH$_3$Br in the United States for dispersive uses restricted by the Montreal Protocol has also varied in recent years. It decreased from 1999 to 2002 but then increased from 2003-2005 owing in part to Critical Use Exemptions (UNEP/MBTOC, 2007). The U.S. methyl bromide consumption in 2003-2005 was 1.3 to 2.8 times higher than consumption in 2002. Global consumption has declined fairly steadily since 1999 and, as a result, the U.S. contribution to global CH$_3$Br reported consumption for uses restricted by the Montreal Protocol increased from 23 ± 4% during 2000-2003 to 36 ± 1% during 2004-2005. Since 2005, amounts approved for Critical Use Exemptions (CUEs) in the United States have declined (UNEP/MBTOC, 2007).

2.1.4.2 UNITED STATES CONSUMPTION WEIGHTED BY GLOBAL WARMING POTENTIAL (GWP)

United States consumption data for ODSs and substitutes (UNEP, 2007) has been combined with the U.S. EPA vintaging model estimates of HFC demand (U.S. EPA, 2007) to assess magnitudes and changes in U.S. consumption of halocarbons weighted by climate-relevant factors. The data suggest large declines in the consumption of ODSs and their substitutes when weighted by 100-year, direct GWPs (Figure

2.3). By 2005, the annual consumption of these chemicals had declined by approximately 2600 Mt CO$_2$-equivalents (87%, Table 2.1) from amounts reported and estimated for 1989. The ozone depletion arising from use of ODSs may have offset some of this warming influence. The magnitude of this offset can be approximated by considering net GWPs that include this indirect effect, though the uncertainties in this indirect influence are large (Box 2.2). With this indirect effect included, U.S. consumption of ODSs and substitutes declined by 81-88% from 1989 to 2005 (1305-3010 Mt of CO$_2$-equivalents). The total U.S. direct GWP-weighted consumption of ODSs and substitutes during 2004-2005 was nearly 400 Mt CO$_2$-equivalent (310-420 Mt CO$_2$-equivalent if the indirect influence associated with ozone depletion is included). The decline in CO$_2$-equivalent consumption has decreased slightly faster in the United States than across the globe; the contribution of the United States to total global ODS production and consumption for regulated, dispersive uses was 30% in 1989 and 21% in 2005 when direct GWP weighting is used. If net GWPs are considered, the U.S. contribution decreased from 24-48% in 1989 to 20-23% in 2005. Whereas in the late 1980s more than 90% of CO$_2$-equivalent U.S. consumption resulted from CFCs, in 2005 more than half of U.S. CO$_2$-equivalent consumption was of HFCs and nearly all the rest was of HCFCs (Figure 2.5).

BOX 2.3: Focus on Methyl Bromide

Methyl bromide is unique among ODSs regulated by the Montreal Protocol for several reasons. First, natural processes emit substantial amounts, in addition to there being significant releases from industrial uses. Emissions arising from human-produced CH$_3$Br accounted for 30 (20-40)% of global emissions during the mid-1990s before industrial production was reduced in response to Montreal Protocol phase-out schedules (Clerbaux and Cunnold et al., 2007). Since 1998, human production for all fumigant-related applications has declined by about 50%.

Second, a substantial fraction of industrial production is for dispersive applications not restricted by the Protocol. These unrestricted uses, primarily in quarantine and pre-shipment (QPS) applications, have increased recently and have led to a slower decline in total global CH$_3$Br production than suggested by UNEP values reported to them for assessing compliance with the Protocol. For example, during 2005 nearly half (43 (36-49)%) of the global, industrially derived emissions of CH$_3$Br were from uses not restricted by the Montreal Protocol (i.e., QPS applications) and, therefore, were not included in the production and consumption data shown by UNEP as reported to them (Box Figure 2.3-1) (UNEP/MBTOC, 2007). Such use is expected to increase in the future (UNEP/MBTOC, 2007). In the United States, QPS consumption increased by about 13% per year, on average, during 2001-2006 (U.S.EPA, 2007), leading to an annual consumption 30 to 80% higher than the annual amounts reported to UNEP during these years.

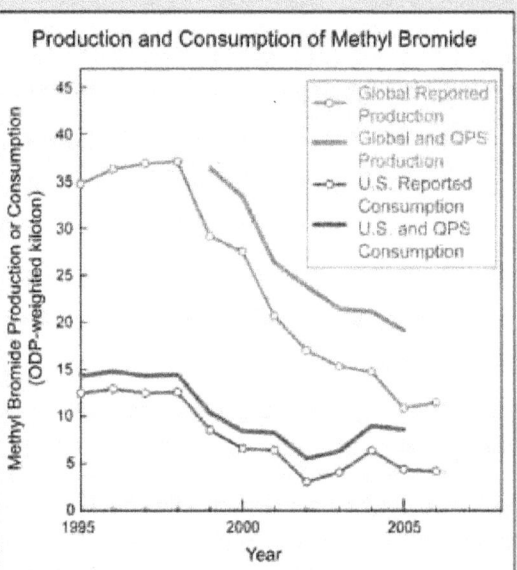

Box Figure 2.3-1 Annual global production and U.S. consumption magnitudes for restricted uses reported to UNEP (UNEP, 2007) (lines with circles), which includes CUE amounts, compared to these reported amounts plus use in QPS applications considered (solid lines) (UNEP/MBTOC, 2007; U.S. EPA, 2007)(feedstock uses not included).

Third, declines in CH$_3$Br production and consumption have also been slowed by exemptions to protocol restrictions for critical uses (Critical Use Exemptions or CUEs) that have allowed substantial continued production and consumption past the 2005 phase-out in developed countries. Enhanced CUEs in the United States have resulted in higher annual consumption of CH$_3$Br and an increased United States/Global consumption ratio during 2004-2005 compared to 2002-2003 (Box Figure 2.3-1).

Despite increases in QPS use and enhanced CUEs in recent years and variability in underlying natural emissions, global atmospheric mixing ratios of CH$_3$Br have declined continuously since 1998 (Clerbaux and Cunnold et al., 2007). While the United States contributed much to this atmospheric decrease through 2002, this U.S. trend reversed in 2003; the atmospheric abundance of bromine attributable to U.S. emissions was higher in 2004-2005 compared to 2002-2003 primarily because of enhanced QPS and CUEs consumption of CH$_3$Br in the United States (Figure 2.14).

BOX 2.4: Focus on Hydrochlorofluorocarbons (HCFCs)

HCFCs were attractive substitutes for CFCs because they have similar properties to CFCs in many applications, but shorter lifetimes, generally fewer chlorine atoms per molecule, and, therefore, lower ODPs and GWPs.

In spite of these attributes, HCFCs still lead to stratospheric ozone depletion and affect climate. Hence, HCFCs are considered only temporary replacements for the most potent ODSs. Production of HCFC-22 causes an additional climate influence through the unintended formation of the byproduct HFC-23, itself a long-lived, potent greenhouse gas.

The temporary nature of HCFC use is reflected in how developed-country consumption totals have changed in recent years (Box Figure 2.4-1). Consumption has declined substantially in developed countries (non-Article 5) and in the United States in response to the HCFC phase-out outlined in the Protocol. Production on a global scale has remained relatively constant over this time, however, as production and consumption in developing countries (Article 5) have increased dramatically.

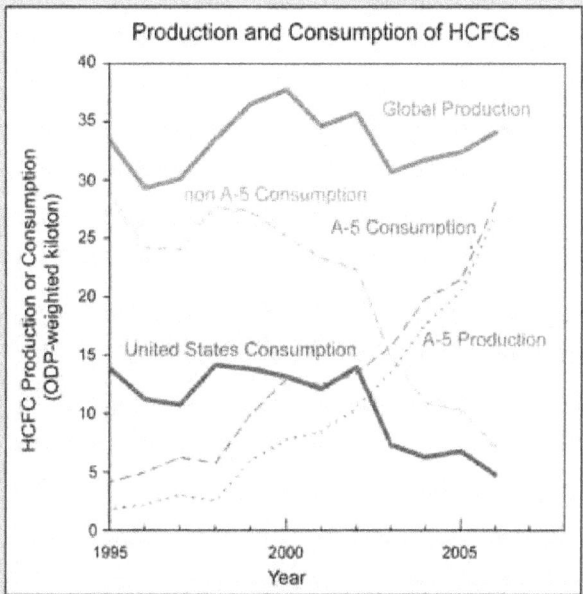

Box Figure 2.4-1 Annual production and consumption totals for HCFCs as reported to UNEP for dispersive and regulated uses, weighted by ODPs (UNEP, 2007). Global production (red line) is compared to U.S. consumption (U.S. Cons.; blue line), consumption in all developed countries (non Article 5; developed countries; red dashed line), and both consumption and production in developing countries (Article 5 country consumption and Article 5 country production; green dashed and dotted lines; developing countries).

U.S. EPA vintaging model estimates suggest that U.S. HCFC annual emissions have increased by about 10% since 2002, despite U.S. reported annual consumption during 2003-2005 being about half of what it was from 1995-2002 (Box Figure 2.4-1). This apparent discrepancy likely arises from the large bank of HCFCs; while HCFC emissions were similar to HCFC consumption in 2005 (~6 ODP-Kt) the HCFC bank was more than ten times larger (Box 2.5). In the U.S. during 2005, HCFC-22, HCFC-142b, and HCFC-141b accounted for 98% of all U.S. HCFC emissions. The remainder was contributed by HCFC-225 (1.2%), HCFC-124 (0.6%), and HCFC-123 (0.3%).

An increased awareness of the influence ODSs have on both climate and stratospheric ozone has led to recent proposals for more stringent HCFC limits to future use by several Parties to the Montreal Protocol, including the United States. The accepted proposal speeds up the production and consumption phase-out schedule for non-Article 5 and Article 5 countries and moves the Article 5 country consumption baseline year forward to 2009-2010 from 2015. This earlier baseline year is expected to reduce Article 5 country consumption beginning in at least 2013, the first year consumption limits would be in force. The potential future implications of this accepted proposal on the evolution of EESC are summarized in Chapter 5.

2.1.5 United States Production and Consumption of ODSs and Substitutes Not Included in Published UNEP Compilations

Production and consumption of ODSs for chemical feedstock purposes and of CH_3Br for QPS applications are not included in UNEP compilations because these uses are not restricted by the Montreal Protocol. While losses from feedstock applications are estimated to be small (0.5%, see Section 2.2.4), most CH_3Br used in QPS applications is emitted to the atmosphere (UNEP/MBTOC, 2007). Furthermore, amounts of CH_3Br used in QPS applications are substantial compared to amounts reported to UNEP for restricted uses and they have increased in recent years. For example, in the United States, annual consumption of CH_3Br in QPS applications during 2001-2006 was 1.8-2.9 Kt, or 57 (±20)% of annual consumption reported by the United States to UNEP for restricted uses; this QPS use had increased by about 13% per year, on average, over this period (U.S. EPA, 2007).

U.S. production data for HFCs are not publicly available either through UNEP, AFEAS, or the U.S. EPA. Estimates of HFC demand and sales, however, are made by the U.S. EPA through its vintaging model (U.S. EPA, 2007). These estimates show how HFC use in the United States has increased by a factor of three over the past decade, when use is weighted by compound-dependent GWPs. HFC use in the United States accounted for about two-thirds of the CO_2-equivalent consumption of ODSs and substitutes in 2005 (Figure 2.5). This vintaging model projects a doubling of CO_2-equivalent HFC use in the United States during 2005-2015 (U.S. EPA, 2007).

2.2 EMISSIONS: OZONE-DEPLETING CHEMICALS AND THEIR SUBSTITUTES

Emissions estimates allow an understanding of how human behaviors influence the atmospheric abundances of ODSs and their substitutes, and how that influence has changed over time as a result of international agreements (such as the Montreal Protocol) and other factors. Only after chemicals become emitted to the atmosphere do they contribute to ozone depletion and

radiative heating of the atmosphere. Nearly all ODSs produced ultimately become released to the atmosphere through direct emission (*e.g.*, use in aerosol cans) or leakage during use or upon disposal. Methyl bromide is an exception, because a substantial fraction that is produced and applied to soils becomes destroyed through hydrolysis and does not reach the atmosphere.

Global emissions can be estimated from production data, knowledge of release rates during production, use, and disposal of ODSs in different use applications, and information on the magnitude of sales for different end uses over time (AFEAS, 2007). Uncertainties can be significant in this "bottom-up" approach— but, in general, emissions are delayed after production with time lags that are application-dependent. Because these estimates rely on the production data considered in Section 2.1 of this chapter, they are not independent of them. Furthermore, restrictions on reporting of production and consumption for ODS and substitutes can substantially influence emission estimates, particularly when a limited number of manufacturers produce a specific chemical.

Independent estimates of global emissions can be derived from an analysis of atmospheric observations. This "top-down" approach provides an important independent check on production and consumption magnitudes reported to UNEP, and is critical for assessing global emissions considering the limitations of the "bottom-up" methodology. The observationally derived emissions are based on the measured change in the global atmospheric burden of an ODS relative to the expected rate of change in the absence of emissions. Accordingly, this calculation incorporates the atmospheric lifetime of the ODS, which is derived from laboratory measurements of destruction rate constants (via photolysis and or oxidation by the hydroxyl radical, OH) and model-derived parameters such as photolytic fluxes, OH abundances, and 3-D distributions of ODS atmospheric mixing ratios. This method of estimating emissions is susceptible to errors in measurement calibration, in estimating the global atmospheric burdens of trace gases in the entire atmosphere from a few measurement locations at Earth's surface, in lifetime, and in the assumption (generally applied) that all observed changes are the result of changes in emissions, not changes in loss rates. Atmospheric measurement techniques have improved over time to the extent that the majority of the uncertainty in this approach for long-lived ODSs is believed to arise from the estimates of lifetime and loss (UNEP/TEAP, 2006).

Global emissions for ODSs have been derived with these different techniques and have been compared and reviewed in past WMO Ozone Assessment Reports (2003; 2007) and in the IPCC/TEAP (2005). Particular discrepancies in bottom-up *versus* top-down emission magnitudes were noted in IPCC/TEAP (2005) for the years since 1990 and were investigated additionally in a special Emissions Discrepancies report (UNEP/TEAP, 2006). In this latter report, the potential for rapid-release applications and time-dependent release functions to influence bottom-up emissions estimates was explored and a more comprehensive analysis of top-down uncertainties was presented. For the compounds studied (CFC-11, CFC-12, HCFC-22, HCFC-141b, and HCFC-142b), the range (± 1 sigma) of emissions estimated with top-down and bottom-up methods overlapped in nearly all years and, therefore, were considered to be consistent estimates (Figure 2.6) (UNEP/TEAP, 2006). The uncertainty ranges are quite large in both approaches, however, such that the mean CFC-11 emissions estimated from these different methods differed generally by a factor of between 1.5 to 2. The overall trends in emissions estimated for these chemicals since 1990 were generally consistent, with the exception being HCFC-142b since 2000. While the bottom-up analysis suggests a rapid decline in emissions of this HCFC over this period, the top-down trends indicate only a small decline.

2.2.1 Global Emissions: Estimates Derived From Atmospheric Observations and Weighted by Ozone Depletion Potentials

Estimates of ODS emissions on a global scale have been derived for the past from a combination of atmospheric observations and industrial estimates (WMO Scenario A1, Daniel and Velders *et al.*, 2007). This emission history indicates substantial declines in total ODP-weighted emissions since 1990. By 2005, annual

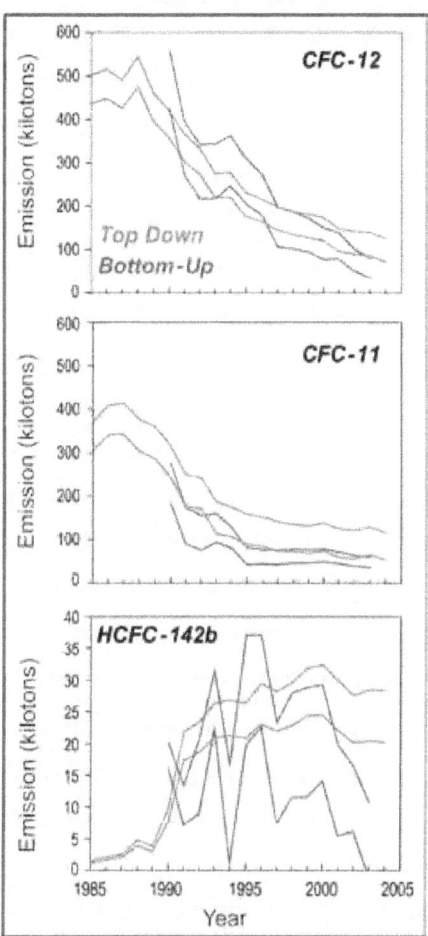

Comparing Emissions Derived with Different Methods

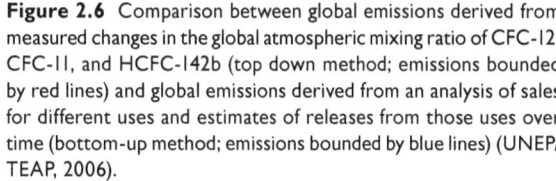

Figure 2.6 Comparison between global emissions derived from measured changes in the global atmospheric mixing ratio of CFC-12, CFC-11, and HCFC-142b (top down method; emissions bounded by red lines) and global emissions derived from an analysis of sales for different uses and estimates of releases from those uses over time (bottom-up method; emissions bounded by blue lines) (UNEP/TEAP, 2006).

emissions had declined nearly 1.1 ODP-Mt from peak emissions in 1988. This corresponds to an 82% decrease in global annual ODP-weighted emissions over this period (Figure 2.7; Table 2.1). Decreases in emissions of CFCs accounted for the majority of this decline (~80%). Decreases in emissions of CH_3CCl_3 and CCl_4 accounted for 6 and 8% of the decline, respectively; emissions decreases in halons and CH_3Br each accounted for 2-3% of the decline. Increases in HCFC ODP-weighted emissions have offset some of the overall decline since 1990; annual HCFC emissions increased from 1.1×10^4 in 1990 to 2.2×10^4 ODP-Tons in 2005. Total global

By 2005, annual emissions of ozone-depleting substances declined by 82% from peak emissions in 1988.

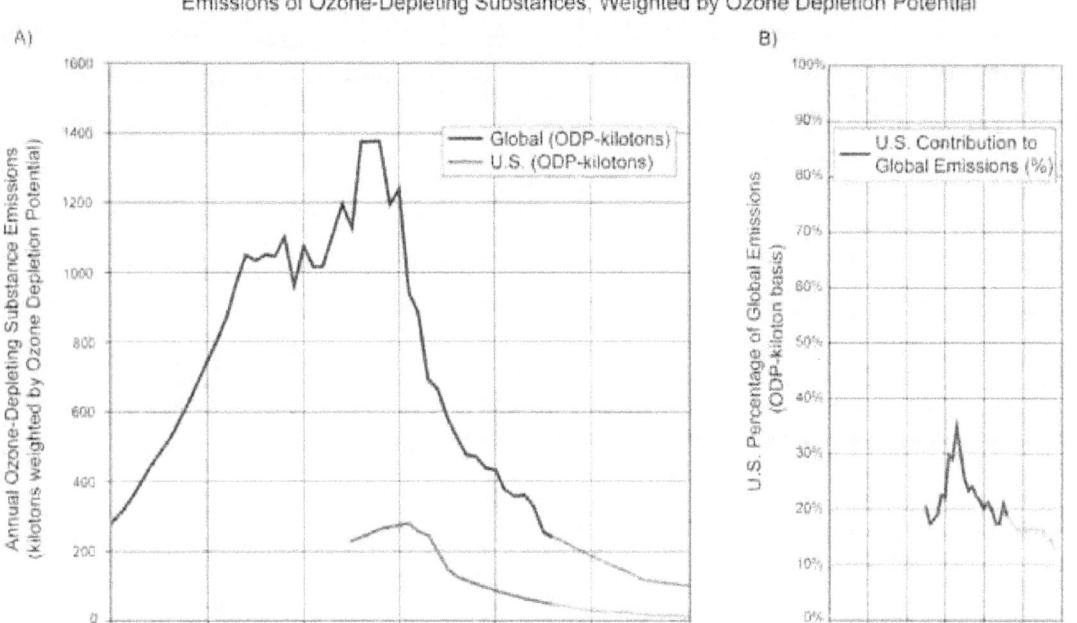

Emissions of Ozone-Depleting Substances, Weighted by Ozone Depletion Potential

Figure 2.7 Panel A: Aggregated emissions of ODSs derived for the entire globe (blue line; Clerbaux and Cunnold *et al.*, 2007; Daniel and Velders *et al.*, 2007) and for the U.S. (red line; U.S. EPA, 2007) over time, weighted by ODP. Lighter lines represent projections into the future. Panel B: The percentage of emissions (weighted by ODP) contributed by the United States to the global total. Global emissions here are derived from atmospheric observations (Clerbaux and Cunnold *et al.*, 2007); U.S. emissions are inferred from a bottom-up analysis of sales data in the U.S. (U.S. EPA, 2007).

emissions of ODSs and substitutes amounted to 2.5×10^5 ODP-Tons in 2005. Because these global emissions estimates are derived from atmospheric observations, they include the influence of all processes releasing ODSs and substitutes to the atmosphere, including releases from non-reported, QPS, Critical Use Exemptions, and all others.

2.2.2 Global Emissions: Estimates Derived From Atmospheric Observations and Weighted by Global Warming Potentials

When the global emission history compiled as the WMO scenario A1 (Daniel and Velders *et al.*, 2007) is combined with global emissions derived for HFCs (Campbell and Shende *et al.*, 2005), the results indicate a substantial decline in total GWP-weighted emissions since the late 1980s when the climate influences of ozone depletion are not included (Figure 2.8). The overall decline in annual emissions amounted to 7270 GWP-Mt by 2005, which corresponds to a 77% decrease from peak global GWP-weighted emissions in 1988 (Table 2.1). With

weighting by net GWPs to include consideration of indirect forcing from ozone changes, the global decline in annual emission is estimated to be between 3600 and 8500 GWP-Mt, or a decline of 77-84%. The decline integrated between 1988 and 2005 amounts to a decrease of over 90 Gt CO$_2$-equivalents compared to constant emissions at 1988 levels. Declines in annual CFC emissions accounted for a decrease of 7900 (5330-8790) GWP-Mt by 2005, but this decline was partially offset by increases in HCFC and HFC annual emissions from 1990 to 2005 of 210 (150-230) and 530 GWP-Mt, respectively (numbers in parentheses represent quantities with a range of net GWP weightings; see Box 2.2).

In 2005, total global emissions of ODSs and substitutes are estimated to have been 2150 GWP-Mt (direct GWP weighting; a range of 675-2600 is calculated with the range of net GWPs). CFCs accounted for 810 (510-910), HCFCs 590 (470-630), and HFCs 625 GWP-Mt of this emission (HFC global emissions for 2005 were interpolated from 2002 estimates and for

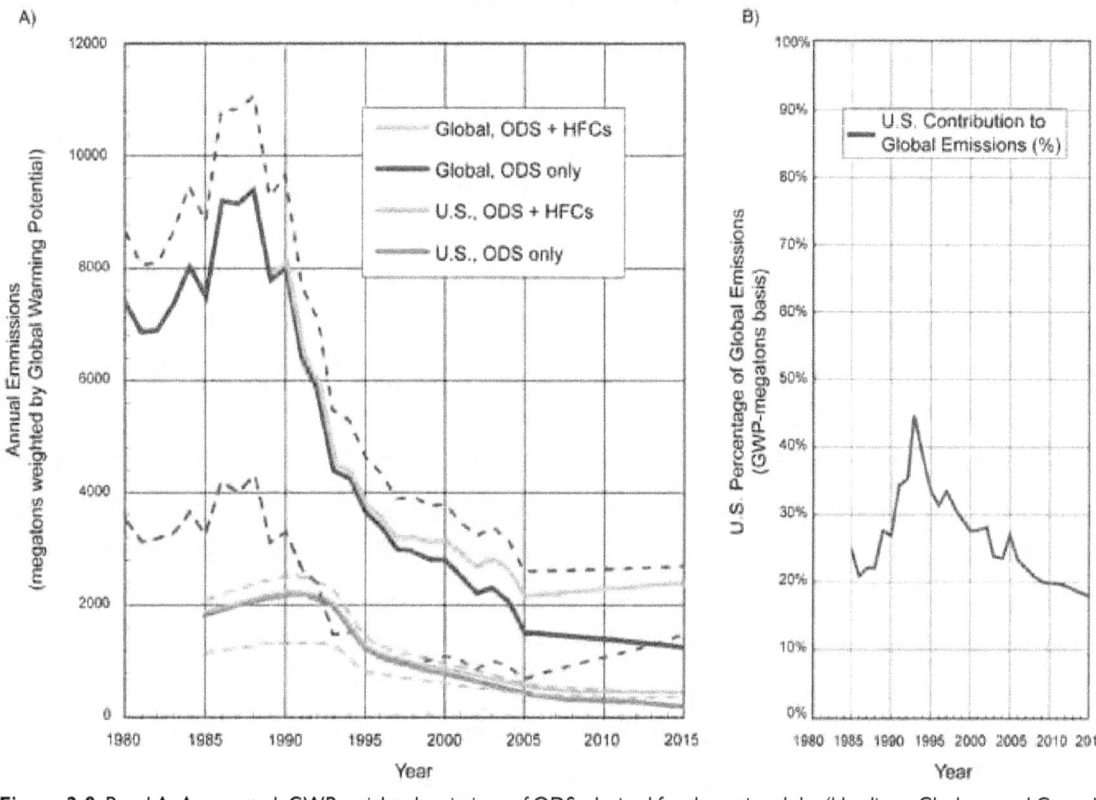

Emissions of Ozone-Depleting Substances, Weighted by Global Warming Potential

Figure 2.8 Panel A: Aggregated, GWP-weighted emissions of ODSs derived for the entire globe (blue lines; Clerbaux and Cunnold *et al.*, 2007; Daniel and Velders *et al.*, 2007) and for the U.S. (red lines; U.S. EPA, 2007). Solid lines represent weighting by direct GWPs (no indirect influences considered), and the lighter colored lines represent the contribution from emissions of HFCs, which were derived on the global scale from Campbell and Shende *et al.*, (2005) and for the U.S. (U.S. EPA, 2007). Dashed blue and red lines indicate the range in overall CO_2-equivalent emissions when the climate influence of stratospheric ozone depletion is included (as indirect GWPs). Panel B: The percentage of emissions (weighted by GWP) contributed by the United States to the global total. Global ODS emissions here are derived from atmospheric observations and global HFC emissions are derived from a combination of atmospheric observations and an analysis of production data (Campbell and Shende *et al.*, 2005; and AFEAS, 2007); U.S. ODS and HFC emissions are inferred from the vintaging model, which is a "bottom-up" analysis of sales and use data in the U.S. (U.S. EPA, 2007). U.S. HFC emissions include those from replacing ODS use and an additional small (~10-20% since 2001) contribution from unintended byproduct emission during HCFC-22 feedstock production.

business-as-usual scenario projections for 2015 [IPCC/TEAP, 2005]).

2.2.3 Global Emissions: The Contribution of Banks and Bank Sizes

"Banks" of ODSs exist where there are reserves of ODSs that potentially could be released at a later date. Though the magnitudes of banks are highly uncertain, the release of ODSs from these banks has become the most important factor in projecting future emissions of many ODSs (*e.g.*, CFCs and halons) for two main reasons. First, the production of CFCs and halons has diminished substantially and is expected to continue to decrease in the future in response to regulations of the Montreal

Protocol; and second, the applications for which CFCs are used today tend to release ODSs only over many years' time. While the continuing production of HCFCs remains important to their future evolution in the atmosphere, the HCFCs banks are currently large enough so that future emissions will also be determined by their size and release rates from them.

The estimated sizes of banks, annual consumption, and annual emissions for the CFCs are shown in Figure 2.9 (see also Box 2.5). The United States' contribution to these values is represented by the lower regions of each bar and amounts to nearly a quarter of global banks in 2005, when ODP weighting

Banks of ozone-depleting substances (ODSs) exist where there are reserves of ODSs that could potentially be released at a later date. The release of ODSs from these banks has become the most important factor in projecting future emissions of many ODSs.

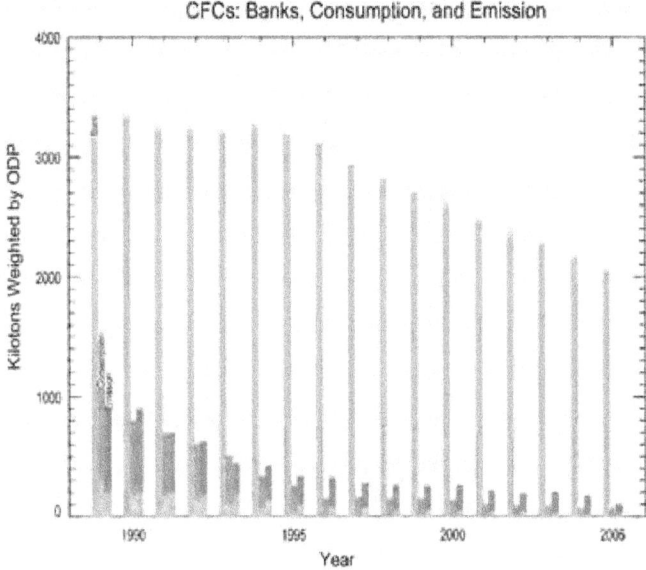

Figure 2.9 Time evolution of CFC banks, annual consumption, and annual emission, weighted by compound-dependent ODPs. The total height of the bars represents global values, while the lower blue/green portions represent the U.S. portions. Global banks are taken from WMO (Daniel and Velders *et al.*, 2007) with reliance on IPCC/TEAP (2005) bottom-up estimates used in combination with annual production and emission estimates. U.S. values are from the U.S. EPA vintaging model analysis of CFCs in the United States.

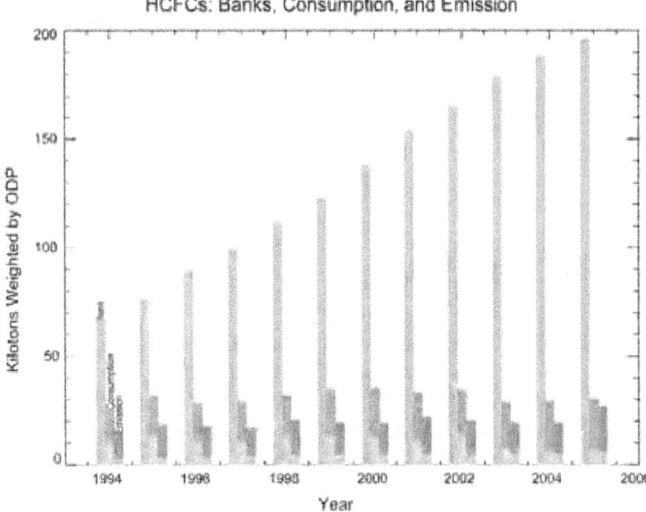

Figure 2.10 Time evolution of HCFC banks, annual consumption, and annual emission, weighted by compound-dependent ODPs. The total height of the bars represents global values, while the lower blue/green portions represent the U.S. portions. Global banks are taken from WMO (Clerbaux and Cunnold *et al.*, 2007; Daniel and Velders *et al.*, 2007) with reliance on IPCC/TEAP (2005) bottom-up estimates used in combination with annual production and emission estimates. U.S. values are from the U.S. EPA vintaging model analysis of HCFCs in the United States.

is considered. Both globally and domestically, the gradual decline of consumption is evident, with the size of the bank remaining as the most important driver of future emission. The importance of the bank is already apparent because the annual emission is substantially larger than the reported annual consumption, with the difference presumably coming from the bank.

The banks of the HCFCs similarly represent an important reservoir that will affect future U.S. and global emissions (Figure 2.10). However, because consumption of HCFCs has not been fully phased out in the developed world and is not yet limited in countries operating under Article 5 of the Montreal Protocol, current global consumption plays a larger relative role influencing current global emission rates than it does for the CFCs (Box 2.5).

Reducing future releases of ODSs from banks would necessitate recovering and destroying some of them. Technical feasibility and the economics of recovery necessarily play important roles in determining which ODS banks could be feasibly recovered and destroyed. Daniel and Velders *et al.* (2007) have evaluated test cases in which the 2007 total global banks of CFCs, HCFCs, or halons were recovered and destroyed in terms of the ozone benefits that could theoretically be achieved as a result. This information can be found in Chapter 5 of this report.

The direct GWP-weighted annual production and emissions are compared to the global bank sizes for CFCs, HCFCs, and HFCs in Figure 2.11. The only HFC considered in this calculation is HFC-134a because there is a lack of information regarding banks of other HFCs.

The decreases in global production, emissions, and bank sizes of the CFCs represent the largest changes and have led to overall decreases in these quantities for the sum of these compounds in a direct GWP-weighted sense. Although global banks for HFC-134a were still a small fraction of total direct GWP-weighted banks in 2005, banks of HCFCs have continued to increase over time and accounted for almost 25% of the total direct GWP-

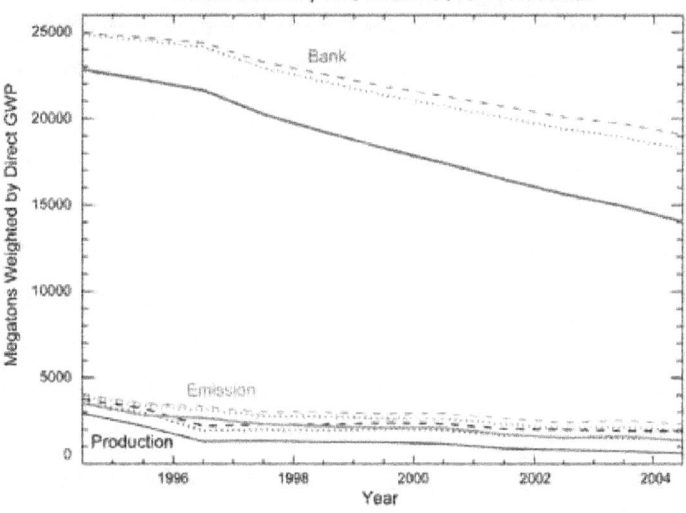

Figure 2.11 Comparison of direct GWP-weighted annual global production and emission with total bank sizes. Solid lines represent contributions of CFCs, dotted lines also include HCFCs, and dashed lines further add the HFC-134a contribution to the bank.

Reducing future releases of ozone-depleting substances from banks would necessitate recovering and destroying some of them.

BOX 2.5: Banks of Ozone-Depleting Substances (ODSs) and Substitute Chemicals

The term "bank" refers to an amount of chemical that currently resides in existing equipment or applications (including, for example, refrigerators, air conditioners, fire extinguishers, and foams) and stockpiles. Banked halocarbons are expected to be released to the atmosphere at some point in the future unless they are recovered and destroyed. Before scientists identified the relationship between chlorine- and bromine-containing halocarbons (now referred to as ozone-depleting substances, or ODSs) and stratospheric ozone, the majority of ODS usage was in fast-release applications like aerosol sprays and solvents. At that time, knowledge of the bank sizes was not critical to an understanding of current or even future projected ODS abundances.

Today the situation is far different. Most ODSs are used in slow-release applications, many of which contain a significant quantity of a halocarbon compared to its current atmospheric abundance; this quantity is in some cases much larger than the amount of halocarbon emitted annually. Accurate knowledge of these bank sizes and rates of halocarbon emissions from banks is now important to the future projections of many halocarbon abundances and to the amount of ozone that these ODSs will destroy.

Estimates of current bank sizes are known to be highly uncertain, though various methods have been used to make these estimates. In ozone assessments prior to 2007 a "top-down" approach was used in which annual changes to bank sizes were determined from the difference between annual production estimates, taken from industry databases or reported amounts to UNEP, and annual emissions, estimated from atmospheric observations and global lifetimes. This method is particularly susceptible to systematic errors in production magnitudes and in the atmospheric lifetime of a chemical; significant errors can arise because the bank size is characterized often by small differences between large production and emission numbers and systematic errors can accumulate over time. A second method, which was discussed in IPCC/TEAP (2005) and is used in the U.S. EPA vintaging model, involves counting the number of application units that use a particular ODS and converting this information to a total bank size by knowing the amount of ODS typically residing in a single unit. This method is often called the "bottom-up" method and is independent of atmospheric lifetime estimates that influence top-down estimates.

Advantages and disadvantages of these methods are discussed in detail elsewhere (IPCC/TEAP, 2005; Daniel and Velders et al., 2007). All have significant uncertainties, and in many cases, they do not agree particularly well. In the most recent WMO ozone assessment report (Daniel and Velders et al., 2007), the bottom-up methodology for estimating the bank, in spite of acknowledged deficiencies, was thought to likely be more accurate than the top-down estimate and was used to project future halocarbon abundances.

BOX 2.5: Banks of Ozone-Depleting Substances (ODSs) and Substitute Chemicals *cont'd*

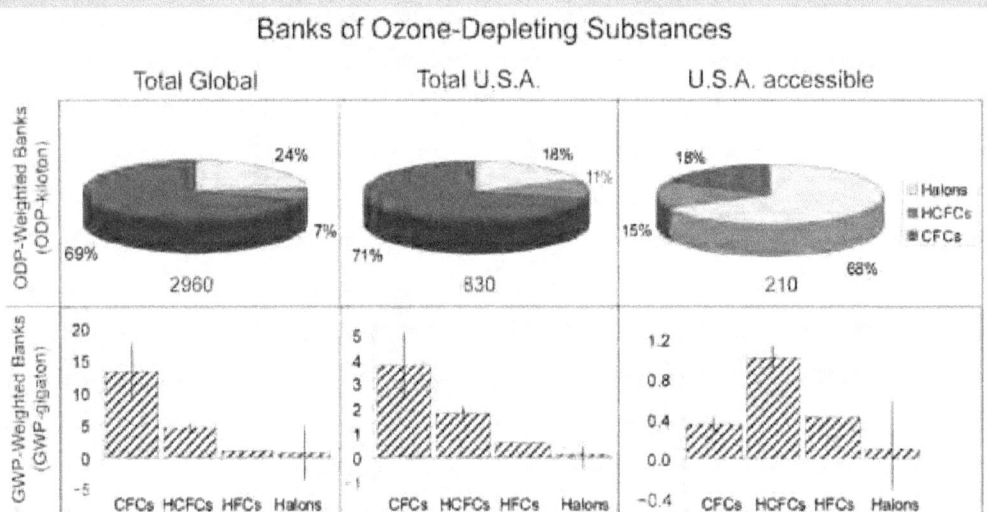

Box Figure 2.5-1 The size and contribution of different compound classes to 2005 banks estimated for the globe, for the U.S., and for the U.S. but classified as accessible (Daniel and Velders et al., 2007; U.S.EPA, 2007). Top row includes bank estimates weighted by compound-dependent ODPs; banks in the bottom row have been weighted by direct (hatched bars) and net (indicated as uncertainties) GWPs (see Box 2.2). Pie charts show relative percentages; units refer to weighted total bank sizes, which are given at the bottom of each box in upper panel. Bar charts in lower panels give weighted absolute quantities; totals in the lower panels are 19 (5-24) GWP-Gt (or Gt CO_2-equivalents) for the global bank, 6.2 (2.9-7.3) GWP-Gt for the U.S. bank, and 1.9 (0.9-2.2) GWP-Gt for the U.S. accessible bank. Note that of the U.S. accessible bank, HCFCs accounted for approximately 1.0 (0.9-1.1) GWP-Gt. Note that halon stockpiles are not included in any of the U.S. bank estimates from the U.S. EPA owing to lack of available data at this time.

The current bank sizes have important scientific and policy implications. Because banks that are not captured will eventually be released into the atmosphere, inaction can result in increased ozone depletion and climate forcing. Banks vary in how technically and cost-efficiently they can be recovered so that emission to the atmosphere is avoided. The U.S. EPA has identified refrigeration, air conditioning, and fire protection banks as accessible and potentially recoverable; other banks may also be recoverable to different extents, albeit with potentially more effort and higher costs.

Sizes and relative contributions of different banks in 2005 weighted by different quantities are displayed in the figure (see Box Figure 2.5-1). When compared to annual emissions in 2005 (Figures 2.7 and 2.8), this analysis suggests that the 2005 total global and total U.S. banks were about 7 to 16 times larger (weighting by ODP or the range of net GWPs). 2005 accessible banks in the U.S. were about four times larger than U.S. annual emissions in that year (ODP or direct GWP weighting). For comparison, CO_2 global emissions from fossil fuel and cement production have been estimated at 27-30 Gt CO_2 in 2005 (IPCC, 2001). Furthermore, while CFCs account for the largest fraction of both global and U.S. banks in 2005 regardless of the weighting considered, they account for a much smaller fraction of the bank classified as accessible in the U.S. (HFCs not included). For 2005, halons accounted for two-thirds of the ODP-weighted, accessible U.S. bank and HCFCs a similar fraction of the GWP-weighted, accessible U.S. bank. The halon contributions to U.S. banks calculated here should be considered underestimates because stockpiles were likely non-zero in 2005.

weighted bank in that year. The 2005 global banks continue to represent an important reservoir in terms of climate forcing, equivalent to about 19 Gt of potential CO_2-equivalent emissions (Figure 2.11). When compared to

the reductions in emissions that already have resulted from compliance with the Montreal Protocol (more than 90 Gt integrated through 2005, direct GWP weighting), they represent a non-zero additional contribution to future climate forcing and ozone depletion (see Section 2.2.1; Velders *et al.*, 2007). The importance of the Montreal Protocol and the quantification of the effect of future policy actions regarding global bank recovery and destruction have also been discussed in WMO (Daniel and Velders *et al.*, 2007), and IPCC/TEAP (2005) reports, and in Velders *et al.*, (2007). Consideration of the indirect forcing influences of ODSs associated with stratospheric ozone declines (and uncertainty in these influences) changes these figures somewhat. When weighted by a range of net GWPs, the 2005 global bank of ODSs and substitutes is estimated to be 5-24 GWP-Gt, comprised of 6-16 GWP-Gt of CFCs, 3-5 GWP Gt of HCFCs, and 0.9 Gt of HCFC-134a (See Box Figure 2.5-1; the uncertainty range stated in these figures represents the range of influences associated with net GWPs; see Box 2.2).

2.2.4 Global Emissions: The Influence of Non-restricted Uses and Other Factors

As indicated in Section 2.1.3 of this chapter, production of ODSs for chemical feedstock purposes is not restricted under the Montreal Protocol. ODP-weighted production of ODSs for feedstock purposes was approximately 1.8 times larger than overall production for dispersive uses reported to UNEP during 2002 (UNEP/CTOC, 2007). Emissions during production of the feedstock chemical are estimated to be 0.5%, which corresponds to 1600 ODP-Tons during 2002 based upon ODSs produced for chemical feedstock purposes, though this estimate does not include emissions that might arise after production. At a rate of 0.5% of production, this emission amounted to less than 1% of total ODP-weighted emission in 2002.

All production of HCFC-22, including feedstock-related production not restricted by the Montreal Protocol, has an added influence on the atmosphere that arises from HFC-23, which is created from over-fluorination as HCFC-22 is produced. Byproduct HFC-23 emissions account for most of the HFC-23 present in the atmosphere today (Clerbaux and Cunnold *et al.*, 2007). Feedstock production of HCFC-22 accounted for approximately one-third of total HCFC-22 production in 2005

(Rand and Yamabe *et al.*, 2005). In a "business-as-usual" scenario regarding HCFC-production for restricted and feedstock purposes, HFC-23 emissions were projected to increase 60% from 2005 to 2015.

Production of CH_3Br for QPS purposes is also not restricted by the Montreal Protocol (Box 2.3). This global production was similar in magnitude to that used for restricted purposes in 2005. The emission rate for CH_3Br in QPS uses is estimated to be 78-90% of the amount produced (UNEP/MBTOC, 2007). Based upon 2005 production for restricted and non-restricted (feedstock and QPS) uses of CH_3Br, we estimate that global emissions from non-restricted applications accounted for nearly half (43 [36-49]%) of all anthropogenic CH_3Br emission during this year. Most of this non-restricted emission (more than 99%) is estimated to arise from QPS uses.

Unlike other restricted ODSs, a substantial amount of CH_3Br emission arises from the natural environment. These emissions arise from the oceans, wetlands, plants, and biomass burning; as a result, humans have little direct control over them. Emissions of CH_3Br arising from industrial production are estimated to have accounted for 30 (20-40)% of total global emissions during the 1990s before industrial production was curtailed (Clerbaux and Cunnold *et al.*, 2007).

The magnitude of variability in non-industrial emissions of CH_3Br on annual and decadal time scales is not well known, and changes in these natural emissions could add to or offset the emission declines brought about by the Montreal Protocol. Despite these uncertainties, atmospheric data (Clerbaux and Cunnold *et al.*, 2007) suggest that global emissions of CH_3Br have declined each year since industrial production was first reduced (1999).

2.2.5 United States Emissions and Banks: Estimates Derived by U.S. EPA Vintaging Models

While global emissions are fairly straightforward to derive from atmospheric measurements of the global background abundances of ODSs and substitutes, provided loss rates are known (Section 2.2), estimating emissions on

In 2005, global emissions from non-restricted applications accounted for nearly half of all anthropogenic CH_3Br emission. More than 99% of this non-restricted emission is estimated to arise from quarantine and preshipment use.

national or regional scales is more difficult. While regional atmospheric monitoring could potentially provide national estimates of emissions, to date such estimates have been sporadic and are based on very few sampling regions (Section 2.2.6). Instead, U.S. emissions have been estimated using "vintaging" models that incorporate data regarding application-specific sales, and leakage rates during and after use of ODSs and substitutes (see Box 2.6 for further description of the vintaging model). With this method the U.S. EPA has estimated annual, U.S. emissions of ODSs and their replacements since 1985 (U.S.EPA, 2007). The data compiled by the U.S. EPA covers industrial production for uses restricted by the Montreal Protocol and for non-restricted uses such as feedstock and QPS applications.

When weighted by chemical-specific ODP values, the U.S. emissions of ODSs and substitute chemicals peaked in 1991 and declined thereafter as a result of limits imposed upon production and consumption by the fully adjusted and amended Montreal Protocol (Figure 2.9). By 2005, total annual U.S. emissions of ODSs and substitute chemicals had declined by 226 ODP-Kt or by 81% (Table 2.1). Emissions have declined less from their peak than consumption or production (Table 2.1) because much of those ODSs are contained currently in in-use foams, fire extinguishers, and cooling devices. United States emissions of ODSs and substitutes are estimated to have been 52.7 ODP-Kt in 2005. Emissions for the substitute HCFCs have increased over this period; U.S. HCFC emissions in 2005 were nearly 6 ODP-Kt, which represents an increase of 3.5 ODP-Kt since 1990. U.S. HCFC emissions in 2005 were predominantly HCFC-22 (87%) and HCFC-141b (7%); other chemicals contributed lesser amounts (HCFC-142b 4%; HCFC-227ca/cb 1%; HCFC-124 0.6%; and HCFC-123 0.3%). Over half (55%) of the U.S. ODP-weighted emissions in 2005 were from CFCs, 25% were from halons, 9% were from CH_3Br, and 11% were from HCFCs.

The contribution of the United States to global ODP-weighted emissions has varied during the period of available data from 18 to 35% (1985-2005; Figure 2.7). When weighted by chemical-specific 100-year direct GWP values,

U.S. emissions estimated by the U.S. EPA suggest a reduction of 1640 GWP-Mt in annual emissions of ODSs and substitutes including HFCs by 2005 compared to 1991 when they were at their peak (a decrease of 74%; Figure 2.8; Table 2.1). When indirect influences are included, a U.S. emissions decline of 900-1880 GWP-Mt is estimated with net GWP weighting (see Box 2.2), which corresponds to a relative decline of 71-75% from peak emissions. The largest decline was for CFCs emissions, which accounted for 90-95% of the direct or net GWP-weighted total annual U.S. emissions in 1985-1995 and only 40-45% of these emissions in 2005. By 2005, direct or net GWP-weighted emissions of HCFCs accounted for about one-third and HFCs one-quarter of total annual U.S. emissions of ODSs and substitutes.

Independent assessments of ODS emissions in the United States are limited. One analysis derived CFC emissions by country during a single year, 1986 (McCulloch et al., 1994). This investigation was based upon consumption data for 1986 compiled by UNEP, AFEAS delineations of use by individual CFCs specific to different geographic regions, and emissions of 86-98% of consumption in each year. These results suggest that emissions in 1986 were within 50% of those estimated by the U.S. EPA for CFC-12, CFC-113, CFC-114, and CFC-115, but a factor of three higher for CFC-11. This discrepancy likely arises because the U.S. EPA analysis suggests that U.S. emissions of CFC-11 were not typical of other nations during this period. In this report we have used the U.S. EPA estimates because they represent the most in-depth and comprehensive analysis of U.S. emissions available. In this analysis, an error range of −25% to +50% is applied to U.S. emission estimates of ODSs after 1985.

As was touched upon in the discussion of ODS banks on global scales, U.S. emissions of many ODSs are dominated currently by slow releases from banks. Bank magnitudes in the United States are estimated with the U.S. EPA's vintaging model, though these estimates likely have large uncertainties. The vintaging model estimates banks as part of its calculation to estimate emissions through an analysis of the number of pieces of equipment in use, the charge size of ODS in the equipment, the loss rate of ODS from these applications, and estimates of how these variables change over time with input from industry. The 2005 banks amounted to seven to sixteen years worth of emissions at rates estimated for 2005, depending upon weighting, or 830 ODP-Kt and 6.2 Gt CO_2-equivalents (direct GWP weighting; Box 2.5) (halon stockpiles not included).

The U.S. EPA vintaging model analysis does include an estimate of how much of the ODS banks are accessible for recovery, where "accessible" refers to ODSs in current air conditioning, refrigeration, and fire protection equipment. While banked halocarbons that are not considered accessible by the U.S. EPA could be recovered and destroyed with appropriate policy measures, market-based incentives, and/or certain technological advances, halocarbons that are not recovered and destroyed will eventually escape into the atmosphere. Banks in the United States that are classified by the U.S. EPA as being accessible amount to approximately 25-30% of the total U.S. bank (ODP, direct or net GWP weighting). Halons account for the majority of the ODP-weighted accessible U.S. bank in 2005, even when stockpiles are not included. HCFCs make up over two-thirds of the accessible GWP-weighted bank. Less than 15% of the total U.S. CFC bank in 2005 is considered accessible (8-14% when net GWP-weighted; ~6% when ODP-weighted) compared to over 30% of the U.S. HCFC bank (55-68% when net GWP-weighted and 37% when ODP-weighted) (Box 2.5).

Comparing these U.S. EPA vintaging model bank size estimates in 2005 with the global banks from WMO (Daniel and Velders *et al.*, 2007) gives that the United States contributed 21 and 26% to the global ODP- and direct GWP-weighted banks, respectively. In this year the U.S. accessible banks accounted for about 5 and 7% of the ODP- and direct GWP-weighted global banks. Future projections from these models suggest that the total U.S. bank will gradually account for less of the global bank over the next decade, shrinking to 14% and 17% of the global ODP- and direct GWP-weighted bank. Similarly, the U.S. accessible bank is projected to decrease to 2 and 4% of the ODP- and direct GWP-weighted global banks. The significance of these banks to integrated EESC

and to ODS recovery times will be addressed in Chapter 5.

Interpretation of these bank comparisons must include consideration of the different assumptions and techniques used to generate the U.S. and the global bank estimates. Although an error analysis has not been performed on either set of numbers, the uncertainties are potentially large, with this uncertainty representing an important gap in our current understanding.

2.2.6 United States Emissions: Derived From Atmospheric Data in Non-remote Areas

Techniques to estimate regional or national emissions of ODSs that are independent of sales data and vintaging models are currently being developed. They rely on high-frequency atmospheric observations (multiple samples per day) in air downwind of source regions. The enhancements observed for ODSs in these air masses can be proportional to emission rates from the upwind source region provided dilution and mixing influences are appropriately considered. These estimates are specific to the region most directly influencing the air reaching a measurement site. Unfortunately, U.S. emissions have been derived using this method with data from only a small number of sites that may not capture regional variations in ODS use and emission rates. The extrapolations are made to the entire United States, for example, based upon population or by reference to enhancements observed in co-measured trace gases whose national emission rates are thought to be better quantified on a national scale (such as carbon monoxide or sulfur hexafluoride).

While the uncertainties associated with this general method can be substantial and estimates have been made for only a small number of years, such an approach offers the only independent test of U.S. emission estimates derived from production and sales data in vintaging models (U.S. EPA, 2007). Estimates of U.S. emissions have been made for selected CFCs, CH_3CCl_3, and CCl_4 over 1996-2003 based on individual studies in California and in the northeast. The results point to a clear decline in U.S. emissions of CFC-12 over this period (Clerbaux and Cunnold *et al.*, 2007). Although U.S. emissions derived for CFCs from these

Halocarbons that are not recovered from banks and destroyed will eventually escape into the atmosphere.

estimates are generally lower than from the U.S. EPA, those for CCl_4 and CH_3CCl_3 are generally higher. On an ODP- or GWP-weighted basis, total U.S. emissions of ODSs derived from observations during 2002-2003 were about half of those estimated by the U.S. EPA's vintaging model analysis. Because it is not known which method is more accurate, no modifications were applied to the history of ODS emissions compiled by the U.S. EPA (2007).

2.3 CHANGES IN THE ATMOSPERIC ABUNDANCE OF OZONE-DEPLETING CHEMICALS AND THEIR SUBSTITUTES

2.3.1 Global Atmospheric Abundances

The influence an atmospheric trace gas has on ozone or climate generally scales with its atmospheric abundance. Atmospheric abundances reflect the integration of past emissions and how persistent a trace gas is in the atmospheric environment (*i.e.*, its global lifetime). A measure of international efforts to minimize the deleterious environmental influences of ODSs and substitutes is found in how successful they are in reducing the atmospheric abundance of these chemicals.

Long-term changes in the global atmospheric abundances of ODSs and substitute chemicals are estimated with different techniques. The atmospheric abundances of the full suite of organic ODSs are generally determined at a small number (less than ten) of remote locations at Earth's surface by independent national and international scientific organizations. These determinations are either made in real time by direct injection of ambient air into on-site instrumentation, or via the analysis of flask samples collected at remote sites and subsequently shipped to a central laboratory. Measurements of the most abundant ODSs (*e.g.*, CFC-12, CFC-11, and HCFC-22) are also made with ground-based infrared solar absorption spectroscopy at selected sites across the globe. The absorption spectroscopy method provides a measure of the total column abundance of these gases above a point on Earth's surface. Measurements are also made with absorption spectroscopy instrumentation onboard satellites. These instruments provide global observations for the most abundant

ODSs. Long-term spectroscopic measurements of these chemicals reveal trends consistent with those observed with ground-based, flask or *in situ* sampling techniques and so are not shown here (Clerbaux and Cunnold *et al.*, 2007).

Because most ODSs and their substitutes have lifetimes of a year or more, they are fairly well-mixed in the atmosphere. As a result, hemispheric and global atmospheric changes can be well captured by measurements at only a few remote sites. Evidence for this can be found in the good agreement noted between global surface mixing ratios derived from the different array of sampling locations and analytical techniques used by these independent organizations. Global surface means derived from these independent laboratories typically agree within a few percent, and often the small discrepancies (typically less than 5%) that do exist for the most abundant ODSs can be attributed to calibration differences (UNEP/TEAP, 2006). Further evidence of this can be found in the consistent measures of atmospheric composition changes provided by spectroscopic total column measurements and those provided by ground-based, *in situ* discreet sample analysis or flask sampling (Clerbaux and Cunnold *et al.*, 2007).

A summary of measured trends for ODSs reveals a wide range of changes in atmospheric mixing ratios for these chemicals and their substitutes, primarily as a result of changes in production and emission brought about by the Montreal Protocol on Substances that Deplete the Ozone Layer (Figure 2.12). As of 2005, the surface mixing ratio and total column burden of the most abundant and long-lived CFC, CFC-12, had begun to decline slowly after reaching a plateau a few years earlier (Clerbaux and Cunnold *et al.*, 2007). Mixing ratios of other ODSs, including CFC-11, CFC-113, CCl_4, CH_3CCl_3, and CH_3Br, have declined persistently over the past five to ten years at rates ranging from −0.5% per year to −18% per year. Halons have been slower to respond to production restrictions, though most data indicate that the atmospheric accumulation of these bromine-containing chemicals has slowed in recent years. Rates of accumulation for halon-1211 and halon-1301 estimated by

Because most ozone-depleting substances and their substitutes have lifetimes of a year or more, they are fairly well-mixed in the atmosphere. As a result, hemispheric and global atmospheric changes can be well captured by measurements at only a few remote sites.

Global Mean Surface Mixing Ratios

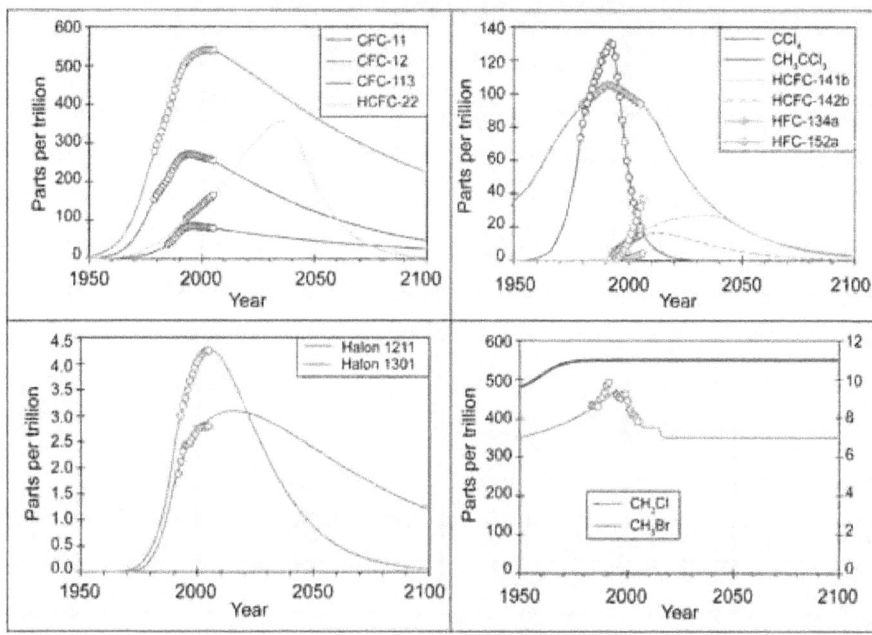

Figure 2.12 Global surface mixing ratios of ODSs and substitute chemicals observed from surface sampling networks (open circles), and as estimated for the past and future in WMO scenario AI (Clerbaux and Cunnold *et al.*, 2007; Daniel and Velders *et al.*, 2007). Past projections are based on histories derived from the analysis of archived air samples, the analysis and modeling of firn-air (air trapped in uncompacted snow in the polar regions) samples, and historic industrial production data.

different laboratories during 2003-2004 range from 0 to 3.2% per year.

Global atmospheric mixing ratios of HCFCs (the chlorine-containing substitutes for CFCs and other ODSs) continue to increase (Clerbaux and Cunnold *et al.*, 2007). The most abundant HCFC, HCFC-22, was present in the background atmosphere at nearly 170 ppt (parts per trillion by mole) in 2005 and has increased fairly steadily at 4 ($\pm$1) % per year for over a decade. Other HCFCs are one-tenth as abundant (or less) in the global atmosphere but increased during 2003-2004 at relative rates similar to HCFC-22. For example, though the global mixing ratio of the least abundant HCFC, HCFC-123, was on the order 0.06 ppt in 2004, its mixing ratio increased at about 6% per year during 2003-2004.

Global atmospheric mixing ratios of HFCs, which are in-kind substitutes that do not contribute any Cl or Br to the atmosphere, have increased quite substantially over the past decade (Clerbaux and Cunnold *et al.*, 2007). HFC-134a is the most abundant of these

substitutes; the global mean surface mixing ratio in the beginning of 2006 was 36 ppt, and it was increasing at a rate of approximately 4.5 ppt per year. The long-lived HFC-23 is the second most abundant HFC; global measured mixing ratios in 2005 were approximately 20 ppt and were increasing. A number of other HFCs have been measured in the remote atmosphere at mixing ratios of a few ppt. After HFC-134a and HFC-23, the HFCs currently emitted in the most significant quantities are HFC-143a and HFC-125.

2.3.1.1 GLOBAL ATMOSPHERIC ABUNDANCES OF OZONE-DEPLETING SUBSTANCES AND SUBSTITUTE CHEMICALS: MEASURED VS. EXPECTED MIXING RATIOS

As of 2005, the initial success of the Montreal Protocol in reducing the threat that ODSs pose to the stratospheric ozone layer is made clear by noting that production declines have led to declining mixing ratios or mixing ratios that are increasing more slowly for all regulated ODSs not considered to be substitutes. The atmospheric response to decreasing production varies for different gases owing to differences

Table 2.2 The most abundant ODSs and substitute chemicals.

Compound	Lifetime (years)	Tropospheric Chlorine or Bromine Contribution (ppt)			EECl Contribution (%)[c]	ODS	ODS Substitute	Regulation[a]
		1985	1995	2005	2005			
CFCs					**45%**			
CFC-11	45	622	808	759	21%	X		M
CFC-12	100	763	1046	1078	18%	X		M
CFC-113	85	114	252	237	5%	X		M
CFC-114	300	25	34	34	0.3%	X		M
CFC-115	1,700	3	8	9	0.1%	X		M
HCFCs					**2.5%**			
HCFC-22	12	61	112	165	1.6%	X	X	M
HCFC-141b	9.3	0	5	35	0.7%	X	X	M
HCFC-142b	17.9	0	6	15	0.2%	X	X	M
HCFC-123	1.3			0.06[b]	0.002%	X	X	M
HCFC-124	5.8			1.7[b]	0.02%	X	X	M
Other Chlorocarbons					**25%**			
CH_3CCl_3	5	324	330	60	1.8%	X		M
CCl_4	26	398	412	376	11%	X		M
CH_3Cl	1	550	550	550	12%[c]	X		None
Bromocarbons					**28%**			
CH_3Br	0.7	8.6	9.5	7.9	15%[c]	X		MA
Halon-1211	16	1.1	3.3	4.2	8%	X		M
Halon-1301	65	0.7	2.3	2.9	3%	X		M
Halon-2402	20	0.4	0.8	0.7	1.4%	X		M
HFCs					**0%**			
HFC-23	270	0	0	0	0		X	K
HFC-125	29	0	0	0	0		X	K
HFC-134a	14	0	0	0	0		X	K
HFC-143a	52	0	0	0	0		X	K
HFC-152a	1.4	0	0	0	0		X	K

[a] "M" represents regulation by the Montreal Protocol, "K" by the Kyoto Protocol, and "MA" is used to show that only the anthropogenic portion of CH_3Br production and consumption is regulated. Lifetimes are from WMO (Clerbaux and Cunnold et al., 2007) and halogen abundances are from WMO Scenario A1 (Daniel and Velders et al., 2007), which are derived from atmospheric observations.
[b] Mixing ratios for these HCFCs are for 2004, not 2005 (Clerbaux and Cunnold et al., 2007).
[c] Bold percentages are calculated for the entire compound class. Methyl chloride and methyl bromide fractions are calculated including natural and manmade components. If only manmade emissions of CH_3Cl and CH_3Br were included (i.e., assuming anthropogenic contributions in 2005 of 0 ppt for CH_3Cl and 1.25 ppt (7.9 - 9.5×0.7) for CH_3Br), the percentages calculated for 2005 abundances would be CFCs: 59%; HCFCs: 3.3%; Other Chlorocarbons: 17%; Other Bromocarbons: 20%; CH_3Br: 3.1%.

in release rates from the applications in which the chemicals were used, and the persistence of the chemicals in the atmosphere. For example, CH_3CCl_3 was used in cleaning applications in which release to the atmosphere followed sales with only a short delay (generally less than one year). This, combined with its relatively short lifetime of approximately five years, resulted in rapid atmospheric decreases once production was curtailed (Figure 2.12). The atmospheric abundance of CH_3CCl_3 has declined since 1998 at near its lifetime-limited exponential rate, which is approximately 20% per year.

Conversely, CFC-11 and CFC-12 were used largely in foam and refrigeration applications in which they only slowly escaped to the atmospheric over decades. This, combined with atmospheric lifetimes of 45 to 100 years, has resulted in only slowly declining atmospheric mixing ratios (Figure 2.12; Table 2.2). The maximum rate of decline in the atmospheric abundance of a chemical with a global lifetime of 100 years, such as CFC-12, is 1% per year and would be observed only if emissions were negligible.

Atmospheric mixing ratios of halons also have been slow to respond to production declines (Table 2.2). This delay is attributable to large banks of chemical in fire-protection installations that are released to the atmosphere during use, servicing, and from leakage, and in the case of halon-1301, its relatively long lifetime.

Carbon tetrachloride is used as a feedstock for production of CFC-11 and CFC-12. As production of these CFCs decreased, so did global emissions and atmospheric mixing ratios of CCl_4. Atmospheric declines have not been as rapid as expected, however, given a lifetime of 26 years. The slower than expected decline suggests the presence of substantial unaccounted emissions (30 to 40 Gg per year since the mid-1990s, or greater than 35% of estimated emissions) or large errors in the estimate of the CCl_4 global lifetime (approximately 26 years; Table 2.2).

The decline in CH_3Br mixing ratios was somewhat faster than expected in response to production declines after 1998, though for this chemical the magnitude of the expected

decline hinges on an accurate understanding of the relative importance of industrial emissions compared to emissions from the natural environment (Clerbaux and Cunnold *et al.*, 2007). The measured decline has been more variable than observed for other ODSs, perhaps because of interannual variability in nonindustrial sources of this chemical, such as biomass burning. Despite these influences over which humans have little direct control, the global mean atmospheric mixing ratio of this chemical through 2006 has decreased each year since 1999, when the gradual phase-out of industrial production and consumption began in developed countries.

Increases in production of ODS substitutes (HCFCs and HFCs) have led to increases in atmospheric mixing ratios for these compounds over the past two decades (Clerbaux and Cunnold *et al.*, 2007).

2.3.1.2 ATMOSPHERIC ABUNDANCES: ON THE ROLE OF VARIATIONS IN LOSS RATES

Atmospheric abundances of ODSs represent a balance between emissions and loss. The Montreal Protocol has resulted in declining emissions of all regulated ODSs not considered to be substitutes. Atmospheric mixing ratios begin to decrease as the natural processes that decompose trace gases in the atmosphere outweigh emissions. For ODSs and their substitutes these natural loss pathways include photolytic destruction primarily in the stratosphere and, for ODSs containing C-H chemical bonds, photochemical oxidation by the hydroxyl radical.

Both loss processes can vary in strength over time because they are influenced by the physical and chemical state of the atmosphere. Accordingly, long-term or short-term variations in rates of photolysis and photo-oxidation have the potential to influence atmospheric mixing ratios in a way that is independent of emission changes brought about by the Montreal Protocol. Estimates of the magnitudes of these loss changes suggest that they are generally small over multi-decadal periods, though it has been suggested that decadal changes in hydroxyl radical abundance can be as large as 15% (Clerbaux and Cunnold *et al.*, 2007). Furthermore, OH reaction rate constants

The atmospheric response to decreasing production varies for different gases owing to differences in release rates from the applications in which the chemicals were used, and the persistence of the chemicals in the atmosphere.

BOX 2.6: On Deriving Atmospheric Abundances from U.S. Emissions

Atmospheric abundances of long-lived ODSs and substitute chemicals can be calculated from an emission history and a simple box model (IPCC/TEAP, 2005; WMO, 2007). A box model includes the influence of emissions and loss rates (or atmospheric lifetimes) to derive atmospheric abundances over time. This common and widely accepted approach is used in this report to estimate atmospheric mixing ratios of ODSs and substitutes arising solely from U.S. emissions of these chemicals. U.S. emission histories are derived over an entire period of ODS use based upon assumptions regarding a potential range of United States/Global emission ratios before 1985 and estimates from a vintaging model analysis thereafter (U.S. EPA, 2007) (Box Table 2.6-1). The U.S. EPA vintaging model is a bottom-up modeling approach that considers market size, amount of ODS in each unit of equipment or application, and ODS substitution trends in order to estimate time-dependent, annual emissions and bank sizes. In the absence of U.S. production or use data for most years before 1985, a number of assumptions were made in order to bracket likely U.S. emissions during these early years. Such assumptions are essential for estimating U.S. contributions to CFC abundances, for example, because much of the pre-1985 emissions of these long-lived chemicals are still present in the atmosphere today. In 1974, as the public became aware of the threat posed to the ozone layer by ODSs, CFC use in fast-release applications in the U.S. was dramatically curtailed. This likely resulted in substantial changes in U.S. emissions of CFCs then. Accordingly, a different approach was used to derive ranges of potential U.S. CFC emissions during 1975-1984 compared to before 1975 (Box Table 2.6-1).

Box Table 2.6-1 Methodology for Deriving Limits to Compound-Specific, Annual U.S. Emissions*:

	Upper Range (emission or emission fraction)	Lower Range (emission or emission fraction)
Pre 1975		
CFCs	[GE]×0.67	[GE]×0.33
Non-CFCs	$GEF_{US/Global (1985-1990)}$ ×1.5	$GEF_{US/Global(1985-1990)}$ ÷1.5
HFC-23	$\alpha \times GPF_{(HCFC-22)US/Global(1985-1990)}$ ×1.2	$\alpha \times GPF_{(HCFC-22)US/Global(1985-1990)}$ ×0.9
1975-1984		
CFCs	interpolate [GE]×0.67 in 1975 to $GEF_{US/Global(1985-1990)}$ ×2 in 1984	$GEF_{US/Global(1985-1990)}$ ÷2
Non-CFCs	$GEF_{US/Global(1985-1990)}$ ×1.5	$GEF_{US/Global(1985-1990)}$ ÷1.5
HFC-23	$\alpha \times GPF_{(HCFC-22)US/Global(1985-1990)}$ ×1.2	$\alpha \times GPF_{(HCFC-22)US/Global(1985-1990)}$ ×0.9
1985-2005		
All HFCs	U.S. EPA (2007) +20%	U.S. EPA (2007) −10%
All others	U.S. EPA (2007) +50%	U.S. EPA (2007) −25%

* where GE = global annual emissions; $GEF_{US/Global(1985-1990)}$ refers to the mean U.S./global emission fraction over the period 1985 to 1990 determined from a ratio of compound-specific emissions from the U.S. EPA (2007) vintaging model and WMO scenario A1 global emissions (Daniel and Velders *et al.*, 2007); and $GPF_{(HCFC-22)US/Global(1985-1990)}$ refers to the mean U.S./global production fraction for HCFC-22 during 1985 to 1990 in data reported by AFEAS (2007) and by the U.S. EPA (2007) and the assumption that HFC-23 emissions were proportional to HCFC-22 production during those years. Alpha (α) indicates the fraction of HFC-23 emitted for a given amount of HCFC-22 produced (~1% by mass, U.S. EPA[2007])

The main uncertainties in calculating atmospheric mixing ratios associated with U.S. emissions are associated with U.S. emission magnitudes and, for some gases, global atmospheric lifetimes. Emission errors stem from uncertainty in the ability of the U.S. EPA vintaging model to accurately capture the mean annual emissions from the many varied applications in which ODSs were used since 1985. Though a chemical-specific uncertainty analysis has not been performed by the U.S. EPA, an uncertainty analysis has been performed on 2005 emissions derived by the U.S. EPA vintaging model for the high-GWP, ODS substitutes as a group (HFCs and PFCs). This analysis, performed with Monte-Carlo techniques, indicated a 95% confidence interval of −9% to +20% on 2005 emissions estimated for this class of compounds.

Compound-specific uncertainties and uncertainties for earlier years are likely to be somewhat larger, though the ODS consumption and production tracking system required for compliance with the Montreal Protocol has added to data reliability.

BOX 2.6: On Deriving Atmospheric Abundances from U.S. Emissions *cont'd*

Given these considerations, we have augmented the uncertainties derived for the high-GWP ODS substitutes by a factor of approximately 2.5 to derive a lower and upper range of −25% to +50% on compound-specific, annual emission estimates derived from the vintaging model during the 1985 to 2005 period.

Limits to United States Emissions of Ozone-Depleting Substances

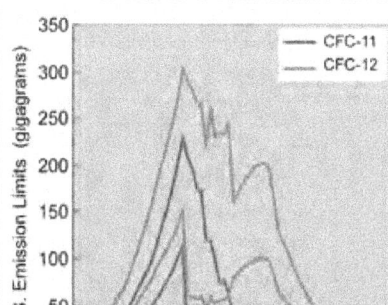
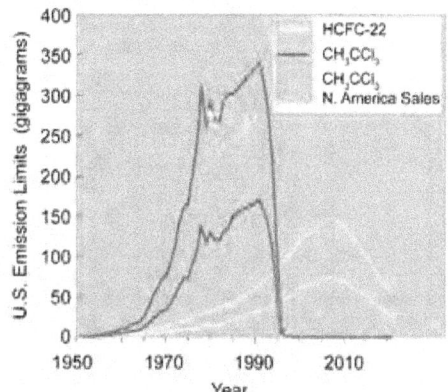

Box Figure 2.6-1 Ranges estimated for U.S. emissions derived from the analysis presented in Box Table 2.6-I; units are Gg, or 10^9 g. These U.S. emissions estimates are compared to North American sales data for CH_3CCl_3 (green points; Midgley and McCulloch, 1999). These data show reasonable consistency within the ranges estimated here for U.S. emissions, considering that the U.S. accounted for approximately 95% of North American consumption of this chemical in the late 1980s (UNEP, 2007) and that releases of CH_3CCl_3 generally occurred within a year after sales. Data to allow similar comparisons for other chemicals are not available.

These approaches have yielded estimates to expected upper and lower ranges to U.S. emissions of ODSs and substitute chemicals (Box Figure 2.6-1).

Errors on calculated atmospheric mixing ratios that are associated with lifetime uncertainties depend upon the use period of an ODS relative to its lifetime. For chemicals that have been emitted for a period that is small compared to their lifetimes (CFC-12 and HFC-23, for example), lifetime uncertainties are relatively small. For other gases such as CH_3Br and CH_3CCl_3 where their use period is long relative to their atmospheric lifetime, lifetime uncertainties are more substantial.

Errors associated with the simple box-model approach are thought to be substantially smaller than the errors already discussed, especially in the case of ODSs whose atmospheric lifetimes are comparable to or much longer than atmospheric mixing times (see, for example, UNEP/TEAP, 2006; Clerbaux and Cunnold *et al.*, 2007).

are temperature sensitive; increases in reaction rates of approximately 10% between OH and CH_4, HCFCs, and HFCs are calculated for a 5°C increase in temperature. Because oxidation by OH is the primary means by which these gases are removed from the atmosphere, compound lifetimes would change similarly. Finally, future changes in large-scale atmospheric circulation patterns have been predicted in some models in response to increased radiative forcing from elevated greenhouse gas abundances and would likely also shorten lifetimes for ODSs where stratospheric losses are relatively important (Section 4.4.1).

2.3.2 The United States Contribution to Global Atmospheric Abundances

Atmospheric mixing ratio histories can be derived from a record of U.S. ODS emissions, estimates of global loss rates (lifetimes), and a simple global box model (Box 2.6). Mixing ratios calculated in this way can be compared to measured and calculated global mixing ratios to estimate the contribution of U.S. emissions to the atmospheric abundance of ODSs and substitute chemicals in the past and future. Though uncertainties in this analysis are large, the results suggest that U.S. emissions of ODSs and substitutes account for between 10

U.S. Contributions to Global ODS Abundances

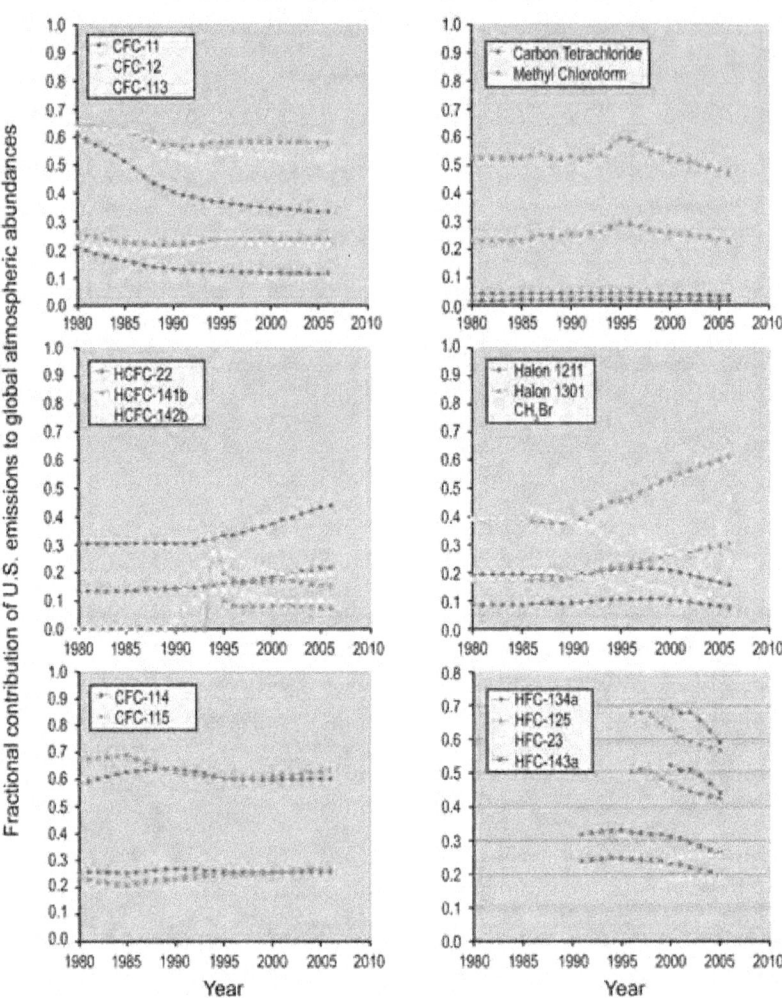

Figure 2.13 Upper and lower ranges to the fractional contributions of U.S. industrial emissions to global atmospheric abundances of ODSs and substitute chemicals over time (see Box 2.6 for the derivation of U.S. emissions and a discussion of uncertainties). For HFC-23, emission from production of HCFC-22 for ODS substitution and feedstock uses is included. For the HFCs, results are only displayed for years when global mixing ratios are greater than 1 ppt. Methyl bromide emissions arising from natural processes are not included, *i.e.*, 146,000 Metric Tons of emission per year (Clerbaux and Cunnold *et al.*, 2007; Daniel and Velders *et al.*, 2007).

and 50% of the global atmospheric abundance measured for most gases in the present-day atmosphere (Figure 2.13). These estimates are most uncertain for the long-lived CFCs because the substantial and poorly constrained emissions that occurred before 1985, when U.S. EPA estimates begin, still contribute significantly to atmospheric abundances today. Uncertainties are smaller for gases having shorter lifetimes and shorter emission histories (*i.e.*, where a higher percentage of total emissions have occurred after 1985).

The smallest fractional contributions of U.S. emissions to present-day mixing ratios are calculated for CCl_4, CFC-11, HCFC-142b, HCFC-141b, and halon-1211. Larger contributions are estimated for some other CFCs, CH_3CCl_3, HCFC-22, halon-1301, CH_3Br, and some HFCs (Figure 2.13).

The largest increases in U.S. contributions in recent years are calculated for CH_3Br, HCFC-22, and halon-1301. The increased fractional contribution of U.S. CH_3Br emissions to its atmospheric abundance arises from increases in U.S. consumption compared to global consumption since 2002 (Section 2.1.4 and Box 2.3). The U.S. fractional contribution to atmospheric mixing ratios of halon-1211, HCFC-141b, CH_3CCl_3 and some HFCs has decreased in recent years (Figure 2.13).

2.4 THE ATMOSPHERIC ABUNDANCE OF AGGREGATED CHLORINE AND BROMINE FROM LONG-LIVED ODSs

Many different chemicals contribute to atmospheric chlorine and bromine, such as sea salt, pool disinfectants, CFCs, and HCFCs. Only those with longer lifetimes (greater than months) and lower water solubilities escape scavenging by aerosols and rain, however, and become efficiently transported to the stratosphere and contribute to ozone depletion (ODSs such as CFCs, HCFCs, halons, CH_3Br, and others). Hence, tropospheric burdens of these long-lived ODSs are closely monitored because they provide a useful measure of changes and amounts of chlorine and bromine being transported to the stratosphere and that will ultimately become available for catalyzing the destruction of stratospheric ozone.

Stratospheric abundances of chlorine and bromine are more difficult to regularly measure with high precision. Such measurements are useful, however, to discern whether amounts and changes in stratospheric chlorine and bromine are well described by the total amount of chlorine or bromine measured in long-lived ODSs at Earth's surface. Stratospheric measurements take advantage of the fact that in the upper stratosphere, nearly all organic compounds have become photo-oxidized and chlorine and bromine exist primarily in only one or two chemical forms (*e.g.*, hydrogen chloride, HCl, and chlorine nitrate, $ClONO_2$; and bromine oxide, BrO, for example). As a result, stratospheric measurements of these few inorganic chemicals provide an integrated estimate of how ozone-depleting halogen levels are changing, and whether or not these changes are consistent with ODS observations in the lower atmosphere.

2.4.1 Atmospheric Chlorine

As a result of the restrictions on production and consumption of ODSs brought about by the Montreal Protocol, the abundance of chlorine measured in long-lived gases has been decreasing in the lower atmosphere since 1995 and has continued to decrease through 2004 (Clerbaux and Cunnold *et al.*, 2007). Global tropospheric chlorine in long-lived chemicals was 3.44 ppb in 2004, or 0.25 ppb below the peak observed in the early 1990s. The rate of tropospheric decline in total chlorine from all regulated ODSs during 2003-2004 was slightly slower than four years earlier, as the influence of CH_3CCl_3 continued to diminish; the mean decline during 2003-2004 was 20 ppt per year (0.6% per year).

Approximately 80-85% of organic chlorine in long-lived trace gases measured in the troposphere is accounted for by gases regulated by the Montreal Protocol (CFC-11, CFC-12, CFC-113, CFC-114, CFC-115, HCFC-22, HCFC-142b, HCFC-141b, HCFC-124, HCFC-123, CH_3CCl_3, CCl_4, and halon-1211). Most of the remaining 15% (or 550 ppt Cl) is accounted for by methyl chloride, a chemical having predominantly (greater than 95%) non-industrial sources. Atmospheric mixing ratios of CH_3Cl have been relatively constant over the past decade, though year-to-year variations on the order of a few percent can be observed at Earth's surface during years with enhanced biomass burning (Clerbaux and Cunnold *et al.*, 2007).

Small additional amounts of chlorine (~2% of the sum from regulated ODSs and CH_3Cl, or 50-80 ppt Cl) may be contributed by short-lived gases such as CH_2Cl_2, $CHCl_3$, C_2Cl_4, and others (Law and Sturges *et al.*, 2007). Tropospheric observations suggest that mixing ratios of some of these gases have decreased in recent years.

Because HCl and $ClONO_2$ have quite low abundances in the lower atmosphere, total column measurements of HCl and $ClONO_2$ from surface-based spectroscopic instruments provide an independent measure of stratospheric chlorine changes (Clerbaux and Cunnold *et al.*, 2007). These measurements show that, after many years of consistent increases, total column inorganic chlorine mixing ratios peaked in the mid-to-late 1990s and have since

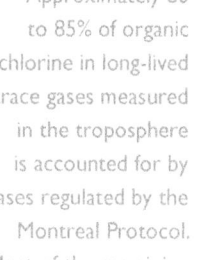

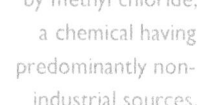

Approximately 80 to 85% of organic chlorine in long-lived trace gases measured in the troposphere is accounted for by gases regulated by the Montreal Protocol. Most of the remaining 15% is accounted for by methyl chloride, a chemical having predominantly non-industrial sources.

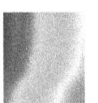

declined. The declines have lagged behind the decreases observed at Earth's surface by a few years because of time lags associated with transporting air in the lower atmosphere to the stratosphere.

Satellite-based spectroscopic instruments also have the potential to measure changes in stratospheric chlorine abundance over time (Clerbaux and Cunnold *et al.*, 2007). To date, however, long-term trends determined from these instruments have added uncertainty from numerous complications related to small unexplained offsets in HCl measured by different instruments, substantial unexplained variability in the longest record (Halogen Occultation Experiment), and relatively short data records for other instruments that offer higher precision (Atmospheric Chemistry Experiment and Microwave Limb Sounder).

Despite these issues, satellite instruments have provided an important independent measure of stratospheric chlorine mixing ratios (Clerbaux and Cunnold *et al.*, 2007). Results from these instruments demonstrate that mixing ratios of chlorine observed in the upper stratosphere

are well explained by measured tropospheric abundances of long-lived ODSs regulated by the Protocol plus a contribution from CH_3Cl of approximately 15%. The scatter among results from different instruments prevents a precise estimate of the contribution of short-lived gases to stratospheric chlorine, but they do suggest it is on the order of a few percent, consistent with the tropospheric observations.

As of 2004, ground-based air sampling results show that CFCs still account for most of the long-lived Cl in the troposphere, (62% in 2004; Clerbaux and Cunnold *et al.*, 2007). The abundance of the three most abundant CFCs has peaked or is decreasing in the troposphere, with declines in CFC abundances accounting for about half of the decline in total tropospheric Cl in 2004 (−9 ppt Cl per year). Methyl chloroform still strongly influences total chlorine trends despite its dramatically reduced atmospheric abundance; in 2004 it accounted for more than half of the observed decline in Cl (−13.5 ppt Cl per year). Carbon tetrachloride has declined fairly steadily at a rate of 1 ppt per year and has accounted for an annual change in tropospheric chlorine of −4 ppt Cl per year during the past

As of 2004, the abundance of the three most abundant CFCs had peaked or was decreasing in the troposphere, with declines in CFC abundances accounting for about half of the decline in total tropospheric chlorine in that year.

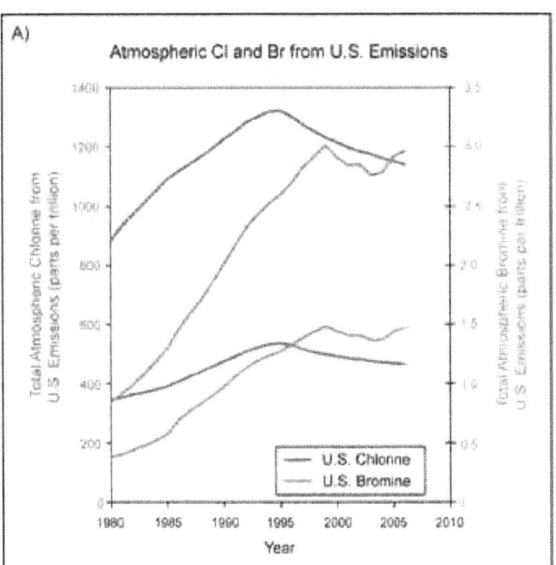

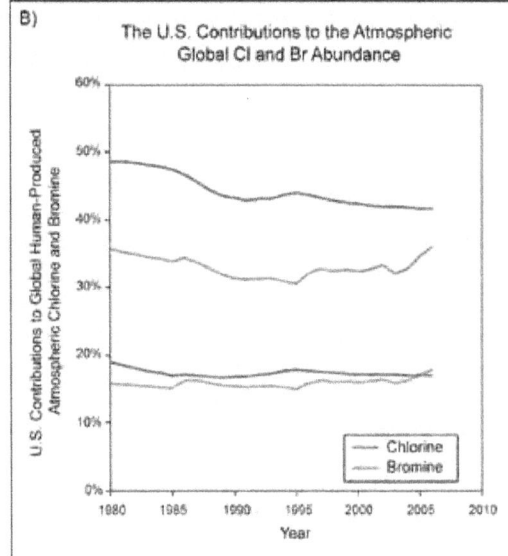

Figure 2.14 A) Estimated ranges of the tropospheric abundance of Cl and Br from U.S. emissions of all regulated ODSs. These ranges are from upper and lower estimates of U.S. emissions (see Box 2.6 for additional information). B) Estimated ranges in the fractional contribution of U.S. emissions to global atmospheric mixing ratios arising from industrial production of ODSs. In both panels, only emissions arising from industrial production of ODSs for restricted uses plus QPS uses of CH3Br were considered; global and U.S. emissions of CH3Cl and CH3Br from natural processes are not included in these calculations. Chemicals included in these estimates of total Cl: CFC-11, CFC-12, CFC-113, CFC-114, CFC-115, HCFC-22, HCFC-142b, HCFC-141b, HCFC-123, HCFC-124, CH3CCl3, CCl4, and halon-1211; and of total Br include: CH3Br, halon-1211, halon-1301, and halon-2402.

decade. In 2004 CH_3CCl_3 accounted for 2% (65 ppt Cl) and CCl_4 accounted for 11% (375 ppt Cl) of all long-lived Cl in the troposphere. (These declines total more than 100% owing to the offsetting increases observed for chlorine from HCFCs.)

HCFCs continue to increase in the atmosphere as they are used as substitutes for CFCs and other ODSs (Clerbaux and Cunnold *et al.*, 2007). They accounted for 6% of total tropospheric Cl in 2004, and chlorine from HCFCs increased at a rate of nearly 8 ppt Cl per year during that year. Though the increase in Cl from HCFCs during 2004 was significantly slower than observed in 1996-2000, near-term projections of production and use, and continued observations since the publishing of the latest WMO Scientific Assessment of Ozone Depletion Report (Clerbaux and Cunnold *et al.*, 2007), show accelerating growth rates since 2004.

Tropospheric chlorine attributable to U.S. emissions of long-lived ODSs also has declined since the early 1990s (Figure 2.14). The U.S. contribution to global tropospheric chlorine from all regulated ODSs (excluding consideration of CH_3Cl) is estimated as being between 17 and 42% in 2005. Despite this large uncertainty range, estimates of the rate of change have smaller uncertainties and suggest that the U.S. relative contribution to global atmospheric chlorine from regulated ODSs has remained fairly constant over the past decade.

2.4.2 Atmospheric Bromine
Bromine in the stratosphere catalyzes the destruction of ozone with a per-atom efficiency that is approximately 60 times that of chlorine (WMO, 2007). As a result, small mixing ratios of stratospheric bromine play an important part in controlling stratospheric ozone abundances. Bromine also differs from chlorine because emissions from regulated uses account for a smaller fraction of the inorganic bromine measured in the stratosphere. Whereas chlorine emissions from uses regulated by the Montreal Protocol accounted for approximately 80-85% of stratospheric chlorine at its peak abundance, emissions of bromine regulated by the Protocol accounted for approximately 50% of the bromine measured in the stratosphere at its

peak abundance of 20-22 ppt (estimated by assuming 30% of 9.5 ppt from CH_3Br as arising from regulated uses, plus 8 ppt Br from halons) (Law and Sturges *et al.*, 2007).

Chemicals containing bromine that are regulated by the Montreal Protocol include halons and methyl bromide, though methyl bromide has both natural and anthropogenic emissions. When both natural and anthropogenic sources of CH_3Br are considered together with the bromine from halons, these chemicals accounted for approximately 80-90% of total bromine reaching the stratosphere in 1998. Surface-based measurements show that total tropospheric bromine from these chemicals peaked in 1998 and has since declined (Clerbaux and Cunnold *et al.*, 2007). By mid-2004, tropospheric bromine from these gases was 0.6 to 0.9 ppt below the peak amount. The decline was entirely a result of declining CH_3Br mixing ratios. By 2004 the tropospheric mean CH_3Br mixing ratio had declined by 1.3 ppt (14%) from its peak in 1998. Although the rate of decline of CH_3Br was variable over this period, global mixing ratios of CH_3Br have declined each year during this period as global production decreased. Global atmospheric mixing ratios of the halons were still increasing slowly in 2004, albeit at slower rates than in earlier years (2004 rates of increase for the halons were less than 0.1 ppt per year). Continued increases in halon mixing ratios arise from continued production allowed in developing nations and slow leakage rates from large banks of halons in developed countries in fire extinguishers that are still in use.

Trends in global mean bromine accounted for by short-lived gases are not easily measured from ground-based stations, because of the high variability observed and the potential for local influences to dominate measured abundances and changes. In these instances, data from firn air (air trapped in uncompacted snow in the polar regions) have improved our understanding of historical changes in the atmospheric abundance of these chemicals. Firn air integrates atmospheric abundances over decadal periods so short-term variations are smoothed, but local influences could mask broader changes. Atmospheric histories of short-lived brominated chemicals such as dibromomethane (CH_2Br_2) and bromoform

Whereas chlorine emissions from uses regulated by the Montreal Protocol accounted for approximately 80-85% of stratospheric chlorine at its peak abundance, emissions of bromine regulated by the Protocol accounted for approximately 50% of the bromine measured in the stratosphere at its peak abundance.

**BOX 2.7: Equivalent Effective Chlorine (EECl) and
Equivalent Effective Stratospheric Chlorine (EESC)**

The threat posed to the ozone layer from ODSs is not directly proportional to the summed mixing ratios of these chemicals in the troposphere. Instead, it depends upon the number of chlorine and bromine atoms contained in the ODSs, how rapidly the ODSs degrade once they reach the stratosphere and liberate ozone-depleting forms of chlorine and bromine, and the abundance of bromine relative to chlorine contained in the mix of ODSs reaching the stratosphere (given that bromine is 60 times more reactive, on average, than chlorine). To account for these influences, indices have been developed to estimate changes in the burden of reactive stratospheric halogen in a simple manner based on observed changes in tropospheric abundances of ODSs.

Equivalent Effective Chlorine (EECl) is one such index used here and elsewhere (WMO, 2007) to quantify overall changes in reactive halogen trends based upon the measured mix of ODSs in the troposphere. The timing associated with EECl changes correspond to the dates those changes were measured in the troposphere.

Equivalent Effective Stratospheric Chlorine (EESC) is a closely related index used to estimate the time evolution of ozone-depleting halogen in the stratosphere. In most past formulations it has differed from EECl only in that it includes a time lag associated with transporting air from the troposphere, where ODS measurements are regularly made, to the stratosphere. EESC is often used to estimate when the cumulative effect of all ODSs on ozone will return to a level attained at some earlier time, often chosen to be 1980, assuming no changes in dynamical, climate, or other non-ODS-related influences (WMO, 2007). Quite different "recovery" times can be calculated for midlatitude EESC and Antarctic EESC in springtime when lag times of three years for midlatitudes and six years for Antarctica are assumed (see Chapter 5 for additional discussion of recovery times). Furthermore, EESC projections for different scenarios of ODS uses have been an important tool for assessing the potential influence of various policy choices on ozone.

Recently, EESC has also been used to improve our understanding of the extent to which changes in ozone abundances may be due to policy restrictions under the currently adjusted and amended Montreal Protocol. Specifically, attempts have been made to identify both a slowing of the declining ozone trends and even a reversal of the decline, and whether these recovery milestones can be attributed to ODS changes.

Despite its usefulness, EESC provides only a rough estimate of the effect of ODSs on stratospheric ozone because it incorporates simplified assumptions regarding mixing processes and degradation rates. Recent efforts to enhance the formulation perhaps provide a more realistic evolution of stratospheric reactive halogen over time and space (Newman et al., 2006, 2007). Other differences in more recent work relating to the calculation of EESC suggest some rather large alterations to estimated ODS recovery times and are currently a source of uncertainty in this analysis.

($CHBr_3$), derived in this way suggest no large long-term changes in atmospheric mixing ratios in polar regions during the past two decades (Law and Sturges et al., 2007).

As was true for chlorine, the integrated influence of changes in the tropospheric abundance of brominated gases can be measured in the stratosphere from airborne, balloon-borne, and satellite instrumentation. These data have been important for quantifying the role non-regulated chemicals play in controlling the abundance of bromine in the stratosphere. They have also demonstrated that the total abundance of Br in the stratosphere has increased over time in a manner that can be explained by the tropospheric mixing ratio changes observed

for halons and CH_3Br, considering lag times associated with air transport (Law and Sturges et al., 2007). A recent study published since WMO (2007) suggests that the accumulation rate of bromine in the stratosphere has slowed in a manner consistent with the trend observed in the troposphere, after considering the time it takes to transport air from the troposphere to the stratosphere (Dorf et al., 2006).

Tropospheric bromine attributable to U.S. emissions of halons and CH_3Br also peaked in 1998-1999 and declined through 2003, as estimated by the U.S. EPA (U.S. EPA, 2007) (Figure 2.14). Since 2003, however, the increased emissions of CH_3Br from Critical Use Exemptions and QPS uses (Box 2.3) have caused the tropospheric abundance of bromine attributable to U.S. emissions to increase. While the contribution of United States to total atmospheric bromine (the industrially derived emissions from regulated uses only) declined throughout the 1980s and early 1990s, it reversed course and actually increased by about 7% from 2000-2005; the U.S. contribution to atmospheric bromine is estimated at between 17 and 35% in 2005 (only halons and anthropogenic CH_3Br considered).

2.4.3 Equivalent Effective Stratospheric Chlorine and Equivalent Effective Chlorine

The combined influence of changes in chlorinated and brominated ODSs on reactive halogen abundances in the stratosphere and on stratospheric ozone can be assessed from aggregate quantities such as EESC and EECl (Box 2.7). These quantities are derived with weighting factors applied to tropospheric ODS abundances to provide a rough estimate of how total reactive halogen abundances are changing or will likely change in the stratosphere based upon observed trends in tropospheric mixing ratios of ODSs.

EECl from measured global surface mixing ratios of regulated ODSs, substitutes, and CH_3Cl peaked in 1994 and has since declined (Figure 2.15). By 2004, EECl had declined by 277 ppt, or 8-9% from the peak. This decline represents about 20% of the decline needed for EECl levels to return to their 1980 levels (Clerbaux and Cunnold et al., 2007). Most of this decline resulted from changes in the atmospheric abundance of the shorter-lived ODSs: CH_3CCl_3 and CH_3Br.

In 2005 CFCs still contributed the most to the atmospheric burden of EECl (45%) from all long-lived chlorinated and brominated chemicals, including those with large natural sources (CH_3Cl and CH_3Br) (Table 2.2). Other chlorinated gases contributed 25%, brominated gases contributed 28%, and HCFCs contributed 2.5% to EECl in 2005 (Table 2.2). These percentages include natural contributions to the atmospheric abundance of CH_3Cl and CH_3Br. When only anthropogenic contributions to 2005 EECl are considered, the relative contribution of CFCs increases (to 59%), that for chlorocarbons

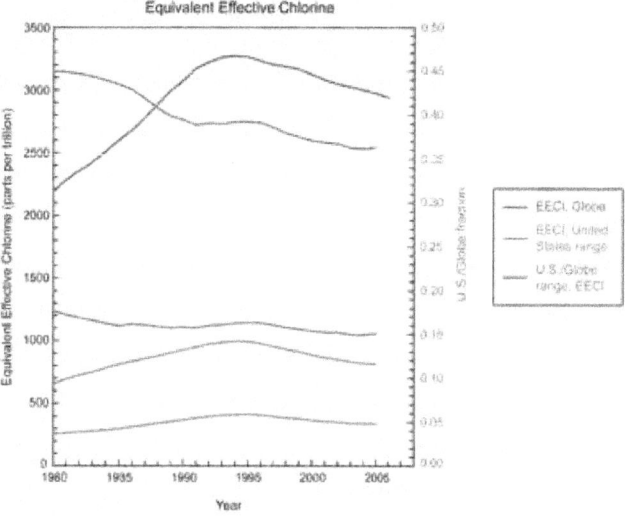

Figure 2.15 EECl from measured global mixing ratios of ODSs (blue line) and from estimates of ODS mixing ratios attributable to U.S. emissions (upper and lower ranges bounded by red lines; see Box 2.6 for discussion of U.S. emissions). While all long-lived ODSs, including CH_3Cl and the natural contribution of CH_3Br, are included in global EECl, natural contributions are not included in the United States/Global fraction or EECl calculated from U.S. emissions alone. Also shown are ranges for the fraction of EECl attributable to U.S. emissions (green lines, right-hand axis).

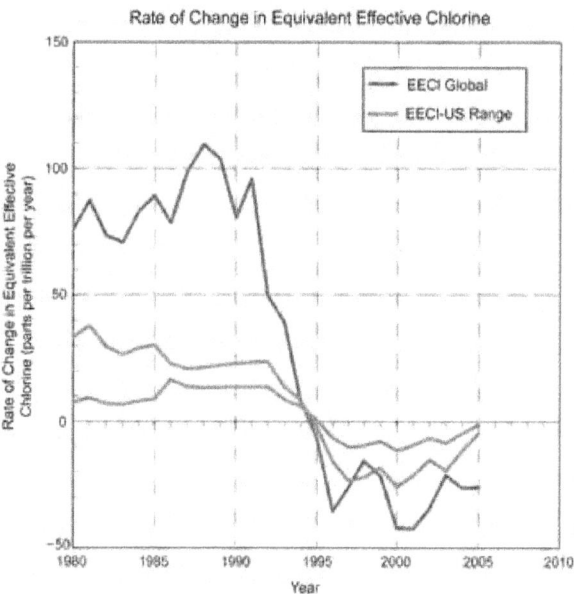

Figure 2.16 Rate of change in global EECl derived from measured global atmospheric mixing ratios of ODSs and substitute chemicals (blue line), and the rate of change in EECl derived from atmospheric mixing ratios calculated from upper and lower bounds on U.S. emissions of ODSs (bounded by red lines; see Box 2.6 for discussion of U.S. emissions). High and low U.S. EECl estimates express the influence of emission uncertainties on these rates (Box 2.6).

decreases (to 17%), that for bromocarbons decreases (to 20%), and that for CH_3Br alone becomes 3.1% (see footnote to Table 2.2 for grouping definitions).

Declines are also calculated for EECl attributable to U.S. emissions of ODSs for regulated uses during 1994-2004, though substantial uncertainty in atmospheric abundances derived from U.S. emissions prevents a precise determination of EECl from U.S. emissions alone. United States emissions of ODSs for regulated uses have accounted for between 15 and 39% of total EECl from regulated chemicals from 1994 to 2004, and between 15 and 36% in 2005.

Despite the added uncertainty U.S. emissions before 1985 add to estimates of the U.S. contribution to Cl, Br, and EECl in today's atmosphere, these uncertainties have a much smaller influence on our understanding of changes in these quantities (Figure 2.16). Global EECl declined fairly consistently since the mid-1990s. EECl from U.S. emissions followed global trends until about 2003, when declines in U.S. EECl slowed substantially. From 2004 to 2005, U.S. EECl declines were substantially smaller than in earlier years, primarily because of the increases in U.S. emissions of brominated gases during these years (primarily CH_3Br) (Figure 2.16).

On uncertainty in the U.S. contribution. Though estimates of the rate of change in EECl attributable to U.S. emissions of ODSs and substitutes are much less dependent upon pre-1985 emission rates, they do rely on the accuracy of emission algorithms of ODSs from in-use applications. Such algorithms are difficult to verify experimentally on national scales (Section 2.2.6).

2.4.3.1 ESTIMATING REACTIVE HALOGEN TRENDS FOR THE MIDLATITUDE STRATOSPHERE

Changes in stratospheric halogen abundance are delayed from those observed in the troposphere because of the time it takes for air to be transported from the troposphere to stratosphere. In addition to this time lag, mixing processes also influence how tropospheric composition changes propagate to the stratosphere. Estimates

of stratospheric halogen trends in the midlatitude stratosphere have been roughly derived with a lag of three years on EECl. The EESC calculated for the midlatitude stratosphere suggests that by 2004, the midlatitude stratospheric halogen burden had declined by approximately 7% from its peak.

2.4.3.2 ESTIMATING REACTIVE HALOGEN TRENDS USING EESC FOR THE ANTARCTIC STRATOSPHERE

Measurements suggest that air found in the lower Antarctic stratosphere during the early springtime has resided in the stratosphere for about six years, or approximately twice as long as it takes to transport air from the troposphere to the midlatitude stratosphere. This influence, combined with the expected slow decline in EESC during the twenty-first century compared to the relatively fast buildup around 1980, suggests that it will take 15-20 years longer for EESC in Antarctica to fall below the 1980s levels than it will for midlatitude EESC to drop similarly (Daniel and Velders *et al.*, 2007) (see also Chapter 5). By 2004, EESC over Antarctica is estimated to have declined from peak levels by only 3%, when estimated simply as a six-year lag of EECl. More detailed analyses of these projections are currently being refined to include mixing effects and a better representation of decomposition rates for individual ODSs and substitute chemicals (*e.g.*, Newman *et al.*, 2006).

2.5 CHANGES IN RADIATIVE FORCING ARISING FROM OZONE-DEPLETING CHEMICALS AND SUBSTITUTES

2.5.1 Changes in Direct Radiative Forcing

As previously noted in this chapter, weighting emissions by 100-year GWPs allows one to compare the integrated radiative forcing from various greenhouse gases, including ODSs, with the intent to gain insight into the resulting climate effects (see Box 2.2). Instantaneous radiative forcing is generally calculated by multiplying the atmospheric mixing ratios of the various GHGs by their radiative efficiencies (Section 2.1.1). While the relationship between changes in radiative forcing and global average temperature vary somewhat among models,

United States emissions of ozone-depleting substances have accounted for between 15 and 39% of total Equivalent Effective Chlorine, an index used to quantify overall changes in reactive halogen trends, from regulated chemicals from 1994 to 2004, and between 15 and 36% in 2005.

radiative forcing remains arguably the best simple metric available to compare the direct climate effect of greenhouse gas abundances. As with "direct" GWPs (Box 2.2), direct radiative forcing represents the forcing of GHGs due to their own absorption of infrared energy and neglects any potential chemical or other feedbacks.

The direct, global radiative forcing due to ODSs and substitutes reached about 0.34 W per m² in 2005 and was still increasing slowly (Figure 2.17). Due to compliance with the Montreal Protocol, however, the recent increase in radiative forcing was much slower than measured in the early 1990s. HFCs contributed a noticeable amount to this increase on a global scale. Radiative forcing from non-HFC ODSs and substitutes changed less than 0.001 W per m² from 2001-2005. For the purpose of putting this total direct forcing into perspective, the amount of radiative forcing due to CO_2, CH_4,

and N_2O in 2005 was approximately 1.66 W per m², 0.48 W per m², and 0.16 W per m², respectively (Forster *et al.*, 2007). While the increase in forcing from ODSs has occurred relatively rapidly, the decrease will be largely limited by the global ODS lifetimes, and will occur more slowly. These future projections will be further discussed in Chapter 5.

The relative direct forcing contributions of classes of ODS chemicals and their replacements are shown in Figure 2.18. The CFCs have contributed between 79 and 86% of the total direct forcing from ODSs since 1980. However, over the last decade the fractional contribution of the CFCs has declined, as have the fractional contributions of CH_3CCl_3 and CCl_4. Increases in HCFC abundances, primarily HCFC-22, and in HFC abundances are counteracting the decline of these other gases. Consideration of the indirect forcing due to ODS-induced ozone depletion alters this figure somewhat, though

The direct, global radiative forcing due to ODSs and substitutes reached about 0.34 W per m² in 2005 and was still increasing slowly. Putting this total direct forcing into perspective, the amount of radiative forcing due to CO_2, CH_4, and N_2O in 2005 was approximately 1.66 W per m², 0.48 W per m², and 0.16 W per m², respectively.

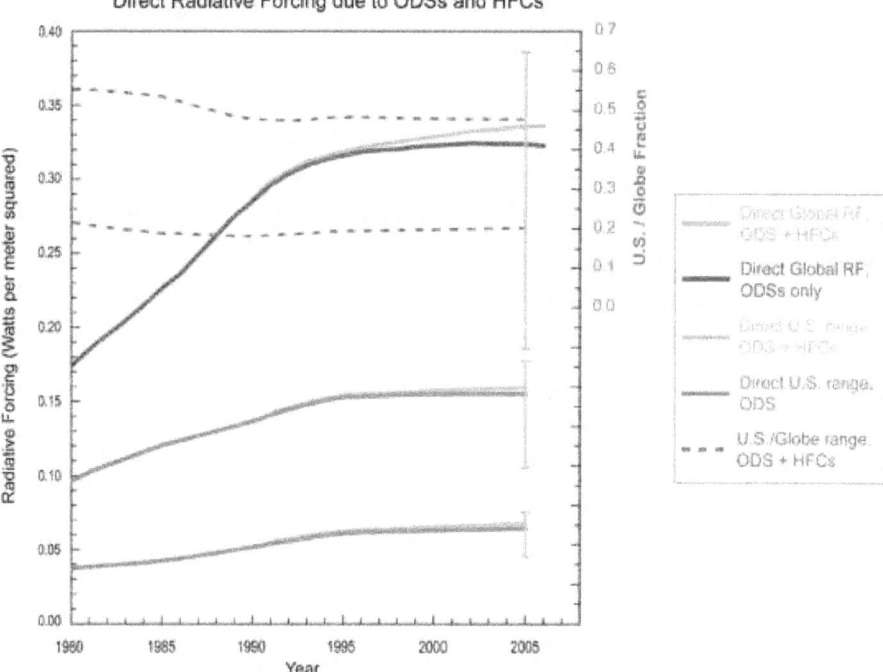

Figure 2.17 Radiative forcing time series arising from changing atmospheric mixing ratios of ODSs and their substitutes. Direct radiative forcing is calculated by weighting global atmospheric mixing ratios of ODSs or a range of ODS mixing ratios attributable to the U.S. (see Box 2.6) by compound-dependent radiative efficiencies for the years 1980-2005. The additional forcing contribution of HFCs to global or U.S. radiative forcing is indicated by the lighter-colored lines. The uncertainties indicated for 2005 include the range of influences (and uncertainty) of the indirect forcing associated with stratospheric ozone depletion. This indirect influence was estimated for the U.S. by considering the range of EECl attributable to the U.S. relative to total global EECl in 2005 (see Box 2.2).

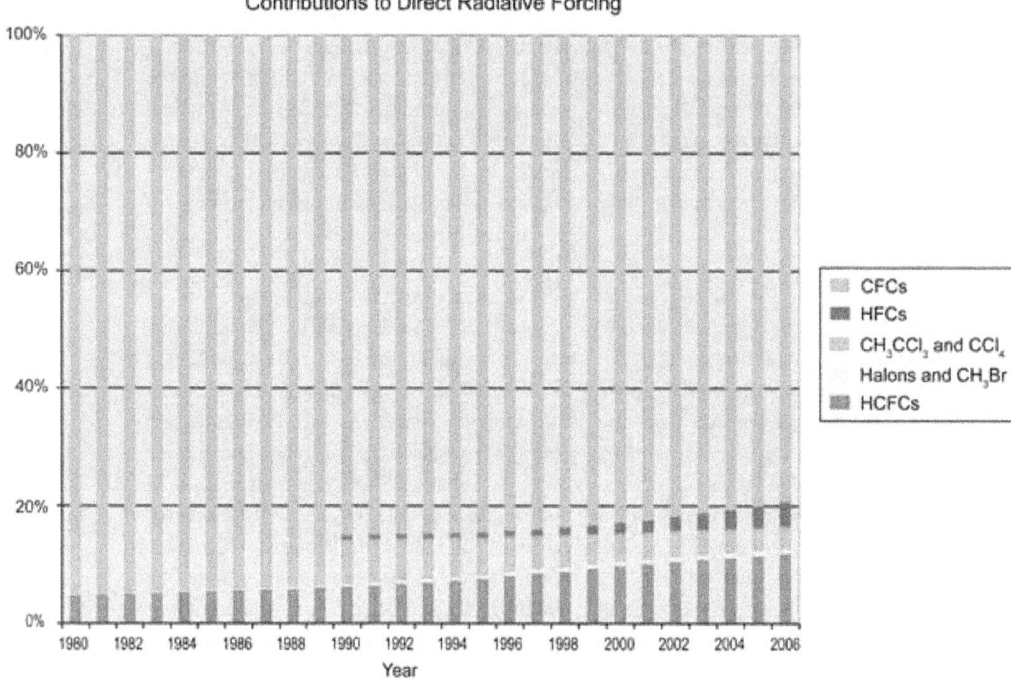

Figure 2.18 Relative contributions of ODS compound classes to global direct radiative forcing. Percentages in 2005 are CFCs=80%, HCFCs = 12%, other ODSs = 5%, HFCs = 4%.

in 2005, for example, the percent contribution of each compound class changes by less than 4% when a mean radiative ozone influence of -0.05 W per m^2 is considered (derived by scaling ozone radiative forcing of -0.05 W per m^2 by the fraction of EECl contributed by each compound class).

2.5.1.1 ESTIMATING THE U.S. CONTRIBUTION TO DIRECT RADIATIVE FORCING CHANGES

Using estimates of atmospheric mixing ratios of ODSs arising solely from U.S. emissions (Section 2.3), the U.S. contribution to the direct global radiative forcing from ODSs and substitutes is calculated to have been between 0.068 and 0.16 W per m^2 in 2005 (Figure 2.17). The error bars are calculated by summing the individual high and low direct forcing estimates for each of the ODSs. Since 1990, the U.S. contribution has accounted for between 19 and 49% of the global direct forcing from ODSs and substitute chemicals. The contributions of the various ODS classes to radiative forcing show the same qualitative behavior for U.S. emissions as for the global emissions that are apparent in Figure 2.18; HCFC contributions are increasing while CFC, CH_3CCl_3, and CCl_4 contributions

are decreasing. We estimate that the United States accounted for about 45% of the increase in direct radiative forcing arising from HCFCs during 2000 to 2006.

2.5.2 Changes in Net Radiative Forcing

As stated in Box 2.2, the direct quantities represent the effect of the ODSs themselves on radiative forcing through their absorption of infrared energy. Their destruction of stratospheric ozone leads to an additional radiative effect, referred to as an "indirect" effect of the ODSs.

In the past, it had been thought that the indirect forcing might have offset as much as 50% of total direct halocarbon radiative forcing, but estimates from different studies varied widely (Ramaswamy *et al.*, 2001). More recently, now that ozone trends are thought to be better quantified, particularly in the radiatively important region near the tropopause, the magnitude of the ozone forcing has been revised to -0.05 ± 0.05 W per m^2 for the changes in stratospheric ozone between 1979 and 1998 and -0.05 ± 0.10 W per m^2 for the changes in stratospheric ozone from preindustrial times to 2005 (Forster *et al.*, 2007). Although revised to a smaller value, the ozone forcing remains potentially significant relative to the direct forcing and continues to be highly uncertain, with even the sign in doubt (Figure 2.17; see Box 2.2). However, it is known that this indirect forcing will gradually become negligible as ozone recovers from depletion due to ODSs (Daniel *et al.*, 1995; see Chapter 5).

The IPCC (2007) estimate for the magnitude of the ozone forcing through 2005 (-0.05 ± 0.10 W per m^2) is added to the global direct ODS forcing in Figure 2.17. As discussed in Box 2.2, there are potential problems related to adding these direct and indirect forcing quantities; nevertheless, such an addition does permit a comparison of the sizes of these globally averaged forcings. Estimates for the U.S. contribution to the indirect forcing are also added to the U.S. direct ODS forcing curves. The U.S. indirect forcing has been calculated by scaling the global ozone forcing range by the U.S. contribution to global EECl. For both the global and U.S. cases, the indirect forcing remains potentially significant but quite uncertain compared to the direct radiative forcing estimates.

2.6 SUMMARY OF FINDINGS RELATED TO THE ROLE OF THE UNITED STATES IN INFLUENCING PAST CHANGES IN PRODUCTION, CONSUMPTION, EMISSIONS, AND MIXING RATIOS OF OZONE-DEPLETING SUBSTANCES AND THEIR SUBSTITUTES

Stratospheric ozone depletion is a global issue because the amount of ozone depletion above the United States, or any other location, results from the global emission of ODSs. A reduction in U.S. ODS emissions leads to reduced ozone depletion above the United States only to the extent to which it reduces global ODS emissions.

In response to restrictions put into place under the Montreal Protocol, U.S. consumption of ODSs for regulated dispersive uses, considered in sum, have declined substantially from peak levels. By 2005, ODP-weighted consumption in the United States had declined by 97-98%,

or nearly 400 ODP-Kt since the late 1980s (UNEP, 2007). These data indicate that the United States accounted for 24 (±2)% of total global production of ODSs during the years of substantial production (1986-1994), and 10 (±2)%, on average, during 2001-2005 (when weighted by ODPs).

United States consumption declines have been slightly faster than phase-out schedules for all developed countries in the adjusted and amended Montreal Protocol for most ODSs. Critical Use Exemptions have resulted in ODS consumption for CH$_3$Br and CFCs above these scheduled allotments in recent years. For example, though CH$_3$Br consumption in developed countries was to have been zero in 2005 and thereafter, U.S. consumption for critical uses was 28% of 1986 baseline consumption during this year (4.4 ODP-Kt) and, in 2006, was 27% of 1986 baseline consumption (4.1 ODP-Kt) (UNEP/MBTOC, 2007). Authorized CUEs for CH$_3$Br consumption in the United States for 2007 were reduced compared to 2005 and 2006 (17% of the U.S. 1986 baseline consumption). U.S. consumption also has continued for critical uses of CFCs despite the 1996 phase-out, though this consumption has been comparably small (less than 1% of 1986 United States baseline CFC consumption in all years since 1996).

The decreases in global and U.S. production and consumption have led to substantial reductions in emissions of most ODSs to the atmosphere. But while global emissions can be derived from observed global atmospheric changes and knowledge of ODS atmospheric lifetimes, U.S. emissions and their changes are more uncertain, as they were derived from vintaging model analyses of sales, use, and release patterns of ODSs (a "bottom-up" analysis; U.S. EPA, 2007). The results suggest that both global and U.S. emissions of ODSs declined overall by 81-82% since the late 1980s when weighted by ODPs (Table 2.1). Furthermore, this analysis suggests that the United States accounted for a decreasing amount of global ODS emissions, from a peak of 35% in 1993 to 20 (±2)% during 2000-2005 (Figure 2.7).

Similar relative contributions of U.S. emissions in 2005 to global atmospheric abundances are calculated for the atmospheric abundance of chlorine (17-42%), bromine (17-35%), and EECl (15-36%) during 2000-2005. EECl arising from U.S. emissions declined every year from 1994 through 2004, but did so much more slowly during 2004 to 2005, largely due to the recent increase in U.S. CH_3Br emissions. U.S. emissions have also resulted in the U.S. accounting for 19-49% of global direct radiative forcing from ODSs in the 2000s.

Weighting ODS emissions by 100-year, direct GWPs allows the magnitude of these emissions to be compared to those of CO_2 to approximate their direct climate effects. Declines of 77 (77-84)% and 74 (71-75)% are calculated for annual global and U.S. GWP-weighted emissions of ODS and substitute chemicals through 2004, or a decline in annual emissions on a global scale of 7270 (3600-8500) Mt CO_2-equivalents and a decline in annual emissions in the United States of 1640 (900-1880) Mt CO_2-equivalents (quantities in parenthesis are calculated with consideration of net GWPs that include radiative forcing and uncertainty associated with stratospheric ozone changes;

see Box 2.2). The U.S. decline alone is likely a large fraction of the global benefit anticipated as a result of adherence the Kyoto Protocol (~2000 Mt CO_2-equivalent emissions; Velders *et al.*, 2007).

APPENDIX 2A

**Table 2A.1 Lifetimes, relative fractional halogen release factor, and Ozone Depletion Potentials
for halocarbons. Reproduced from Daniel and Velders et al. (2007).**

Halocarbon	Lifetime (years)	Relative Fractional Release Factor[a]	Semi-Empirical ODP	ODP in Montreal Protocol
Annex A-I				
CFC-11	45	1	1.0	1.0
CFC-12	100	0.60	1.0	1.0
CFC-113	85	0.75	1.0	0.8
CFC-114	300	0.28±0.02[b]	1.0	1.0
CFC-115	1700		0.44 [†]	0.6
Annex A-II				
Halon-1301	65	0.62	16	10.0
Halon-1211	16	1.18	7.1[c]	3.0
Halon-2402	20	1.22	11.5	6.0
Annex B-II				
Carbon tetrachloride	26	1.06	0.73	1.1
Annex B-III				
Methyl chloroform	5.0	1.08	0.12	0.1
Annex C-I				
HCFC-22	12.0	0.35	0.05	0.055
HCFC-123	1.3	1.11	0.02	0.02
HCFC-124	5.8	0.52	0.02	0.022
HCFC-141b	9.3	0.72	0.12	0.11
HCFC-142b	17.9	0.36	0.07	0.065
HCFC-225ca	1.9	1.1	0.02	0.025
HCFC-225cb	5.8	0.5	0.03	0.033
Annex E				
Methyl bromide	0.7	1.12	0.51	0.6
Others				
Halon-1202	2.9		1.7[d]	
Methyl chloride	1.0	0.80	0.02	

[†] Model-derived values, WMO (2003).

[a] From WMO (2003), Table 1-4, except for the value for CFC-114. For the EESC calculations in Section 1.8 of WMO (2003), slightly different relative fractional release factors were used by mistake for the halons.

[b] From Schauffler et al., 2003.

[c] The ODP of halon-1211 should have been reported as 5.3 in the previous Assessment (WMO, 2003), but was incorrectly reported as 6.0 due to a calculation error.

[d] WMO (2003), with adjustment for updated α value.

Table 2A.2 Direct Global Warming Potentials for selected gases. Reproduced from Daniel and Velders et al. (2007).

Industrial Designation(s) or Common Name	Chemical Formula	Radiative Efficiency[b] (W per square m per ppbv)	Lifetime (years)	Global Warming Potential for Given Time Horizon		
				20 years	100 years	500 years
Carbon dioxide	CO_2	$1.41 \times 10^{-5\,c}$		1	1	1
Nitrous oxide	N_2O	3.03×10^{-3}	114[d]	289	298	153
Chlorofluorocarbons						
CFC-11	CCl_3F	0.25	45	6,730	4,750	1,620
CFC-12	CCl_2F_2	0.32	100	10,990	10,890	5,200
CFC-13	$CClF_3$	0.25	640	10,800	14,420	16,430
CFC-113	CCl_2FCClF_2	0.30	85	6,540	6,130	2,690
CFC-114	$CClF_2CClF_2$	0.31	300	8,040	10,040	8,730
CFC-115	$CClF_2CF_3$	0.18	1700	5,310	7,370	9,990
Hydrochlorofluorocarbons						
HCFC-21	$CHCl_2F$	0.14	1.7	530	151	46
HCFC-22	$CHClF_2$	0.20	12.0	5,160	1,810	549
HCFC-123	$CHCl_2CF_3$	0.14	1.3	273	77	24
HCFC-124	$CHClFCF_3$	0.22	5.8	2,070	609	185
HCFC-141b	CH_3CCl_2F	0.14	9.3	2,250	725	220
HCFC-142b	CH_3CClF_2	0.20	17.9	5,490	2,310	705
HCFC-225ca	$CHCl_2CF_2CF_3$	0.20	1.9	429	122	37
HCFC-225cb	$CHClFCF_2CClF_2$	0.32	5.8	2,030	595	181
Hydrofluorocarbons						
HFC-23	CHF_3	0.19[e]	270	11,990	14,760	12,230
HFC-32	CH_2F_2	0.11[e]	4.9	2,330	675	205
HFC-41	CH_3F	0.02	2.4	323	92	28
HFC-125	CHF_2CF_3	0.23	29	6,340	3,500	1,100
HFC-134	CHF_2CHF_2	0.18	9.6	3,400	1,100	335
HFC-134a	CH_2FCF_3	0.16[e]	14.0	3,830	1,430	435
HFC-143	CH_2FCHF_2	0.13	3.5	1,240	353	107
HFC-143a	CH_3CF_3	0.13	52	5,890	4,470	1,590
HFC-152	CH_2FCH_2F	0.09	0.60	187	53	16
HFC-152a	CH_3CHF_2	0.09	1.4	437	124	38
HFC-227ea	CF_3CHFCF_3	0.26[e]	34.2	5,310	3,220	1,040
HFC-236cb	$CH_2FCF_2CF_3$	0.23	13.6	3,630	1,340	407
HFC-236ea	CHF_2CHFCF_3	0.30	10.7	4,090	1,370	418
HFC-236fa	$CF_3CH_2CF_3$	0.28	240	8,100	9,810	7,660
HFC-245ca	$CH_2FCF_2CHF_2$	0.23	6.2	2,340	693	211
HFC-245fa	$CHF_2CH_2CF_3$	0.28	7.6	3,380	1,030	314
HFC-365mfc	$CH_3CF_2CH_2CF_3$	0.21	8.6	2,520	794	241
HFC-43-10mee	$CF_3CHFCHFCF_2CF_3$	0.40	15.9	4,140	1,640	499

Industrial Designation(s) or Common Name	Chemical Formula	Radiative Efficiency[b] (W per square m per ppbv)	Lifetime (years)	Global Warming Potential for Given Time Horizon		
				20 years	100 years	500 years
Chlorocarbons						
Methyl chloroform	CH_3CCl_3	0.06	5.0	506	146	45
Carbon tetrachloride	CCl_4	0.13	26	2,700	1,400	435
Methyl chloride	CH_3Cl	0.01	1.0	45	13	4
Bromocarbons						
Methyl bromide	CH_3Br	0.01	0.7	17	5	1
Halon-1201	$CHBrF_2$	0.14	5.8	1,380	404	123
Halon-1211	$CBrClF_2$	0.30	16	4,750	1,890	574
Halon-1301	$CBrF_3$	0.32	65	8,480	7,140	2,760
Halon-2402	$CBrF_2CBrF_2$	0.33	20	3,680	1,640	503
Fully fluorinated species						
Sulfur hexafluoride	SF_6	0.52	3200	16,260	22,810	32,600
Trifluoromethylsulfur-penta-fluoroide	SF_5CF_3	0.57	650-950	13,120-13,180	17,540-17,960	20,060-22,360
Perfluoromethane	CF_4	0.10[e]	50000	5,210	7,390	11,190
Perfluoroethane	C_2F_6	0.26	10000	8,620	12,200	18,180
Perfluoropropane	C_3F_8	0.26	2600	6,310	8,830	12,450
Perfluorobutane	C_4F_{10}	0.33	2600	6,330	8,850	12,480
Perfluorocyclobutane	$c-C_4F_8$	0.32	3200	7,310	10,250	14,660
Perfluoropentane	C_5F_{12}	0.41	4100	6,510	9,150	13,260
Perfluorohexane	C_6F_{14}	0.49	3200	6,620	9,290	13,280
Perfluorodecalin	$C_{10}F_{18}$	0.56[f]	1000	5,500	7,510	9,440
Halogenated alcohols and ethers						
HFE-125	CHF_2OCF_3	0.44	136	13,790	14,910	8,490
HFE-134	CHF_2OCHF_2	0.45	26	12,190	6,320	1,960
HFE-143a	CH_3OCF_3	0.27	4.3	2,630	756	230
HCFE-235da2	$CHF_2OCHClCF_3$	0.38	2.6	1,230	349	106
HFE-245fa2	$CHF_2OCH_2CF_3$	0.31	4.9	2,280	659	200
HFE-254cb2	$CH_3OCF_2CHF_2$	0.28	2.6	1,260	359	109
HFE-7100 (HFE-44-9)	$CH_3OC_4F_9$	0.31	5.0	1,390	404	123
HFE-7200 (HFE-56-9)	$C_2H_5OC_4F_9$	0.30	0.77	200	57	17
HFE-245cb2	$CH_3OCF_2CF_3$	0.32	5.1	2,440	708	215
HFE-347mcc3	$CH_3OCF_2CF_2CF_3$	0.34	5.2	1,980	575	175
HFE-356pcc3	$CH_3OCF_2CF_2CHF_2$	0.33	0.93	386	110	33
HFE-374pc2	$CH_3CH_2OCF_2CHF_2$	0.25	5.0	1,930	557	169
	$CH_3OCF(CF_3)_2$	0.31	3.4	1,200	343	104

Industrial Designation(s) or Common Name	Chemical Formula	Radiative Efficiency[b] (W per square m per ppbv)	Lifetime (years)	Global Warming Potential for Given Time Horizon		
				20 years	100 years	500 years
HFE-43-10pccc124 [a]	$CHF_2OCF_2OC_2F_4O-CHF_2$	1.37	6.3	6,320	1,870	569
	$(CF_3)_2CHOH$	0.28	2.0	764	217	66
HFE-236ca12	$CHF_2OCF_2OCHF_2$	0.66	12.1	8,040	2,820	859
HFE-338pcc13	$CHF_2OCF_2CF_2O-CHF_2$	0.87	6.2	5,070	1,500	456
Species whose lifetimes have a high uncertainty						
Nitrogen trifluoride	NF_3	0.21[e]	740	13,370	18,000	21,270
Perfluorocyclopropane	$c-C_3F_6$	0.42	>1000	>12,700	>17,340	>21,800
HFE-227ea	$CF_3CHFOCF_3$	0.40	11	4,540	1,540	468
HFE-236ea2	$CHF_2OCHFCF_3$	0.44	5.8	3,370	989	301
HFE-236fa	$CF_3CH_2OCF_3$	0.34	3.7	1,710	487	148
HFE-245fa1	$CHF_2CH_2OCF_3$	0.30	2.2	1,010	286	87
HFE-329mcc2	$CHF_2CF_2OCF_2CF_3$	0.49	6.8	3,060	919	279
HFE-338mcf2	$CF_3CH_2OCF_2CF_3$	0.43	4.3	1,920	552	168
HFE-347mcf2	$CHF_2CH_2OCF_2CF_3$	0.41	2.8	1,310	374	114
HFE-356mec3	$CH_3OCF_2CHFCF_3$	0.30	0.94	355	101	31
HFE-356pcf2	$CHF_2CH7_2OCF_2CHF_2$	0.37	2.0	931	265	80
HFE-356pcf3	$CHF_2OCH_2CF_2CHF_2$	0.39	3.6	1,760	502	153
	$CHF_2OCH(CF_3)_2$	0.41	3.1	1,330	379	115
	$-(CF_2)_4CH(OH)-$	0.30	0.85	254	72	22

Note: Values are calculated for a CO_2 mixing ratio of 378 ppm, compared with 370 ppm in IPCC/TEAP (2005) and WMO (2003), which tends to increase all GWPs.

---

[a] Referred to as H-Galden 1040x in WMO/UNEP ozone assessments prior to WMO (2007).
[b] All values not otherwise noted from IPCC/TEAP (2005).
[c] See Section 8.2.3 of Daniel and Velders et al., 2007.
[d] This value is an adjustment time that includes feedbacks of emissions on the lifetime.
[e] See Table 8-3 of Daniel and Velders et al., 2007.
[f] From Shine et al., 2005.

Table 2A.3 Direct and indirect GWPs for a 100-year time horizon

Gas	Direct GWP[a]	Indirect GWP[b]	
		Low	High
CFC-11	4,750	−3,790	1,263
CFC-12	10,890	−2,160	720
CFC-113	6,130	−2,530	843
CH_3CCl_3	146	−643	214
CCl_4	1,400	−3,630	1,210
HCFC-22	1,810	−286	95
HCFC-123	77	−83	28
HCFC-124	609	−120	40
HCFC-141b	725	−667	222
HCFC-142b	2,310	−362	121
HCFC-225ca	122	−93	31
HCFC-225cb	595	−156	52
CH_3Br	5	−2,150	717
Halon 1211	1,890	−40,280	13,430
Halon 1301	7,140	−49,090	16,360
Halon 2402	1,640	−62,000	20,670

[a] Uncertainties associated with direct GWPs are estimated to be 35% at the 2σ level.

[b] Indirect GWPs are affected by numerous uncertainties as discussed in the text. Here, the only contribution to the indirect GWP range considered is uncertainty in the ozone forcing through 2005 (−0.05±0.1 W per m²; IPCC, 2007). The 'Low' GWP assumes an ozone forcing of −0.15 W per m², while the 'High' is calculated for +0.05 W per m². Both indirect values are calculated by scaling the indirect GWPs in WMO (2007) by the ratio of the appropriate ozone forcing used here to the forcing assumed in WMO (2007), which was −0.15 W per m².

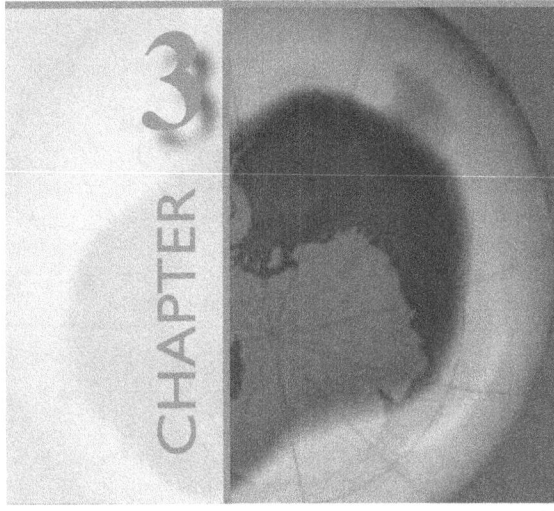

Ozone and UV Observations

Convening Lead Authors: Paul A. Newman, NASA;
Jay R. Herman, NASA

Lead Authors: Richard Bevilacqua, NRL; Richard Stolarski, NASA;
Terry Keating, U.S. EPA

CHAPTER 3

KEY ISSUES

As atmospheric concentrations of ozone-depleting substances change as a result of implementation of international policies, concentrations of stratospheric ozone and levels of ultraviolet radiation reaching the Earth's surface should also change. However, ozone concentrations and ultraviolet (UV) levels are affected by other natural and human processes as well. To understand whether the international policies are working, we must be able to determine changes in stratospheric ozone and ground-level UV and separate out the effects of ozone-depleting substances (ODS) changes and the effects of other factors.

Stratospheric ozone depletion is a global problem that has its most profound effects in the polar regions. However, the processes that drive stratospheric ozone depletion in the polar regions are somewhat different than those that drive depletion in the rest of world. Therefore, the impact of ODS changes may be different in polar regions than over the midlatitude United States.

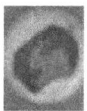

In this chapter, we briefly review the observations and current understanding and uncertainties in long-term trends in atmospheric ozone and ground-level UV radiation to address the following questions:

- What is the current state of ozone in the stratosphere in the Earth's midlatitudes and over the polar regions?
- What do the observations indicate about the abundances and trends of stratospheric ozone over the United States and elsewhere?
- How do midlatitude ozone levels and the processes that drive them differ from ozone levels and driving processes in the polar regions?
- What is the trend in the occurrence, depth, duration, and extent of the Antarctic ozone hole?
- What is the state of stratospheric ozone depletion in the Arctic region?
- How well do we understand the chemical and meteorological processes that determine stratospheric ozone concentrations in the polar regions and midlatitudes?
- How have UV radiation levels at the Earth's surface in the United States and elsewhere changed as a result of changes in stratospheric ozone?

KEY FINDINGS

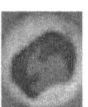

- Total global ozone has remained relatively constant over the last four years (2002-2006). Northern midlatitude ozone reached a minimum in 1993 because of forcings from the Mt. Pinatubo eruption and the solar cycle minimum, and has increased somewhat since then. Southern midlatitude ozone decreased until the late 1990s, and has been constant since. There are no significant ozone trends over the tropics.

- Ozone over the continental United States has followed the behavior of ozone for the entire northern midlatitude region; a decrease to a minimum in 1993, and an increase since then (see previous bullet).

- Ozone depletion in the upper stratosphere has closely followed the trends in chlorine. The slow down of the negative (or decreasing) trend is attributed to the leveling off of chlorine in this region of the stratosphere.

- Over the last decade (1995-2006), the Antarctic ozone hole has not worsened. Most Antarctic ozone hole diagnostics show losses leveling off after the mid-1990s. Saturation of ozone loss inside the ozone hole due to complete ozone destruction over a broad vertical layer plays the major role in this leveling off. This complete ozone destruction over a deep vertical layer is modulated by year-to-year dynamical variations. Antarctic ozone hole diagnostics showed an increase of ozone levels in some recent winter years (e.g., 2002, 2004), but these increases resulted from higher levels of dynamical forcing which warmed the Antarctic stratosphere, and not decreases in Equivalent Effective Stratospheric Chlorine levels. In contrast, the Austral spring of 2006 had below average dynamical forcing resulting in below average Antarctic temperatures, causing the 2006 Antarctic ozone hole to be one of the largest on record.

- Arctic spring total ozone values over the last decade were lower than values observed in the 1980s. In addition, spring Arctic ozone is highly variable depending on dynamical conditions. For current halogen levels, anthropogenic chemical loss and variability in ozone transport are about equally important for year-to-year Arctic ozone variability. Colder-than-average vortex conditions result in larger halogen-driven chemical ozone losses. Warmer-than-average vortex conditions result in smaller halogen-driven chemical ozone losses. Variability of temperatures and ozone transport are correlated because they are both driven by dynamic variability.

- Erythemal irradiance (which is a weighted combination of UVA and UVB based on skin sensitivity) over the United States increased roughly by 7% when the ozone minimum was reached in 1993 and is now about 4% higher than in 1979.

- Ground-based measurements of UV irradiance can detect UV trends related to ozone change when data from only days with clear sky are used by correcting for aerosol scattering and absorption using measured aerosol data.

- UV irradiance estimated from satellite data are usually 10% to 30% too high because satellite algorithms neglect the effects of absorbing aerosols.

- UVB irradiance trends can be estimated directly from satellite-measured ozone changes since regional cloud cover and aerosol loadings have not undergone large changes since 1979 except for a short period after the June 1991 Mt. Pinatubo eruption.

- Increased adverse human health effects associated with excessive UV exposure have been observed in Australia, where there are lower ozone amounts and less cloud cover, compared with similar latitudes in the United States.

3.1 INTRODUCTION

Ozone is a trace constituent of the atmosphere, with maximum volume mixing ratios of about 10 to 12 molecules per million air molecules (*i.e.*, 10 to 12 ppm). Figure 3.1 (top) shows the annually averaged, longitude-averaged ozone distribution.

The total amount of ozone (*i.e.*, the vertical integral of ozone density from the surface to space) is highest in the mid-to-high latitudes. The bottom panel of Figure 3.1 shows the total ozone integrated from the top panel. In midlatitudes, ozone density is highest in the lower stratosphere between 12 and 25 km (Figure 3.1 middle panel). While the maximum of the ozone mixing ratio (Figure 3.1 top panel) is highest in the tropics at 32 km, the total column ozone is highest in the midlatitudes, not in the tropics (illustrated in the bottom panel).

The distribution of ozone mixing ratios (Figure 3.1 top), density (Figure 3.1 middle), and total ozone (Figure 3.1 bottom) is controlled by the photochemical production, catalytic destruction, and transport. The basic circulation (shown as the yellow streamlines in the upper two panels of Figure 3.1) is known as the Brewer-Dobson circulation (Shepherd, 2007). This Brewer-Dobson circulation carries air into the stratosphere in the tropics near 16 km, leading to very low ozone in the tropical lower stratosphere as the low ozone air is carried upward from the troposphere. The poleward and downward flow of ozone from the tropics produces the midlatitude maximum in both hemispheres.

As air rises in the tropical stratosphere, ozone is produced when molecular oxygen (O_2) is split by solar ultraviolet radiation (hv) to form oxygen atoms (O) that combine with O_2 in the presence of a third air molecule (M) to form O_3.

$$O_2 + hv \rightarrow O + O \qquad (1a)$$
$$2\,(O + O_2 + M \rightarrow O_3 + M) \qquad (1b)$$
Net: $3\,O_2 + hv \rightarrow 2\,O_3$

This solar production of ozone leads to very high ozone concentrations in the mid-stratosphere in the tropics (near 32 km).

Average Ozone Distribution

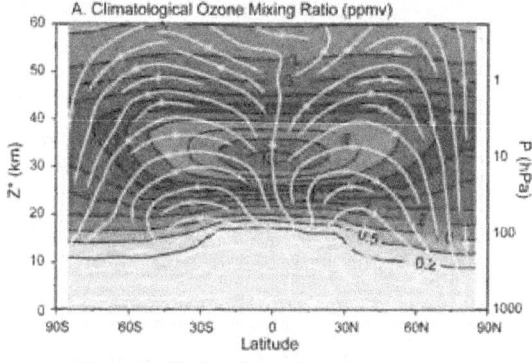

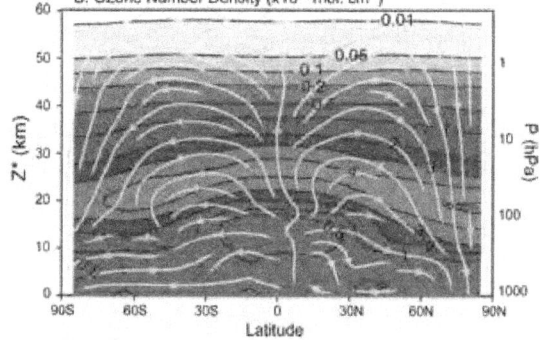

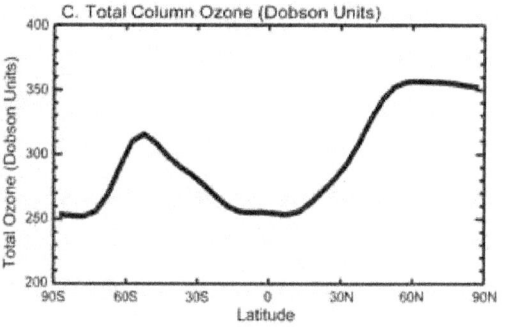

Figure 3.1 Annual longitudinal averaged ozone mixing ratios (top), ozone density (middle), and annual longitudinal averaged total ozone (bottom). Top panel units are parts per million (ppm); middle panel units are molecules per cm³; bottom panel units are Dobson Units (DU). One DU is equal to a column amount of 2.69×10^{16} molecules per cm² or about 0.01 mm of pure ozone at standard temperature and pressure. Equivalently, an ozone number density of 10^{12} molecules per cm³ is equivalent to 3.7 DU per kilometer (km). The bottom panel is the vertical integral of the middle panel. The annual average flow field stream lines are shown in the top and middle panels. The rising motion in the tropical stratosphere and sinking motion in the polar region is known as the Brewer-Dobson circulation. Adapted from McPeters *et al.* (2007).

Ozone is destroyed when it reacts with oxides of nitrogen, hydrogen, chlorine, bromine, or oxygen atoms in catalytic reactions to reform molecular oxygen.

$$O_3 + X \rightarrow O_2 + XO \qquad (2a)$$
$$XO + O \rightarrow O_2 + X \qquad (2b)$$
$$O_3 + h\nu \rightarrow O_2 + O \qquad (2c)$$
Net: $2\ O_3 + h\nu \rightarrow 3\ O_2$

Here, X represents the catalysts chlorine atoms (Cl), bromine atoms (Br), and the oxides of nitrogen (nitric oxide, NO) and hydrogen (hydroxyl, OH), while $h\nu$ represents the absorption of solar ultraviolet light to photochemically break a chemical bond of ozone. The net effect of the catalytic cycle is to destroy two ozone molecules while regenerating the catalytic agent. All of these catalysts are highly reactive free radicals, meaning they have an unpaired electron, which tends to attach to other molecules in order to form a chemical bond. Since these reactions have an initial energy barrier to reaction, warmer temperatures will speed up this catalytic cycle, and cooler temperatures (as predicted to occur by recent climate models) will slow down this ozone loss cycle. In Figure 3.1, ozone decreases above 32 km as this ozone destruction begins to dominate over the ozone production.

The source gases for the ozone-destroying catalysts are compounds such as chlorofluoro-carbons, CFCs (chlorine), halons and methyl bromide (bromine), nitrous oxide (nitrogen), and methane (hydrogen) (see Chapter 2 for a complete discussion of these source gases). The relative contributions of the oxides of hydrogen (HO_x), oxides of chlorine (ClO_x), and oxides of nitrogen (NO_x) reactions can be found in Figure 1.11 of IPCC/TEAP (2005). As the air rises in the stratosphere, the catalytic agents are liberated from the source gases by both the UV radiation and chemical reactions.

The catalytic reactions that cause stratospheric ozone decreases are principally those involving chlorine and bromine. These chlorine and bromine compounds are from halogen species such as CFCs and halons. These species are inert in the troposphere, but are carried into the stratosphere by the slow rising circulation (Figure 3.1, top panel). As they ascend in the stratosphere, the halogen species are broken down by UV radiation or oxidation, releasing chlorine and bromine to catalytically destroy ozone. The rate of catalytic destruction of ozone is limited by the conversion of the chorine and

bromine oxides to reservoir compounds such as hydrochloric acid (HCl), chlorine nitrate, ($ClONO_2$), and bromine nitrate ($BrONO_2$). These chlorine and bromine species are eventually returned to the troposphere, where they are removed in wet processes.

These ozone catalytic cycles involve oxygen atoms (O), and thus operate most rapidly in the mid-stratosphere of the tropics and the midlatitudes, where the concentration of oxygen atoms increases with increasing altitude. Oxygen atom concentrations increase with altitude because their loss slows as the density of O_2 and M ($O + O_2 + M \rightarrow O_3 + M$) decreases with altitude. Maximum halogen-catalyzed ozone loss at midlatitudes occurs around an altitude of about 40 km (just above the peak ozone concentrations), where these oxygen atoms are more abundant. While fractional ozone loss peaks near 40 km for a stratosphere unperturbed by cold temperatures (about 8 to 10% of the naturally-occurring ozone at that altitude), the contribution of ozone loss at 40km to the fractional loss in the total column is small, since ozone density falls off rapidly above the 20-25 km layer (Figure 3.1, middle panel).

Ozone depletion in the polar lower stratosphere involves different chemistry than described above. During winter, the lower stratosphere over the poles is characterized by air that the Brewer-Dobson circulation has carried poleward and downward from the upper stratosphere and mesosphere (Figure 3.1, top panel), extremely low temperatures (less than 200 K ($-73°C$)), and a circumpolar jet stream that isolates the air over the polar regions from midlatitude influence (the polar vortex). These extremely cold and isolated conditions enable polar stratospheric clouds (PSCs) to form (Crutzen and Arnold, 1986; Toon *et al.*, 1986). The ozone loss occurs in two steps. First, heterogeneous chemical reactions occur on the surfaces of the PSC particles, liberating chlorine from the two reservoir species (HCl + $ClONO_2$ [on PSCs]$\rightarrow$ Cl_2 + HNO_3) (Solomon *et al.*, 1986). Second, two principal chlorine and bromine catalytic reactions that do not involve oxygen reactions (Equation 1a) produce rapid depletion:

The catalytic reactions that cause stratospheric ozone decreases are principally those involving chlorine and bromine.

$$ClO + ClO + M \rightarrow ClOOCl + M \qquad (3a)$$
$$ClOOCl + hv \rightarrow 2\ Cl + O_2 \qquad (3b)$$
$$2\ [\ Cl + O_3 \rightarrow ClO + O_2\ ] \qquad (3c)$$
Net: $2\ O_3 + hv \rightarrow 3\ O_2$

$$BrO + ClO + hv \rightarrow Br + Cl + O_2 \qquad (4a)$$
$$Br + O_3 \rightarrow BrO + O_2 \qquad (4b)$$
$$Cl + O_3 \rightarrow ClO + O_2 \qquad \text{see } (3c)$$
Net: $2\ O_3 + hv \rightarrow 3\ O_2$

Equation (4a) represents a sequence of reactions that together lead to the products shown. Again, hv represents the absorption of solar light to photochemically break the chemical bonds, and M represents any air molecule, typically nitrogen (N_2) or oxygen (O_2), which carries away the excess energy of the reaction. In contrast to the intense UV necessary to photolyze oxygen molecules in (1a), the reactions (3b and 4a) require only visible light. These two catalytic cycles account for all but a few percent of the polar ozone loss, which occurs in the lowermost stratosphere (12 to 24 km altitude). This effect is strongest in the Antarctic stratosphere where the stable polar vortex allows the nearly complete destruction of ozone between about 12 and 22 km altitude each spring, forming the Antarctic ozone hole (see the low ozone amounts in Figure 3.1, bottom panel). The principal ingredients for large ozone losses in the polar regions are: (1) cold temperatures (< 195 K or −78°C) for the formation of PSCs; (2) high concentrations of chlorine and bromine; and (3) visible light for photolyzing both molecular chlorine (Cl_2) and the ClO dimer (ClOOCl).

The dramatic seasonal ozone losses occur over Antarctica during the Austral spring August-October period (with more than 50% of the total column ozone depleted) and to a smaller extent over the Arctic during the Boreal spring February-March period. The difference in hemispheres has to do with the contrast between the presence of polar stratospheric clouds and the timing of the break up of the polar vortex in the two polar regions. First, PSC extent is much greater in the Antarctic due to colder stratospheric temperatures than in the Arctic. Thus, molecules to participate in the two catalytic cycles involving chlorine and bromine atoms are much more abundant in the Antarctic. Second, the Arctic vortex breaks up and warms

at an earlier time in spring than the Antarctic, shutting off the ozone loss.

In the midlatitudes, ozone destruction can take place locally or ozone-depleted air may be transported from polar regions. During periods following major volcanic eruptions, the sulfur injected into the stratosphere can lead to enhanced aerosols in the lower stratosphere. The surfaces of these aerosols promote the conversion of reservoir compounds of chlorine and bromine back to catalytically active oxides that increase ozone destruction.

The solar UV radiation that reaches the Earth's surface is strongly screened by ozone. The UV radiation important for biological processes is described by two bands, UVA (315 to 400 nanometers, nm) and UVB (280 to 315 nm). In a cloud-free atmosphere, both UVA and UVB are scattered by both molecules (Rayleigh scattering) and aerosols, while UVB is also significantly absorbed by ozone. Ozone absorption increases rapidly with decreasing wavelength, which is why there is little detectable radiation below 280 nm at the Earth's surface. For a given sun angle, the relationship of percent UV increase to percent ozone decrease is proportional to the ozone absorption. Human exposure to UV radiation has both negative (*e.g.*, skin cancer and eye cataracts) and positive (*e.g.*, Vitamin D production) effects. The negative effects of UV overexposure are the major reason for concern over stratospheric ozone decreases. In addition to changes in ozone, long-term changes in the amount of aerosols and cloud cover affect exposure at the surface to all UV wavelengths.

The following sections of this chapter briefly review the observed trends in ozone and ground ultraviolet radiation levels and discuss our current understanding of the processes that determine these levels. For each of these issues, the polar regions will be discussed separately from the low and midlatitudes because of the fundamentally different issues associated with those regions.

The dramatic seasonal ozone losses occur over Antarctica during August to October and to a smaller extent over the Arctic in February and March.

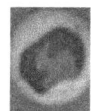

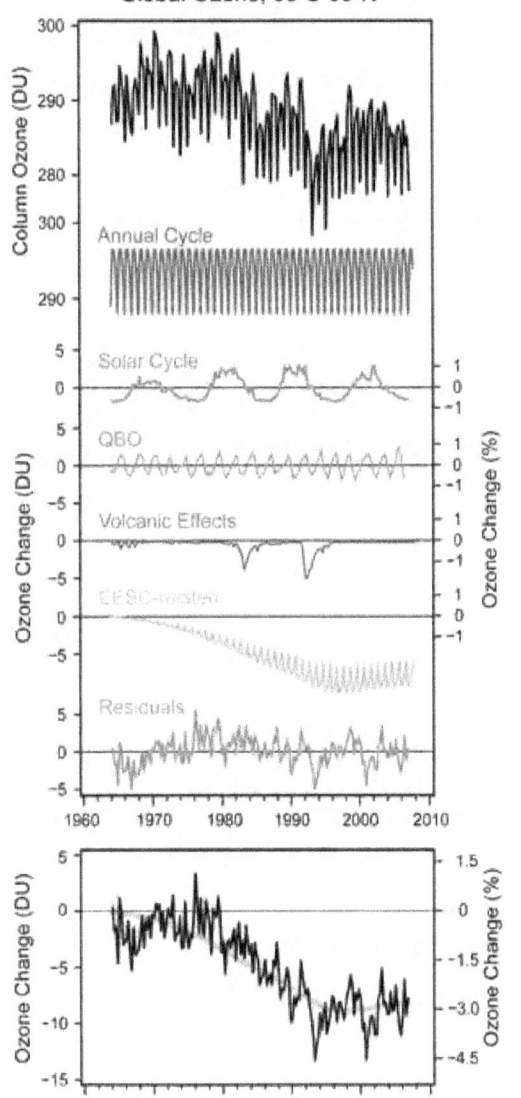

Global Ozone, 60°S-60°N

Figure 3.2 Top panel: Ozone observations for 60°S to 60°N estimated from ground-based data and individual components that comprise ozone variations (Dobson Units or DU). Bottom panel: Ozone deviations after removing annual cycle (blue line), solar cycle (red line), quasi-biennial oscillation or QBO (magenta line), and volcanic effects (green line) from original time series. Seasonal variations in the Equivalent Effective Stratospheric Chlorine (EESC) related component (up and down variations in orange line) are also removed. The thick orange line in the bottom panel represents the annual average EESC component derived from the regression model. See Box 3.1 for additional details.

3.2 OZONE

In this chapter we briefly review the most recent observed trends in observations of total ozone (Section 3.2.1) and ozone vertical distributions (Section 3.2.2). We then discuss our current understanding and recent findings related to the chemical and meteorological or dynamical processes that affect ozone (Section 3.2.3).

3.2.1 Total Ozone Observations
3.2.1.1 GLOBAL OZONE (EXCLUDING POLAR REGIONS)

After nearly two decades of decrease, the column amount of ozone at midlatitudes of the Northern and Southern Hemispheres has been relatively stable over the last decade. Polar ozone is considered in more detail in Section 3.2.1.3 below. We can integrate over the globe to get a simple measure of the recent changes in the ozone layer (Figure 3.2). The global mean total column ozone values for 2002-2005 were approximately 3% (about 10 Dobson Units or DU) below 1964-1980 average values. The 2002-2005 values are similar to the 1998-2001 values and this indicates that, overall, ozone is no longer decreasing. Several global datasets confirm this conclusion, although differences of up to 1% between annual averages exist between some individual sets (WMO, 2007).

Total column ozone over the tropics (25°S to 25°N) remains essentially unchanged. Total ozone trends in this region for the period 1980-2004 are not statistically significant, consistent with earlier assessments (Figure 3.4, WMO, 2007).

The behavior of ozone at midlatitudes in the Northern Hemisphere during the 1990s was different from that in the Southern Hemisphere during the same period. The Northern Hemisphere shows a minimum around 1993 resulting from forcings from the Mt. Pinatubo eruption and the solar cycle minimum, followed by an increase. The Southern Hemisphere shows an ongoing decrease through the late 1990s, followed by relatively constant levels (Figure 3.3). The average for the period 2002-2005 of total ozone at midlatitudes in each hemisphere is similar to the average for the previous four years, 1998-2001. Ozone in the southern midlatitudes remains about 5.5% below its 1964-1980 average, while ozone in the northern midlatitudes remains about 3% below (Figure 3.3).

BOX 3.1: Estimating Ozone Trends

Isolating the ozone response to manmade ozone-depleting substances from natural variations in the ozone, such as seasonal changes or volcanic perturbations, is accomplished using a statistical time series regression analysis. The top panel of Figure 3.2 (black line) shows total ozone time series from ground-based measurements taken over the period 1964 to 2006 and averaged seasonally and over the 60°N-60°S area (87% of the Earth's area). The observations are statistically modeled as a linear combination of the known individual processes that cause ozone to vary. In this analysis (following Fioletov *et al.*, 2002), the regression model used is:

$$O_3(t) = \mu + \text{seasonal cycle} + \alpha \cdot EESC + \beta \cdot QBO + \gamma \cdot Solar + \delta \cdot Volcano + noise$$

Here, μ, α, β, γ, and δ are constants estimated such that the model (terms on the right hand side) best matches the observed ozone time series. The mean (μ) and seasonal cycle are calculated directly from the ozone data from 1979 to 1987 (blue line in Figure 3.2). Equivalent Effective Stratospheric Chlorine (EESC, see Chapter 5) is used to represent anthropogenic trace gases that react with ozone (orange line in Figure 3.2). The magenta line shows the quasi-biennial oscillation (QBO). The QBO is a variation in stratospheric winds with a period of about 26 months that is represented using equatorial radiosonde wind observations (Reed *et al.*, 1961). The solar term is represented using the 10.7 cm radio flux measured at Ottawa, Canada (red line). The volcanic term is derived from stratospheric aerosol observations (dark green). The noise term includes all variations required to make the model exactly equal the observed ozone (grey). The coefficients (μ, α, β, γ, and δ) are estimated by a mathematical regression that minimizes the noise term.

The bottom panel highlights the ozone changes due to chlorine and bromine (*i.e.*, EESC) with the natural forcings (seasonal cycle, QBO, solar, and volcano) removed. This line is the original observations with only the annually-averaged EESC-related time series (smoothed orange line) and the residual noise term remaining (grey line).

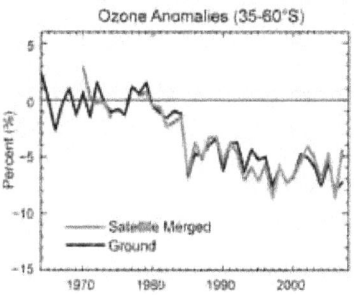
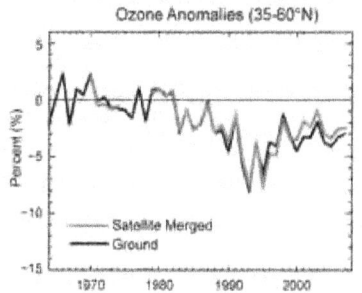
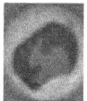
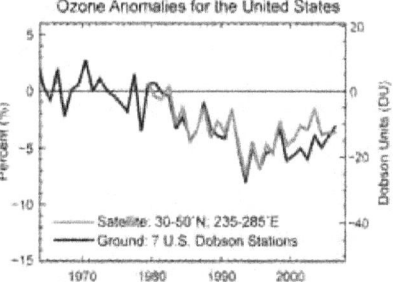

Figure 3.3 Top: deseasonalized, annually averaged, area-weighted total ozone deviations from satellite (red) and ground stations (black) for the latitude bands 35°N to 60°N (left) and 35°S to 60°S (right). Anomalies were calculated with respect to the time average for the period 1964-1980. Updated from Fioletov *et al.* (2002) and WMO (2003). Bottom: Average total ozone over the United States from the TOMS/SBUV series of satellite instruments (red), and seven ground stations in the United States. Both time series are plotted relative to the 1964-1980 mean of the ground-station data. Updated from Stolarski and Frith (2006).

Average March Total Column Ozone

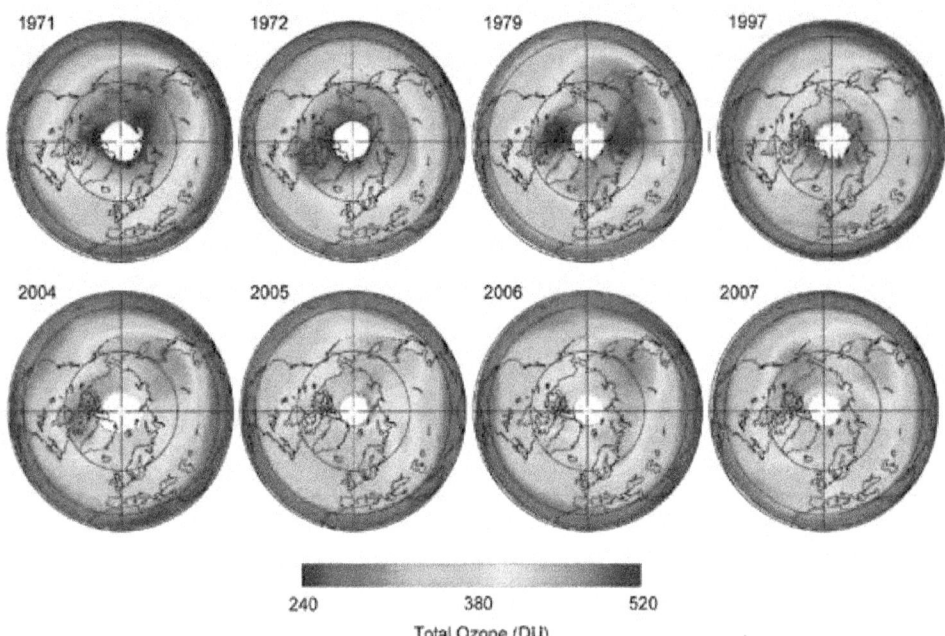

Figure 3.4 March monthly averaged total ozone. The 1971 and 1972 images are from the Nimbus-4 BUV instrument, the 1979 is from the Nimbus-7 TOMS satellite instrument, the 1997 and 2004 images are from the Earth Probe TOMS, and the 2005, 2006, and 2007 images are from the Aura Ozone Monitoring Instrument (OMI). This figure is updated from Figure 4-6 of WMO (2007).

Releases of ozone-depleting substances in the United States affect global ozone levels, and releases across the globe affect the United States, because of the long lifetimes of CFCs and their mixing, or spread, around the world.

Total ozone over the United States tends to parallel the entire Northern Hemisphere because these levels are driven by the response to the worldwide chlorine and bromine releases and by hemispheric-scale transport processes (Figure 3.3, bottom). Releases of ozone-depleting substances in the United States affect global ozone levels, and releases across the globe affect the United States, because of the long lifetimes of CFCs and their mixing, or spread, around the world. Total ozone over the United States is shown in the bottom panel of Figure 3.3. The total ozone changes are similar to ozone over the entire northern midlatitudes (compare to top right panel). The minimum value was reached shortly after the eruption of Mount Pinatubo and was about 5 to 8 percent below the amounts present prior to 1980 (as a result of the volcanic aerosol effect). The ozone increase since 1993 has diminished the ozone deficit to about 2 to 5 percent below the pre-1980 amounts. The average for the last four years (2002-2005) is essentially the same as the previous four years.

3.2.1.2 POLAR OZONE

Significant ozone depletion has occurred in the polar regions over the last few decades as a result of anthropogenic halogen-containing compounds. The ozone loss chemistry, as described in Chapter 1 (also WMO, 2007 and references therein), begins with very cold temperatures that lead to the formation of PSCs. Chlorine is rapidly converted from inactive to reactive forms on the cold aerosol surfaces. The Antarctic ozone hole is the most extreme manifestation of this phenomenon. Reactive chlorine is released within the stratospheric polar vortex beginning in the winter darkness. In August through September, when sunlight has returned to the Antarctic, halogen photochemistry rapidly destroys ozone. Some ozone loss is also observed in the June-August period at the edge of the polar vortex (Roscoe *et al.*, 1997). Ozone loss maximizes by the late September to early October period, after which temperatures warm, ozone loss ceases, the polar vortex breaks up, and high ozone air from midlatitudes mixes in, rapidly filling in the ozone hole (typically in the November-December period).

In this section, we illustrate trends in total ozone for both the Arctic and Antarctic. The ozone content in the polar lower stratosphere is dependent on background chemical conditions, temperatures, transport, and dynamics. The Arctic polar stratosphere shows large interannual variability, while the Antarctic is more stable because the Antarctic polar vortex is more stable. This section discusses the behavior of polar ozone over the last few decades. Section 3.2.1.2.1 focuses on the Arctic, while 3.2.1.2.2 shows the Antarctic.

3.2.1.2.1 Arctic total ozone

Arctic total ozone has had a substantial downward trend since the 1970s with slightly higher values over the last ten years than in the previous six years. Figure 3.4 displays a series of March polar averages for selected years from 1971 to 2007 (updated from Figure 4-6 in WMO, 2007). The 60°N latitude circle generally encloses the region of ozone depletion, but in some years (e.g., 2005) the vortex and low ozone region are displaced from the pole, extending somewhat southward of 60°N. Nevertheless, Arctic ozone for recent March averages is low compared to the observations prior to 1980 (shown in the upper row of Figure 3.4).

The springtime average total ozone values in the Arctic poleward of 63°N latitude (upper line) are shown in Figure 3.5, in comparison with the average total ozone for the years 1970-1982 (gray horizontal line). The difference between the observed values and the 1970-1982 average indicates the combined changes in ozone due to chemistry and dynamics. In the last ten years Arctic column ozone is higher than the low values of the mid-1990s, except in the cold and chemically active winter of 1999/2000, when a large decrease of 63°-90° Northern Hemispheric total ozone was observed (Rex *et al.*, 2002).

The record-cold winter of 2004/2005 led to very large ozone losses (Manney *et al.*, 2006; Rex *et al.*, 2006; Singleton *et al.*, 2007; Goutail *et al.*, 2005; Feng *et al.*, 2007a). However, this large loss

showed a less pronounced impact on the March polar average total ozone. Although Northern Hemisphere polar column ozone averages are a general indicator of Arctic ozone depletion and trends (WMO, 2007), the chemical loss can oftentimes be masked by the 63-90°N polar averaging. For example, the 2005 March average had a strong influence of dynamics. Vortex fragments moved outside the 63°-90°N and the total ozone showed a distinct minimum near 60°N (Figure 3.4). This created a higher value relative to other recent cold winters even though chemical ozone loss in the lower stratospheric vortex in mid winter of 2005 was as high as or higher than ozone loss in other recent cold winters.

3.2.1.2.2 Antarctic total ozone

In the Southern Hemispheric (SH) polar region, very large ozone depletions in the Austral spring have led to extremely low ozone values over Antarctica during October, the "ozone hole" (Figure 3.5, bottom line). Figure 3.6 displays a series of Antarctic total ozone images (values shown in Figure 3.5 are averaged from these images). A comparison of the moderate values of total ozone over

Arctic total ozone has had a substantial downward trend since the 1970s, with slightly higher values over the last ten years than in the previous six years.

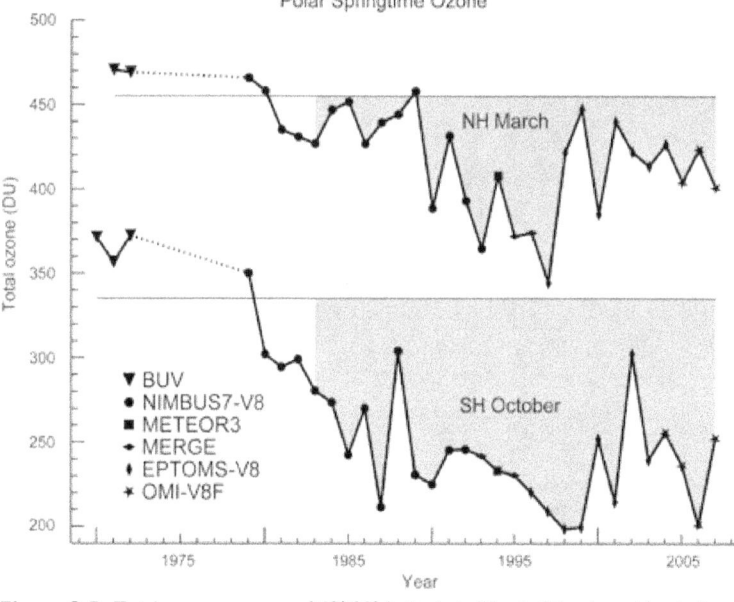

Figure 3.5 Total ozone average of 63°-90° latitude in March (Northern Hemisphere) and October (Southern Hemisphere). Symbols indicate the satellite data that have been used in different years. The horizontal gray lines represent the average total ozone for the years prior to 1983 for the Northern Hemisphere and Southern Hemisphere. The grey shading shows the contribution of chemical ozone destruction and natural variations. Updated from Figure 4-7, WMO (2007).

Average October Total Column Ozone

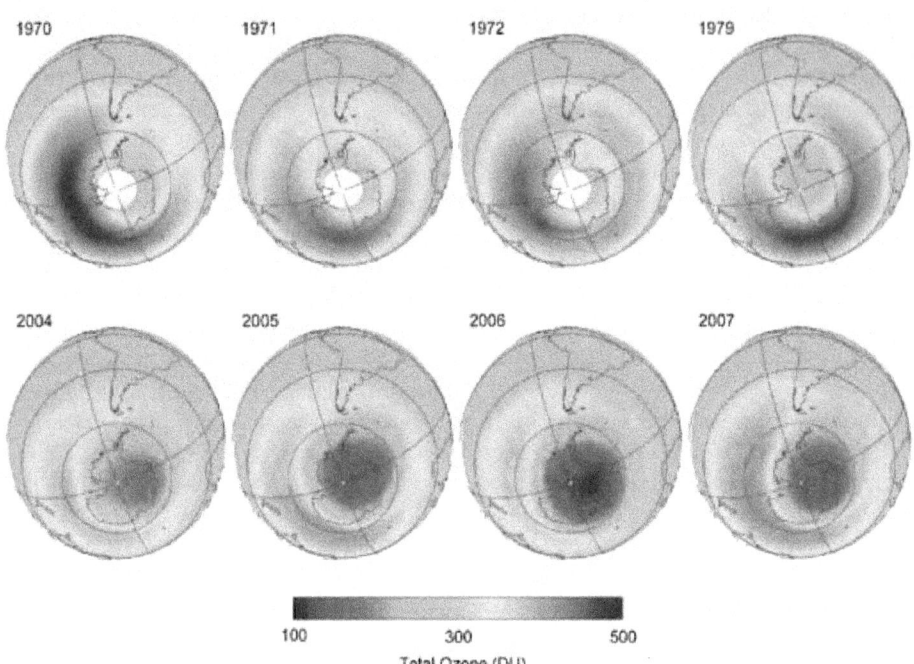

Figure 3.6 Satellite observations of October monthly averaged total ozone. The 1970, 1971, and 1972 images is from the Nimbus-4 BUV instrument, the 1979 image is from the Nimbus-7 TOMS instrument, and the 2004, 2005, 2006 and 2007 images are from the Aura OMI instrument.

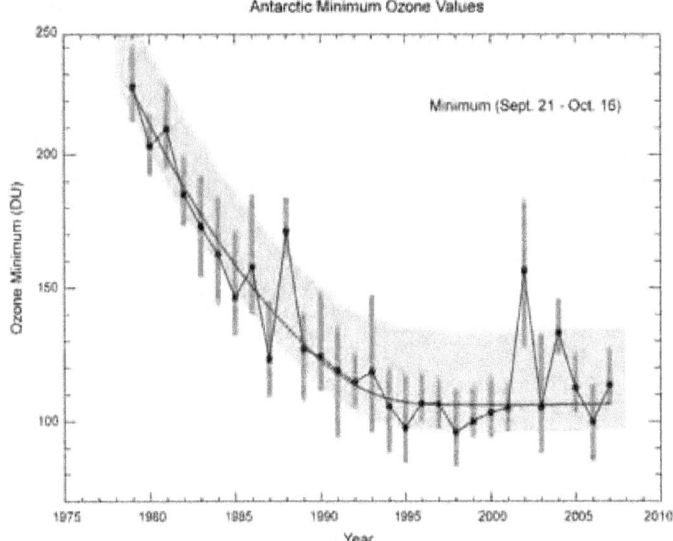

Figure 3.7 The minimum ozone values over Antarctica are averaged for the period from 21 September to 16 October (black dots). The vertical grey bars indicate the range of ozone values used in the average. The blue line shows the fit to these ozone values as was shown in Newman *et al.* (2004), and now using EESC, as derived in Newman *et al.* (2006) (also Box 2.7 in Chapter 2). The EESC has a mean age of 5.5 years, an age spectrum width of 2.75 years, and a bromine-scaling factor of 60. The fit is quadratic in EESC. The background lighter grey shading shows the expected variation of minimum ozone values between warm (upper side = +10°C) and cold years (lower side = −10°C). This figure was generated using TOMS and OMI total ozone. Updated from Figure 4-8 of WMO (2007).

Antarctica in the early years (1970s, top row) to the reduced values over Antarctica in the last two decades (bottom row) illustrates the Antarctic ozone hole. In Figure 3.5, the years from 2000 to 2005 showed an increase in polar column ozone averages compared to 1998 and 1999. The interannual variations in ozone depletion observed from 2001 to 2005 primarily result from variations in the dynamics (*i.e.*, stratospheric weather variations), and have not been caused by changes in Equivalent Effective Stratospheric Chlorine (EESC). See Box 2.7 of Chapter 2 for a definition of EESC and see Chapter 5 for more discussion on its usage. Since the early 1990s, total loss of ozone occurs in the lowermost stratosphere inside the polar vortex in September and October (Solomon *et al.*, 2005). Estimates of EESC inside the vortex reached a value of about 3.2 parts per billion by mole (ppb) in 1990 and peaked in early 2001 at about 4.0 ppb (Newman *et al.*, 2007). Hence, the EESC concentrations since the early 1990s have exceeded those necessary to cause total loss. The Antarctic ozone hole, therefore, has had low sensitivity to moderate decreases in EESC and the unusually small ozone holes in some recent years (*e.g.*, 2002 and 2004) are strongly attributable to a dynamically driven warmer Antarctic stratosphere.

Various metrics that capture different aspects of the Antarctic ozone hole are used to describe the severity of ozone depletion, such as Antarctic ozone hole area, ozone minimum, ozone mass deficit, and profile shape (Section 3.2.2.2). The polar average from 63-90°S tends to exaggerate dynamical fluctuations (Figure 3.5). Figure 3.7 displays the Antarctic ozone hole minimum values averaged for the period 21 September to 16 October. Because the Antarctic ozone hole chemical losses peak in late September, the average minimum ozone columns in this period provide a very useful metric for the depletion severity. Again, this figure shows a clear decrease from 1979 to the mid-1990s, with particularly low values in the mid- to late 1990s. Following Newman *et al.* (2006), we have added a statistical fit of these metrics (blue line) to a quadratic function of Antarctic EESC. The fit shows how ozone levels have responded to chlorine. In addition to the fit to chlorine, the figure also includes a background grey shading that shows the expected natural variation of the ozone minimum values for warmer than average years (+2σ = 10 K, upper part) and colder years (−2σ = −10 K, lower part). The 2002 minimum value stands out because it was the warmest year on record. The minimum ozone values in 2002 and 2004 were higher than the expected values (the blue line) because of the warmer temperatures.

3.2.2 Vertical Distribution of Ozone
3.2.2.1 GLOBAL

In addition to the polar regions, the upper stratosphere also shows clear evidence for ozone destruction due to increasing chlorine compounds. Measurements from both the Stratospheric Aerosol and Gas Experiment (SAGE I+II) and Solar Backscatter Ultraviolet (SBUV[/2]) satellite instruments show significant declines in upper stratospheric ozone from 1979 through 2004 (Figure 3.8). The net ozone decrease over the 1979 to 1995 period was approximately 10-15% over midlatitudes, with smaller but significant changes over the tropics (Figure 3-7 in WMO, 2007). During the last decade, upper stratospheric ozone has remained relatively constant. Available independent Umkehr ground-based optical

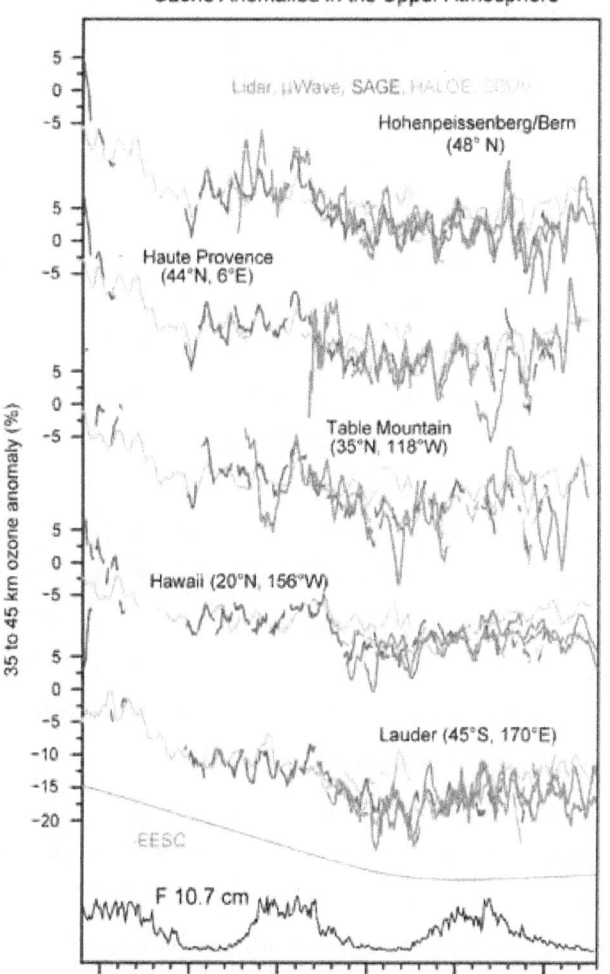

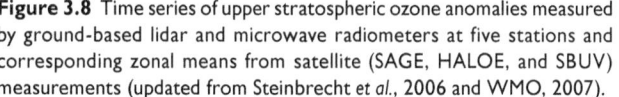

Figure 3.8 Time series of upper stratospheric ozone anomalies measured by ground-based lidar and microwave radiometers at five stations and corresponding zonal means from satellite (SAGE, HALOE, and SBUV) measurements (updated from Steinbrecht *et al.*, 2006 and WMO, 2007).

remote sensing, Light Detection and Ranging (lidar) remote sensing, and microwave ozone measurements confirm these findings.

The bulk of column ozone is found in the lower part of the stratosphere (Figure 3.1). The evidence shows that lower stratospheric ozone declined over the period 1979-1995, but has been relatively constant with significant variability over the last decade. Figure 3.9 shows the vertical profile of ozone trends in midlatitudes of the Northern (left panel) and Southern (right panel) Hemispheres. The trends are actually fits to EESC ($\Delta O_3 = \alpha \bullet \Delta EESC$, see Box 3.1) that is converted to a percentage per

The Antarctic
ozone hole first
began to develop
in the early 1980s
and reached its
current full extent
by the mid-1990s.

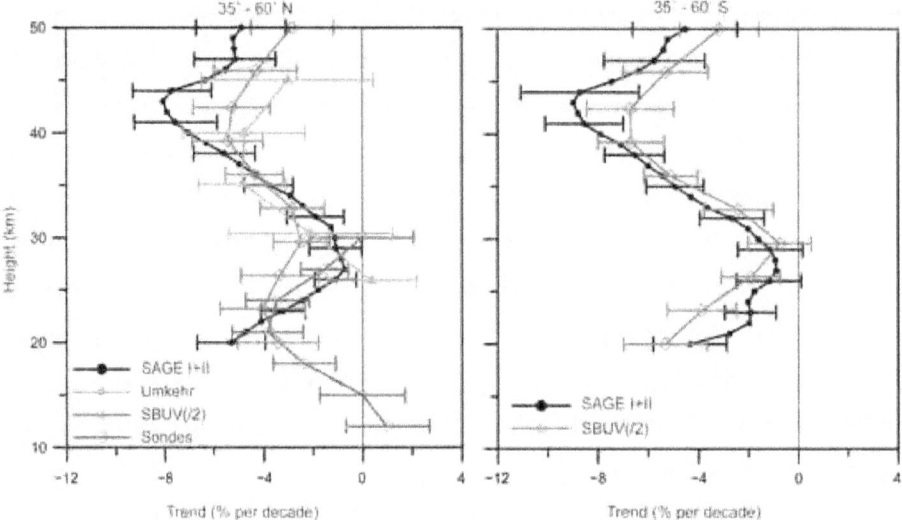

Midlatitude Vertical Ozone Trends

Figure 3.9 Vertical profile of ozone trends over northern and southern midlatitudes estimated from ozonesondes, Umkehr, SAGE I+II, and SBUV(/2) for the period 1979-2004. The trends were estimated using regression to an EESC curve and converted to % per decade using the variation of EESC with time in the 1980s. The trends were calculated in geometric altitude coordinates for SAGE and in pressure coordinates for SBUV(/2), sondes, and Umkehr data, and then converted to altitude coordinates using the standard atmosphere. The two standard deviation error bars are shown.

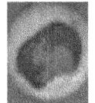

decade by scaling the α coefficient with the linear 1 ppb change of EESC observed during the 1980s. Measurements by SAGE I+II and SBUV(/2) showed declines of 7 to 9% (or 10 to 15% cumulative by 1995) between 40 and 45 km altitude (Figure 3.9).

These midlatitude ozone decreases are not linear, and did not continue in the last decade. This non-linear trend has been accounted for by using the ozone regression against the EESC time series and then converting to % per decade using the variation of EESC with time in the 1980s. At lower altitudes, between 12 and 15 km, in the Northern Hemisphere, a strong decrease in ozone was observed from ozonesonde data between 1979 and 1995, followed by an overall increase from 1996 to 2004, leading to no net long-term decrease at this level. These changes in the lowermost stratosphere have a substantial influence on the column because most of the ozone resides in the lowermost stratosphere.

3.2.2.2 POLAR

The Antarctic ozone hole first began to develop in the early 1980s and reached

its current full extent by the mid-1990s (Hofmann *et al.*, 1997; Solomon *et al.*, 2005). The most complete record of the morphology of the Antarctic ozone hole vertical structure is found from the balloon-borne ozonesonde

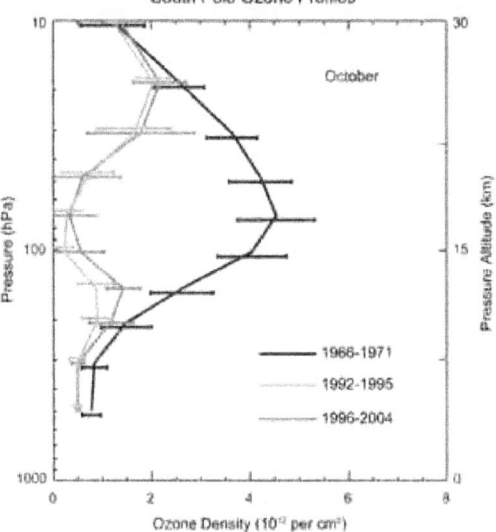

South Pole Ozone Profiles

Figure 3.10 Observations of the October average ozone profiles measured at the South Pole in different time periods; prior to the Antarctic ozone hole (1966-1971), after the Mt. Pinatubo eruption when aerosol abundances were enhanced in (1992-1995), and current conditions (1996-2004). Adapted from Solomon *et al.* (2005).

measurements at the South Pole, which extend back to the mid-1960s. Figure 3.10, from Solomon *et al.* (2005), uses the South Pole ozonesonde data to delineate the Antarctic ozone hole region relative to the pre-ozone hole conditions of the 1970s. The altitude range of the Antarctic ozone hole has been very stable in the 1990s. In the vicinity of the lower edge of the Antarctic ozone hole (10 to 14 km), Figure 3.10 shows that ozone abundances were lowest in the 1992-95 time period. This is presumably the result of increased ozone loss resulting from the enhanced aerosol loading after the Mt. Pinatubo eruption (Hofmann *et al.*, 1997; Solomon *et al.*, 2005).

Also of interest is the ozone variability near the top edge of the Antarctic ozone hole. Ozone abundances in this layer between 18 and 22 km may provide an early indication of Antarctic ozone hole recovery (Hofmann *et al.*, 1997). However, as discussed further below, the higher abundances in the 2001-2004 period have been attributed to meteorological variations rather than to ozone recovery (*e.g.*, Hoppel *et al.*, 2005). During 2002-2004, the temperature in the 20-22 km region tended to be warmer than average from mid-August through September, resulting in fewer PSCs which inhibited ozone loss (Hoppel *et al.*, 2005). The most extreme manifestation of this inhibited ozone loss occurred in 2002. As described in Section 3.2.3.1.1, in September of that year the first documented Antarctic major warming event took place (Roscoe *et al.*, 2005). Major warmings are defined as reversals of both the vortex flow and the temperature gradient in the middle stratosphere; these events are relatively common in the Arctic, but had not been previously observed in the Antarctic. In 2002, anomalously high ozone levels and temperatures extended down to 15 km.

The Antarctic ozone hole generally behaves in a regular fashion, since the Antarctic winter stratosphere is consistently cold, with a stable, isolated vortex and an abundance of PSCs each winter. As discussed in Section 3.2.3.1.1, the Arctic winter stratosphere exhibits much more variability. Compared to the Antarctic, the Arctic is generally warmer with fewer PSCs (Fromm *et al.*, 2003 Figure 3-13). Periods of cold temperatures with elevated reactive chlorine

also tend to persist for shorter lengths of time in the Arctic and these cold regions are generally not concentric with the Arctic polar vortex, but are frequently centered roughly in the region between Greenland and Norway. Thus, ozone levels in the Arctic lower stratosphere exhibit a large amount of variability, which is well correlated with temperature. This is primarily the result of the fact that in the Arctic lower stratosphere the average temperature is very near the PSC formation threshold temperature. Therefore, in cold winters, PSCs tend to be very abundant and large halogen-catalyzed ozone depletion occurs, whereas in warm winters PSCs are very infrequent and little chemical ozone depletion occurs (Rex *et al.*, 2004). This is illustrated later in Section 3.2.3.1.1 in Figure 3.12, which shows a very good correlation between the volume of air with temperatures cold enough to be capable of forming PSCs and the chemical loss of ozone in the lower stratosphere.

A particular problem with regard to assessing trends in polar ozone loss is that the distribution and variation of stratospheric ozone are controlled by both transport processes and photochemical processes. Ozone trends resulting from changes in atmospheric halogen loading must be separated from trends resulting

> Compared to the Antarctic, the Arctic is generally warmer with fewer polar stratospheric clouds (PSCs), which play a role in the depletion of the ozone layer.

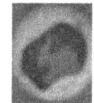

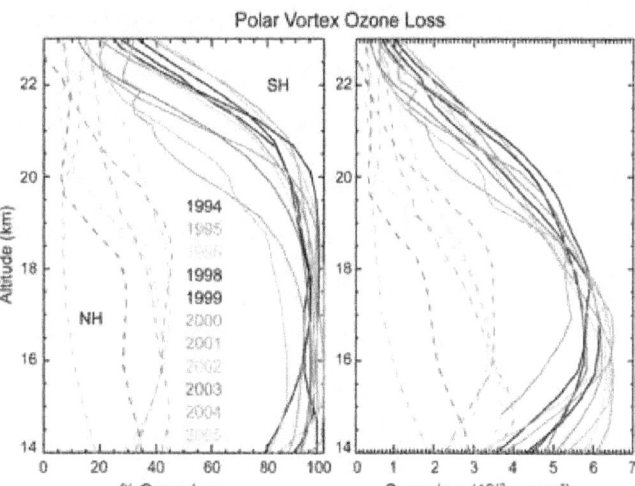

Figure 3.11 Ozone loss estimates from the Polar Ozone and Aerosol Measurement (POAM II & III). Southern Hemisphere October 5 estimates (solid lines) and Northern Hemisphere March 10 estimates (dashed lines) are based on the "vortex average technique" described in WMO (2007) and Hoppel *et al.* (2002, 2003). Estimates are shown only for Northern Hemisphere winters which had a relatively persistent, isolated vortex from January 1 - March 10.

from transport variations. Instruments measure ozone abundances and their variations, but do not directly measure ozone photochemical loss. Isolating the photochemical ozone change in the Arctic is more complicated than in the Antarctic because of the much larger degree of dynamical variability. Several different methods have been developed for isolating photochemically driven ozone change from transport-driven change. For cold Arctic winters (in which there is measurable loss), ozone loss derived from each of these methods now agree fairly well (*e.g.*, WMO, 2007, Figure 4-11). Therefore, we now have a fairly reliable record of ozone chemical loss for all Antarctic winters, and for cold Arctic winters, dating back to the mid-1990s. As an example, Figure 3.11 shows vertical profiles of photochemical loss derived from Polar Ozone Aerosol and Measurement (POAM II & III) measurements, for both the Arctic and Antarctic, during the 1994-2005 time period (Hoppel *et al.*, 2003). Ozone loss in the Antarctic ozone hole was fairly stable in the 1990s, with nearly complete loss in the 14 to 19 km altitude range. Figure 3.11 shows that the anomalously high ozone levels in the upper region of the Antarctic ozone hole in 2001 through 2004 were the result of reduced ozone chemical loss. In contrast to the Antarctic, the ozone loss profiles for the Arctic are highly variable with peak losses of almost 50% (losses up to approximately 60% have been reported by other analyses [Rex *et al.*, 2004; WMO, 2003; WMO, 2007]).

3.2.3 Processes That Affect Ozone
3.2.3.1 TRANSPORT AND DYNAMICS
Stratospheric ozone levels are strongly influenced by both transport and the temperatures of the stratosphere. In this section, we will summarize the influence of dynamical processes on ozone levels. First, there is the direct influence of winds that carry ozone-enriched air from the photochemical production region into other regions, thereby increasing ozone. Second, the opposite process can occur where winds carry ozone-depleted air into other regions, thereby decreasing ozone (*e.g.*, from the Antarctic ozone hole into the midlatitude stratosphere). Third, the radiatively and dynamically driven local temperature can influence ozone by affecting catalytic loss reaction rates.

This section is divided into two subsections. The first subsection discusses the influence of dynamics on polar ozone, while the second subsection addresses the influence of dynamics on midlatitude ozone.

3.2.3.1.1 Polar
Variability in the dynamical conditions in the troposphere/stratosphere system results in variability of ozone transport and temperatures in the polar stratosphere. Previous World Meteorological Organization (WMO)/ United Nations Environment Programme (UNEP) assessments have shown that, on short timescales, interannual variability in polar ozone chemistry is mainly driven by temperature variability, which in turn is the result of variable dynamical conditions. The combined effect of dynamically-induced variability in both chemistry and transport is the main driver of interannual variability of the abundance of ozone in the polar stratosphere.

As described in the Introduction (Section 3.1), the air in the polar lower stratosphere is transported downward from the upper stratosphere and mesosphere over the course of the winter period by the Brewer-Dobson circulation. The Brewer-Dobson circulation is driven by large-scale atmospheric waves that propagate upward from the troposphere. Figure 3.1 shows this poleward and downward circulation in the annual average. This upper stratospheric air has on average been in the stratosphere for five to six years since entering the stratosphere at the tropical tropopause. In the absence of polar ozone destruction, this air would be characterized by relatively high ozone concentrations. Furthermore, because the air has been in the upper stratosphere and exposed to intense solar UV, the organic chlorine and bromine compounds have been almost completely converted to inorganic forms that can participate in ozone loss processes.

The chemical ozone loss processes precipitated by the presence of halogens are initiated by the formation of PSCs in the extremely cold polar lower stratosphere (Crutzen and Arnold, 1986; Toon *et al.*, 1986). PSCs provide a surface upon which heterogeneous (not gas-phase, but at the surface between a solid/liquid and a gas) reactions take place that convert comparatively

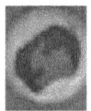

The Antarctic ozone hole is more severe in colder than average years, while less severe in warmer than average years.

unreactive chlorine reservoirs into ones that are exceedingly reactive in sunlight. While the chlorine and bromine levels in the stratosphere directly cause ozone loss, year-to-year variation of the chemically driven polar ozone loss is directly tied to the temperature by a modulation of polar stratospheric clouds and transport.

A number of studies have shown that the Antarctic ozone hole is more severe in colder than average years, while less severe in warmer than average years (Newman and Randel, 1988; WMO, 1989). In the Arctic, Rex et al. (2004) quantitatively related the volume of polar stratospheric clouds (V_{PSC}) to the chemical ozone loss estimated from ozonesondes (extended to the Antarctic by Tilmes et al., 2006). Figure 3.12 shows ozone loss plotted against V_{PSC} for the years from 1992 to the present (the latest data are available through 2007). This 1992-2007 period has high chlorine and bromine levels (Chapter 2). For the coldest Arctic winters, the volume of air with temperatures low enough to support polar stratospheric clouds (V_{PSC}) increased significantly since the late 1960s (Rex et al., 2006). The cooling of the lower stratosphere is much larger than expected from the direct radiative effect of increasing greenhouse gas concentrations. The reason for the change is not clear and it could be due to long-term natural variability or an unknown dynamical mechanism.

The year-to-year variation of spring temperatures in the polar stratosphere is primarily driven by year-to-year variability of planetary waves that propagate upward from the troposphere to the stratosphere. The relationship of waves to stratospheric ozone was recognized by a number of early investigators who saw large increases of total ozone following major stratospheric warmings (London, 1963).

The large variability of polar total ozone shown in Figures 3.5 and 3.7 is directly tied to the variations in the levels of the planetary waves (Randel et al., 2002). The Southern Hemisphere winter of 2002 provides an excellent example of a year with extremely high levels of planetary waves propagating into the stratosphere. The planetary wave forcing of the stratosphere is estimated from the eddy heat flux (a cross

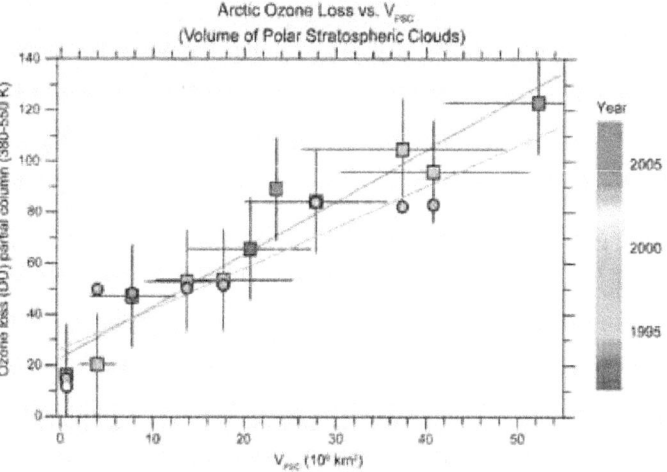

Figure 3.12 Scatter plot of vortex-average chemical loss of column ozone (ΔO_3, calculated over the range 380 to 550 K) versus V_{PSC} inferred from ozonesonde observations for the 1991/1992 to 2006/2007 Arctic winters (update from Rex et al., 2004). Colored squares and the red fit line show results based on ozonesonde analyses; colored circles and the green fit line show results from tracer correlation studies based on HALOE data (update from Tilmes et al., 2006). Adapted from Rex et al. (2006, 2004), and Tilmes et al. (2006).

correlation of the north-south wind and the temperature) at an altitude of 16 km in the 45-75°S zone (see Andrews et al. (1987) for a more complete description of the wave driving of the stratosphere by the troposphere). In September 2002, a major warming had a dramatic impact on total ozone, splitting the Antarctic ozone hole into two pieces (Stolarski et al., 2005). Meteorological conditions in 2002 showed that the early winter was already unusually disturbed (Hio and Yoden, 2005; Newman and Nash, 2005; Allen et al., 2003). There were several significant wave events from May to October that each warmed the stratosphere by a few degrees until the major warming in late September. Several models reproduced the chemistry and dynamics of this 2002 warming, revealing the direct impact of tropospheric waves on Antarctic ozone levels (Manney et al., 2005; Ricaud et al., 2005; Konopka et al., 2005; Grooß et al., 2005; Sinnhuber et al., 2003; Feng et al., 2005).

3.2.3.1.2 Midlatitude dynamic and transport effects on ozone

The influence of transport and dynamics on the midlatitude lower stratosphere (16 to 30 km) and lowestmost stratosphere (8 to 16 km) principally occurs through the Brewer-Dobson circulation and through mixing processes. While photochemistry plays an important role

The lifetime of ozone in the midlatitude lower stratosphere (16 to 30 km) is more than 100 days. Therefore, transport through the atmosphere plays a very important role in determining midlatitude ozone levels.

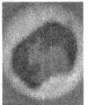

Heterogeneous
reactions on polar
stratospheric clouds
(PSCs) convert
the comparitively
unreactive chlorine
reservoirs into
chlorine that is very
reactive in sunlight.

for ozone in the midlatitudes, the lifetime of ozone in the lower stratosphere is long (greater than 100 days), and hence, transport plays a very important role in determining ozone levels. In the upper stratosphere, dynamically or radiatively forced temperature changes can have a large effect on ozone loss rates by modifying the catalytic loss processes. Dynamically forced ozone changes in the lower stratosphere occur because of:

- interannual and long-term changes in the strength of the stratospheric Brewer-Dobson circulation (Figure 3.1), which is responsible for the winter-spring buildup of extratropical ozone (*e.g.*, Fusco and Salby, 1999; Randel *et al.*, 2002; Weber *et al.*, 2003; Salby and Callaghan, 2004a; Hood and Soukharev, 2005) and;
- changes in tropospheric circulation, particularly changes in the frequency of local nonlinear synoptic wave forcing events, which lead to the formation of extreme ozone minima ("mini-holes") and associated large increases in tropopause height and horizontal mixing (Steinbrecht *et al.*, 1998; Hood *et al.*, 1997, 1999, 2001; Reid *et al.*, 2000; Orsolini and Limpasuvan, 2001; Brönnimann and Hood, 2003; Hood and Soukharev, 2005; Koch *et al.*, 2005).

The effects of dynamics on ozone trends and variability are extremely difficult to quantify. This difficulty is caused by the relationship between the strength of the Brewer-Dobson circulation, the wave mixing processes, and the position and strength of the polar vortex. As is well recognized, the propagation of planetary scale waves from the troposphere into the stratosphere drives the Brewer-Dobson circulation, while at the same time the breaking of these waves irreversibly mixes air latitudinally. The estimation of ozone advection is further confused by the need to multiply the transport "variables" by the ozone horizontal and vertical gradients. This effect of the ozone gradient is mainly evident in two regions: the mixing of lower stratospheric ozone depleted air from the polar latitudes to the midlatitudes during the spring period, and the mixing of air from the tropical upper troposphere (with very low ozone amounts) into the midlatitude lowermost stratosphere.

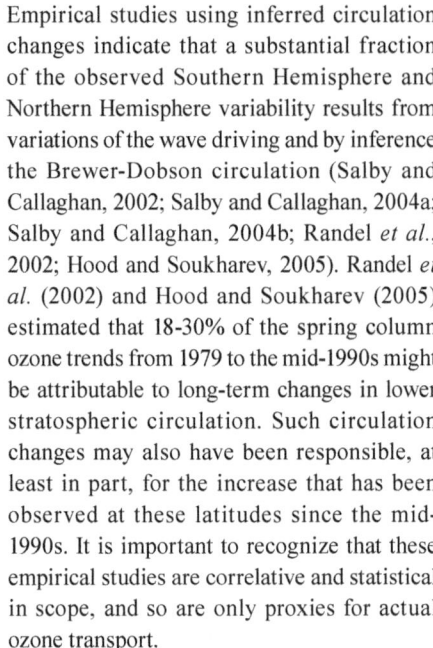

Empirical studies using inferred circulation changes indicate that a substantial fraction of the observed Southern Hemisphere and Northern Hemisphere variability results from variations of the wave driving and by inference the Brewer-Dobson circulation (Salby and Callaghan, 2002; Salby and Callaghan, 2004a; Salby and Callaghan, 2004b; Randel *et al.*, 2002; Hood and Soukharev, 2005). Randel *et al.* (2002) and Hood and Soukharev (2005) estimated that 18-30% of the spring column ozone trends from 1979 to the mid-1990s might be attributable to long-term changes in lower stratospheric circulation. Such circulation changes may also have been responsible, at least in part, for the increase that has been observed at these latitudes since the mid-1990s. It is important to recognize that these empirical studies are correlative and statistical in scope, and so are only proxies for actual ozone transport.

Estimates of the dynamically induced contributions to ozone interannual variability and trends can be derived by using chemical transport models (CTM) driven by observed temperature and wind fields (Hadjinicolaou *et al.*, 1997; 2002; 2005). Using the SLIMCAT three-dimensional (3-D) CTM, Hadjinicolaou *et al.* (2005) found that about one-third of the observed ozone trend from 1979 to the mid-1990s could be explained by transport-related changes. In addition, Hadjinicolaou *et al.* (2005) also found that all of the midlatitude "increase" (see the period from the mid-1990s to 2004 in top right panel of Figure 3.3) could be explained by transport alone, and not by halogen decreases. However, the interannual variation discrepancies between CTMs and observations are large, making it difficult to place much weight on CTM results to attribute long-term transport changes.

The midlatitude ozone is influenced by polar loss via air mass mixing after the polar vortex breakup in early spring. Using regression analysis, Dhomse *et al.* (2006) concluded that this mechanism is one of the main factors responsible for the recent increase in Northern Hemisphere total ozone.

3.2.3.2 CHEMISTRY
3.2.3.2.1 Polar chemistry

Heterogeneous reactions on PSCs convert the comparatively unreactive chlorine reservoirs hydrochloric acid (HCl) and chlorine nitrate ($ClONO_2$) first to chlorine gas (Cl_2) in the long, dark polar night. As soon as the Sun first appears over the horizon in the Antarctic spring in August each year, the Cl_2 photolyzes (breaks apart into chlorine atoms in the presence of sunlight, $Cl_2 + h\nu \rightarrow 2\ Cl$) and Cl atoms react with ozone to make chlorine monoxide (ClO) (see reaction 3c in Section 3.1). These reactions are often called "chlorine activation," since the chlorine compounds are converted from comparatively unreactive forms to much more photochemically reactive forms. At high concentrations, ClO reacts both with itself (reaction 3a forms the ClO dimer, dichlorine peroxide [ClOOCl], a reaction that actually proceeds faster at lower temperatures) and with the analogous bromine monoxide, BrO (see reaction 4a). Almost all of the rapid ozone loss in the Antarctic spring is attributed to catalytic cycles formed from the reaction of ClO with itself (reactions 3) and with BrO (reactions 4) (Frieler et al., 2006).

Thus, stratospheric chlorine levels provide the fundamental driver for polar ozone loss, since chlorine is involved in the principal catalytic cycles responsible for polar ozone loss. Beyond this basic understanding, however, the calculated chemical loss rates of polar ozone are still quantitatively uncertain. Questions remain to be resolved on the photolysis rate of the ClOOCl (Equation 3b) and the balance between ClO and ClOOCl in the Antarctic stratosphere and the atmospheric abundance of bromine. Higher levels of bromine would improve the comparison between theory and observation for Arctic and Antarctic loss rates, but the exact sources of the extra bromine are somewhat uncertain.

From *in situ* aircraft measurements, Stimpfle et al. (2004) suggested that the ClO dimer cycle (reactions 3) may be a more efficient process for polar ozone loss than previously thought (Frieler et al., 2006), and good overall consistency between *in situ* observations of ClO and the ClOOCl and model calculations can be achieved if it is assumed that ClOOCl photolyzes faster than assumed in WMO (2003). However, one recent laboratory study of the absorption cross-section of ClOOCl does not support this. It indicates that ClOOCl may photolyze (Equation 3b) slower than previously understood (Pope et al., 2007). However, this slower photolysis rate results in severe underestimates by photochemical models of observed O_3 depletion rates and observed ClO levels, and hence poor representations of the severity of polar ozone losses (von Hobe et al., 2007). Current models (without Pope et al., 2007) reproduce the basic features of the Antarctic ozone hole and Arctic ozone losses using previous laboratory recommendations for photochemical parameters (e.g., WMO, 2003; WMO, 2007). Clearly, more work will be required to understand this discrepancy.

Recent measurements show that bromine exists in the stratosphere at higher concentrations than is found in most 3-D models (WMO, 2007 and references therein). Profiles of BrO measured in the Arctic vortex suggest that inorganic bromine levels may be 3 to 8 parts per trillion (ppt) by volume larger than the amount of bromine carried to the stratosphere by methyl bromide (CH_3Br) and halons alone (Canty et al., 2005; Frieler et al., 2006). Although still uncertain, the additional 3-8 ppt of bromine is probably derived from very short lived (VSL) species containing bromine that enter the stratosphere at the tropical tropopause (WMO, 2007). Considering that the BrO + ClO cycle is now estimated to contribute up to half of total chemical loss of polar ozone, using the more efficient ozone loss by the ClO dimer cycle, this observation indicates the BrO + ClO catalytic cycle is likely to be a more efficient ozone loss process than considered in WMO (2003). Hence, bromine may play a more important role in polar ozone depletion than previously thought.

Polar stratospheric clouds are critically important in ozone photochemistry primarily through two processes: chlorine activation and denitrification. The chlorine heterogeneous reactions and sunlight lead to chlorine activation, while removal of nitric acid (HNO_3) occurs as PSCs fall out of the lower stratosphere and remove nitrogen, or denitrify, that air. Satellite observations of aerosols and clouds

Bromine may play a more important role in polar ozone depletion than previously thought.

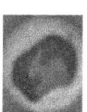

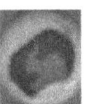

in the polar stratosphere began with NASA's Stratospheric Aerosol Monitor (SAM) II in 1978, and continued nearly uninterrupted to 2005. These measurements used solar occultation in the visible and shortwave infrared portion of the electromagnetic spectrum. In addition to SAM II, other instruments included the NASA series of Stratospheric Aerosol and Gas Experiment (SAGE) I-III and the Naval Research Laboratory's Polar Ozone and Aerosol Measurement (POAM) II-III.

As noted in Section 3.2.3.1.1, V_{PSC} is a key parameter for estimating ozone loss. It is important to recognize that V_{PSC} actually represents the volume of temperatures cold enough to form PSCs, not the actual PSC volume over the polar region. Nevertheless, temperatures are directly related to PSC occurrence frequency (Steele *et al.*, 1983). The long-term PSC statistics are presented in Figure 3.13 (Fromm *et al.*, 2003). Here the PSC frequency (the number of profiles with a PSC divided by the number of profiles inside the polar vortex) for entire winter seasons is shown. In the Antarctic, PSCs are more frequent than the Arctic (Fromm *et al.*, 2003). There are large interannual variations in Antarctic PSC frequency but no obvious long-term trend. In the Arctic, as described in Section 3.2.2.2, stratospheric temperatures exhibit large

Models predict that the decline in ozone outside the polar regions should have ceased and that the next few decades should show the beginning of a recovery from the maximum depletion.

variability and the average temperature is close to the PSC formation threshold temperature. In warm Arctic winters, little or no PSC activity is evident (for example, in the winter of 1984/1985). However, even in the coldest Arctic winters, PSCs only reach a 25% frequency.

3.2.3.2.2 Global and midlatitude chemical processes

As in the polar regions, halogen increases (chlorine and bromine) have been the principal driver of ozone depletion over the past few decades in the midlatitudes. There is good overall agreement between observed long-term changes in ozone outside of the polar regions and model simulations that include the effects of increasing halogens. The models generally reproduce the observed ozone changes as a function of altitude, latitude, and season, confirming our understanding that halogen changes are the main drivers of global ozone changes (WMO, 2007). These models predict that the decline in ozone should have ceased and that the next few decades should show the beginning of a recovery from the maximum depletion. This is supported by the statistical fit of globally averaged ozone observations with Equivalent Effective Stratospheric Chlorine (EESC), a quantity that peaked in the late 1990s (Figure 3.2).

The explosive eruption of Mt. Pinatubo in 1991 injected large quantities of sulfur into the stratosphere (Trepte *et al.*, 1993). The sulfur-enhanced stratospheric sulfate aerosols provided significantly more surfaces that could support heterogeneous chemical reactions, thus converting a higher fraction of stratospheric chlorine to catalytically active forms. The impact of aerosols on midlatitude ozone was greatest in the early 1990s after the eruption of Mt. Pinatubo in 1991 (Figure 3.3). The observed decrease in Northern Hemisphere column ozone in 1993 agrees with chemical dynamical models that include these effects (WMO, 2003, 2007). The same models predict that the aerosols from Mt. Pinatubo should have produced a significant decrease in ozone over midlatitudes of the Southern Hemisphere, but no effect has been seen in either satellite measurements or ground measurements at stations such as Lauder, New Zealand.

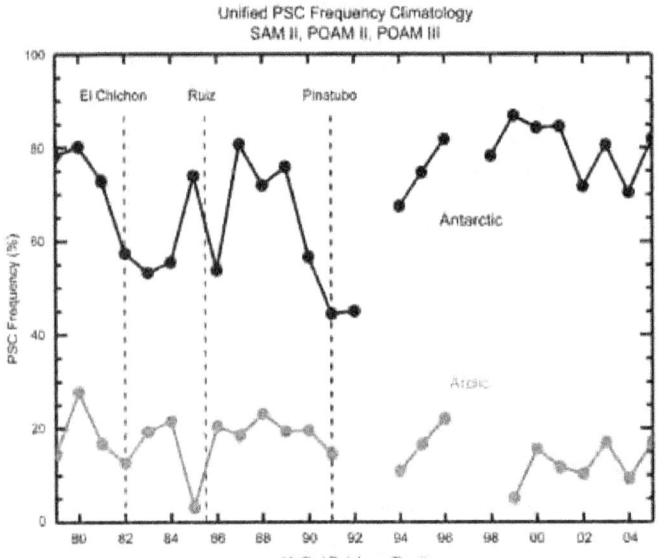

Figure 3.13 PSC frequency for the entire winter season. The frequency is calculated as the number of profiles with a PSC divided by the number of profiles inside the polar vortex. See Fromm *et al.* (2003) for details.

The inclusion of additional inorganic bromine (Br_y) from very short-lived substances (VSLS) in models leads to larger ozone destruction at midlatitudes, compared with studies including only long-lived bromine source gases (*e.g.,* Salawitch *et al.,* 2005; Feng *et al.,* 2007b). The enhanced ozone loss occurs in the lower stratosphere via interactions of this bromine with anthropogenic chlorine. Midlatitude ozone loss is primarily enhanced during periods of high aerosol loading. The impact on long-term midlatitude ozone trends (1980-2004), assuming constant VSLS Br_y, is calculated to be small because aerosol loading was low at the start and end of this time period.

The profile of upper stratospheric ozone trends from 1980-2004 is generally consistent with our understanding of gas-phase chlorine chemistry as the cause of declining ozone, modulated by changes in temperature and other gases such as methane (WMO, 2007). However, global dynamical-chemical models have not demonstrated that they can simultaneously reproduce realistic trends in all relevant parameters, although observations over the full time period are limited (Eyring *et al.,* 2006). Chemical models without interactive radiation obtain ozone changes that peak at about 14% for 1980-2004 (in altitude coordinates), consistent with SAGE observations.

Our ability to reproduce observed past changes in the Northern Hemisphere is better than that for the Southern Hemisphere. Two-dimensional (2-D) models show large model-model differences in the Southern Hemisphere due to different treatments of the Antarctic ozone loss and how it is spread to the midlatitudes. Three-dimensional CTMs are inherently better at simulating the polar regions and this leads to smaller model-to-model differences. These CTMs, however, still do better at reproducing long-term changes in the Northern Hemisphere than in the Southern Hemisphere (WMO, 2007). This ongoing disagreement between model-observation comparisons in the Northern versus the Southern Hemisphere indicates that we do not yet have a full understanding of the combined chemical and transport processes controlling ozone changes at midlatitudes.

3.3 ULTRAVIOLET RADIATION AT THE EARTH'S SURFACE

3.3.1 Background (Factors Controlling UV Surface Irradiance)

The amount of UV radiation reaching the Earth's surface is controlled by several key factors including cloud cover, aerosols, and amount of atmospheric ozone (with most of the ozone being in the stratosphere). Ozone and cloud cover are the most important atmospheric components limiting the amount of UVB (280 to 315 nm) radiation able to reach the ground. Clouds and scattering aerosols reduce UV radiation at all wavelengths by reflecting a fraction of UV energy back to space, whereas ozone absorbs a fraction of the UV radiation only in the 280 to 340 nm range, with more absorption at shorter wavelengths than at longer wavelengths. Under special conditions, clouds can locally increase UV from 1% to 10% by cloud edge reflections. Extremely heavy cloud cover (black thunderstorm) can decrease UV almost 100%. Radiation with wavelengths shorter than 280 nm does not reach the surface in significant amounts because of absorption by the atmosphere (O_3 and O_2). Air pollution is an additional factor that can affect UV reaching the surface through the absorption and scattering by aerosols and absorbing trace gases such as tropospheric O_3 and nitrogen dioxide (NO_2). UV radiation at the surface is generally highest near the equator following the seasonally changing sub-solar point (latitude between $\pm 23°$), where stratospheric ozone is a minimum and the solar zenith angle (SZA) is the smallest. Larger amounts of UV radiation are seen at high altitude sites, especially those with predominantly dry and clear weather and large surface reflectivity (*e.g.,* from snow or ice cover). Understanding, modeling, and measuring the factors affecting the amount of UV radiation reaching the Earth's surface is important, since increases in UV radiation affect human health adversely through skin cancer (Diffey, 1991), eye cataracts (Taylor, 1990), and suppression of the immune system (Vermeer *et al.,* 1991), and positively through increased Vitamin D production (Grant, 2002; Holick, 2004). Changes in UV radiation also have important effects on ecosystem biology (Smith *et al.,* 1992; Ghetti *et al.,* 2006).

Reductions in ozone lead to increases in ultraviolet (UVB) radiation at the Earth's surface. Higher UVB radiation can increase skin irritation.

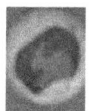

Both theory and observations (Figures 3.14 and 3.15) show that reductions in ozone lead to increases in UV erythemal radiation and UVB at the Earth's surface. Erythemal radiation is a weighted average of UVA (315 to 400 nm) and UVB used as a measure of skin irritation caused by exposure to sunlight (McKinlay and Diffey, 1987). The UV erythemal irradiance data shown in Figure 3.14 was obtained under clear-sky conditions at Mauna Loa, Hawaii, and shows the measured inverse relationship between ozone change and UVB radiation, which is the dominant portion of erythemal radiation. The relation to UV index and the units for

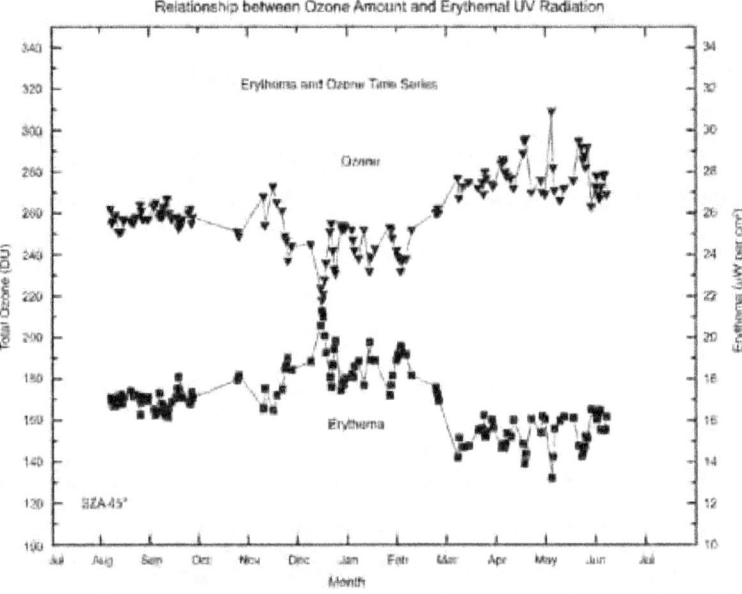

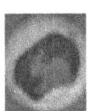

Figure 3.14 Measured erythemal irradiances (lower curve) from an ultraviolet spectroradiometer at SZA 45° compared with total ozone (upper curve) for 132 clear mornings during July 1995 to July 1996 at Mauna Loa Observatory (19.5°N, 155.6°W, 3.4 km), showing the inverse relationship between erythemal UV and ozone amount (adapted from WMO, 1999). (UV index 10 = 25 μW per cm² = 250 mW per m²)

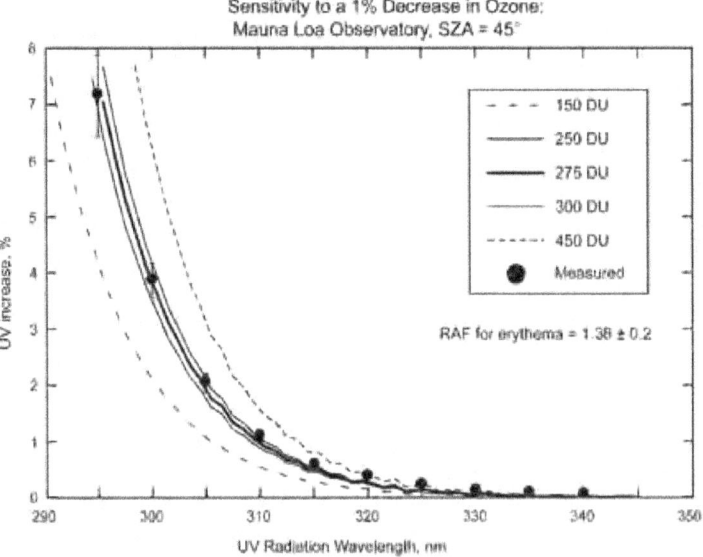

Figure 3.15 Validation of Equation 1 using the measured (dark circles) changes in ozone and UV irradiance from Mauna Loa, Hawaii, shown in Figure 3.14 (adapted from WMO, 1999).

irradiance and dose are discussed in Appendix 3B.

Increases in 280 nm to 340 nm UV radiation caused by decreases in ozone are easily estimated using radiative transfer calculations. For clear-sky conditions, the changes can also be accurately estimated using a simple relation between ozone and irradiance given in Equation 1:

$$dF/F = -d\Omega/\Omega \ \alpha\Omega \ \sec(\theta) = -d\Omega/\Omega \ (RAF) \quad (1)$$

where the quantity $\alpha\Omega \ \sec(\theta)$ is known as the Radiation Amplification Factor (RAF).

The relationship is derived from the standard Beer's Law of irradiance F attenuation in an absorbing atmosphere, $F = F_o \exp(-\alpha\Omega \ \sec(\theta))$, where Ω is the ozone column amount in Dobson Units (DU, equal to milli-atm-cm), α is the ozone absorption coefficient (in cm^{-1}), θ is the solar zenith angle, and F_o is the irradiance at the top of the atmosphere (Madronich, 1993). An example to show the magnitude of the RAF as a function of wavelength is shown in Figure 3.16 for $\theta = 45°$ and $\Omega = 330 \ DU = 0.33$ atm cm. The RAF method accurately estimates UV irradiance change compared to clear-sky radiative transfer (Herman *et al.*, 1999b). For example, radiative transfer shows that a 1% decrease in O_3 produces a 2.115% increase in 305 nm irradiance, while the RAF method estimates a 2.064% increase ($\Omega = 375 \ DU$, $\theta = 30°$). Changes in measured erythemal irradiance are approximated very accurately using Equation 1 with an RAF = 1.38 when the ozone amount changed by 1% (Figure 3.14, $\Omega = 275 \ DU$, $\theta = 45°$). For most conditions, erythemal irradiance change with ozone change behaves roughly the same as 308 nm irradiance.

The RAF approximation is useful for mid-day during the spring, summer, and autumn at most latitudes. During summer solstice, Equation 1 applies up to 83° latitude. In the presence of constant attenuation by cloud cover or scattering aerosols, Equation 1 still

approximately gives the fractional change in irradiance for a change in ozone amount.

Fioletov *et al.* (1997) reported an extensive analysis of UVB irradiance and its dependence on total ozone. The analysis provides an empirical wavelength-by-wavelength measure of the increase of UVB irradiance for a 1% decrease of total ozone. These values were found to be essentially the same for clear and cloudy conditions (except for very heavy clouds) and are in good agreement with model results for longer wavelengths and moderate SZA.

UV radiation reaching the Earth's surface varies on all time scales, from seconds to seasons. Hourly to daily changes, *i.e.*, the short-term variations, are mostly due to cloud cover changes and aerosols, and to ozone in the UVB range. The extent of cloud cover also causes changes on daily and monthly time scales as the weather changes. In today's atmosphere, the multi-year variations are controlled principally by changes in stratospheric ozone, changes in the extent of cloud cover, and other longer-term changes such as in the amount of aerosol and

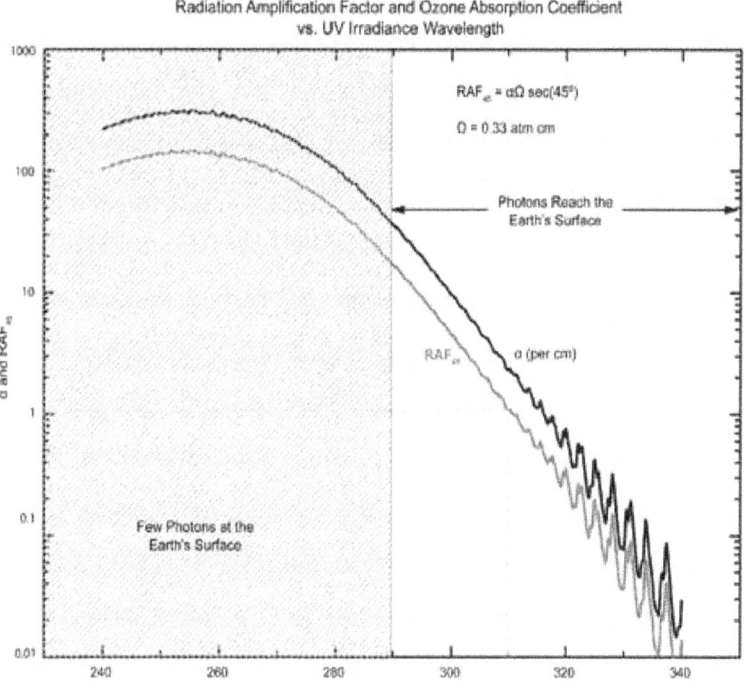

Figure 3.16 Ozone absorption coefficient α (cm^{-1}) and the Radiation Amplification Factor RAF_{45} for a solar zenith angle SZA = 45° and ozone amount of 330 DU ($\Omega = 0.330$ atm cm). Note that at 310 nm the RAF_{45} is approximately 1, so that a 1% increase in O_3 would produce a 1% decrease in 310 nm irradiance.

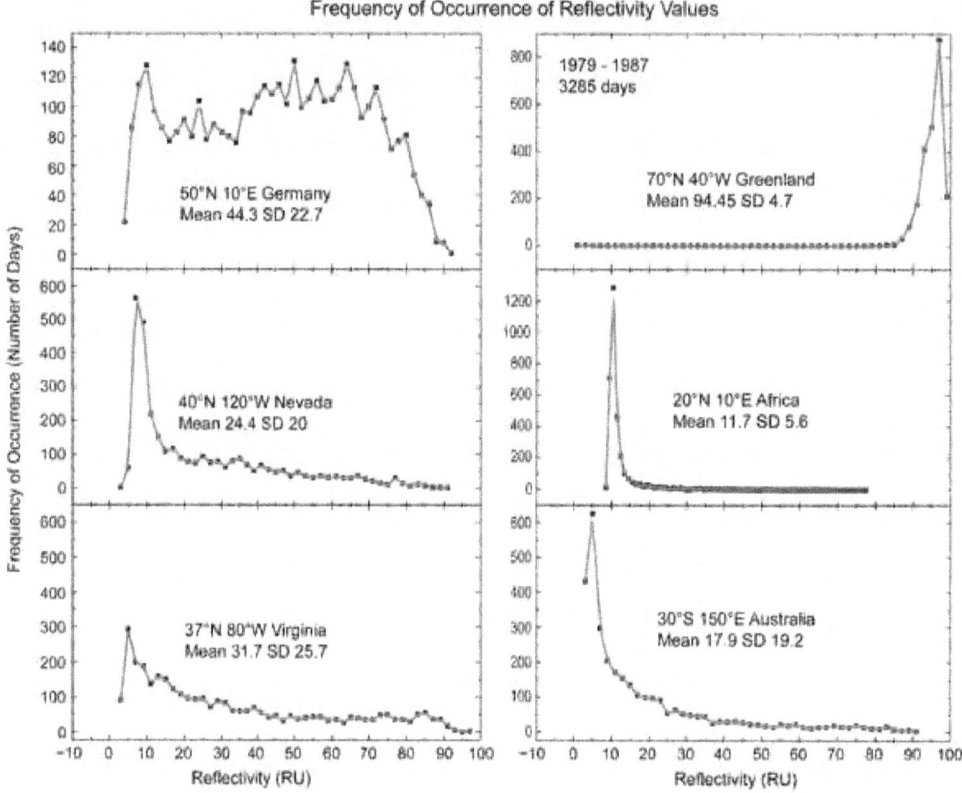

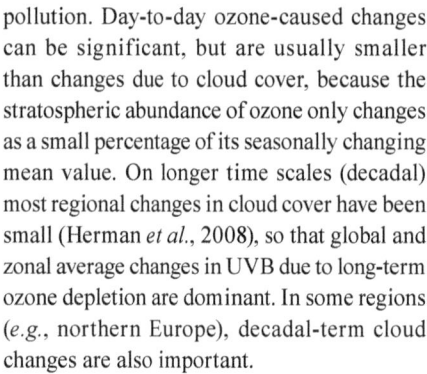

Figure 3.17 Frequency of occurrence of reflectivity values from 1979 to 1987 (3285 days) for six different locations. The mean and standard deviation (SD) are in RU (1 RU = 1%). Based on Herman *et al.* (2001a).

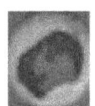

pollution. Day-to-day ozone-caused changes can be significant, but are usually smaller than changes due to cloud cover, because the stratospheric abundance of ozone only changes as a small percentage of its seasonally changing mean value. On longer time scales (decadal) most regional changes in cloud cover have been small (Herman *et al.*, 2008), so that global and zonal average changes in UVB due to long-term ozone depletion are dominant. In some regions (*e.g.*, northern Europe), decadal-term cloud changes are also important.

Ozone data from Nimbus-7/TOMS, obtained during June for the entire 5° longitudinal zone centered at 40°N, shows that the ozone amount can vary by 50 DU about the mean value of 350 DU, or $d\Omega/\Omega = \pm 0.14$. The day-to-day June ozone variation is obtained from figures similar to those shown in Herman *et al.* (1995). Using an average noon SZA for June of about 23° and an ozone absorption coefficient for 305 nm α = 4.75 cm^{-1} yields a typical 305 nm irradiance change $dF/F = -d\Omega/\Omega \ \alpha\Omega \ \sec(\theta) = \pm 0.14$ * 4.75 * 0.35 * 1.09 = ± 0.25. In other words, for

clear-sky conditions, the 305 nm irradiance typically changes by $\pm 25\%$ during June just from to day-to-day ozone changes. As will be discussed later, the day-to-day variability of clear-sky 40°N UV June irradiance is about three times larger than the change caused by the long-term June decrease in ozone from 1980 to 2007 ($d\Omega/\Omega \sim -0.04$).

Identification of long-term (decadal) changes from ground-based measured surface UV radiation due to stratospheric ozone depletion can be accomplished if the data are filtered to remove the effects of clouds. Trend detection from ground-based measurements under all sky conditions, though appealing and relevant, has many difficulties. This is primarily because the surface UV is highly variable, as noted above, due to factors such as cloud cover and aerosols, and because the stratospheric ozone depletion has been rather small (less than 10%) over the past decades, with the exception of high latitudes (greater than 60°).

Other factors, such as Rayleigh scattering and surface reflectivity, affect the magnitude of measured or theoretically estimated UV irradiance. However, these factors do not significantly affect the short- or long-term changes in irradiance, since their changes are small. Hourly or daily changes in Rayleigh scattering follow the small changes in atmospheric pressure, which usually are less than 2%. There have been no long-term changes in mean atmospheric pressure. The UV surface reflectivity, R_G, is small (3 Reflectivity Units, RU, to 10 RU, where 1 RU = 1%) and almost constant with time except in regions seasonally or permanently covered with snow or ice. Based on radiative transfer studies, clear-sky atmospheric backscattering to the surface contributes less than 0.2 R_G to the measured UV irradiance, which is quite small for most ice/snow-free scenes.

3.3.1.1 REDUCTION OF UV BY CLOUDS

A measured daily cycle of UV reaching the surface will show large UV irradiance reductions from clear-sky conditions as clouds pass over a site. These reductions are frequently in excess of those caused by measured ozone changes from climatological values for wavelengths longer than 305 nm. In general, the effect of clouds is to reduce the UV amount at all wavelengths reaching the Earth's surface. The average amount of UV radiation reduction by clouds can be estimated from the Lambert Equivalent cloud reflectivity R, which varies significantly between locations (Figure 3.17). The operational definition of R is given in Appendix 3A.

Satellite data (Figure 3.17) show that the most commonly occurring values of R are about 3 to 5 RU greater than the surface reflectivity representing haze or very sparse cloud cover. Central Europe, represented by Germany, is quite different from North American sites in that the most frequent values are around 10 RU (127 days out of 3285 days) and around 50 RU (128 days), with almost the same number of days (80 to 128 days) having 10 to 70 RU. Greenland is another extreme, where the reflectivity is always high because of the ice cover. Nevada and Virginia are similar, except that Nevada has a lower average reflectivity representing less cloud cover. An extreme case is represented by Australia, where the average reflectivity (due to cloud cover) is very low and cumulative UV exposure is high compared to the same latitude in the United States.

Satellite observations of reflected UV indicate that reflectivities for typical midlatitude cloud-covered scenes have a wide range of values, which can reach 90 RU over high altitude cloud tops that occur most frequently in the tropics. Under snow-free conditions, the surface reflectivity R_G is usually between 2 RU and 4 RU, reaching about 10 RU in the Libyan Desert and similar small areas (*e.g.*, Andes Mountain high deserts). Area-averaged clear-sky UV surface irradiance is then approximately reduced as a linear function of the cloud plus aerosol reflectivity, which can be written in terms of effective transmission (Krotkov *et al.*, 2001), $T \approx (1 - R)/(1 - R_G)$, with local values occasionally exceeding clear-sky irradiances by about 10% because of reflections from the sides of clouds. Midlatitude UV irradiance reductions caused by clouds range up to 50%, which is larger than the day-to-day 305 nm UV variability caused by ozone (25%), and comparable to the change at 300 nm.

Snow-covered scenes cannot be distinguished from cloud cover by observations in the UV wavelengths. Because of this, the use of reflectivity to estimate the amount of UV radiation at the surface in the presence of snow is likely to be in error. For example, the very high reflectivity values observed in Greenland (Figure 3.17) are almost independent of the cloud cover.

Long-term changes in regional cloud and aerosol reflectivity must be considered when estimating long-term changes in UV irradiance. However, for most populated regions of the Earth, long-term (decadal) cloud and aerosol scattering changes have been shown to be small even where they are statistically significant (Herman *et al.*, 2001b, 2008). Local values of aerosol amounts and absorption are currently estimated from the widely distributed AERONET network of ground-based sunphotometers (Holben *et al.*, 2001).

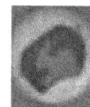

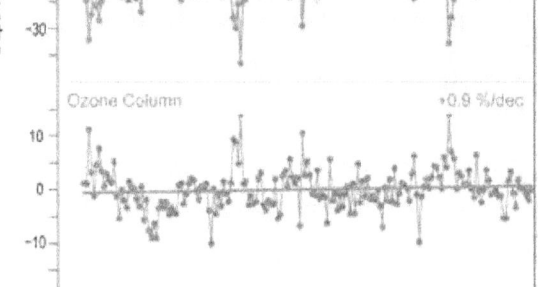

Figure 3.18 Combined effects of ozone, aerosols, and other absorbing components on UV radiation. Long-term variability in monthly mean solar spectral irradiances at 324 nm (upper panel) and at 305 nm (middle panel) measured at Thessaloniki, Greece, under clear skies at 63° solar zenith angle, shown as departures from the long-term (1990-2006) averages. The lower panel shows the corresponding departures in the ozone column of 375 DU (adapted from WMO, 2007).

3.3.1.2 UV ABSORPTION

The amount of UV reaching the surface can also be affected by air pollution, *i.e.,* absorption by aerosols (black carbon, dust, and smoke), tropospheric O_3, NO_2, and other gases. These can cause reductions in UV of up to 10% in polluted sites, but with much higher reductions occurring in certain highly polluted cities, *e.g.,* occasionally in Los Angeles and frequently in Beijing. Nitrogen dioxide causes small reductions mainly to UVA since its absorption cross-section peaks near 410 nm, but is still significant at 330 nm. Aerosols of most types have much weaker wavelength dependence and affect UV and visible radiation at all wavelengths. Pollution abatement, especially in highly polluted regions, can decrease the atmospheric reflectivity and absorption, which has the effect of increasing the amount of UV and visible light reaching the ground.

3.3.1.3 ESTIMATING UV TRENDS: GROUND-BASED

Instrumental requirements for making long-term UV irradiance measurements are well understood in terms of calibration and stability for both spectrometers and broadband radiometers. While useful work can still be done with broadband instruments, much more information can be derived from high spectral resolution spectrometers (*e.g.,* the global network of Brewer spectrometers represented in the United States by the NOAA-EPA network of single-grating Brewers <http://www.esrl.noaa.gov/gmd/neubrew>, the National Science Foundation/Biospherical network, and at NASA by a modified double-grating Brewer (Cede *et al.,* 2006). Long-term surface UV spectral irradiance measurements must be carefully made and analyzed to preclude variations due to clouds that could be mixed into UV trend estimates, or whose variability can mask the detection of small changes. If ground-based data are filtered for cloud-free observations, then UVB changes caused by changes in ozone amount are easily observed in multi-year data records. Aerosols and other forms of pollution can also produce apparent changes in UV irradiance that mask the effect of ozone changes. These can be taken into account if measurements are made simultaneously in the UVB range (*e.g.,* 305 nm) and outside of the ozone absorbing range (*e.g.,* 324 nm). The lack of ability to separate aerosol and pollution effects from ozone-induced changes limits the usefulness of broadband instruments (300 to 400 nm) for understanding the observed irradiance changes.

Radiometric and wavelength calibration of spectrometers used for trend estimates must be carefully maintained to detect the relatively small changes caused by ozone and aerosols. Making accurate spectral measurements is quite difficult, since the natural UV spectrum at the ground changes by several orders of magnitude from 300 to 400 nm. A slight wavelength misalignment can cause significant errors in the measured UVB irradiance amount. Wavelength misalignment is less important for integrated quantities such as the erythemal irradiance.

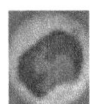

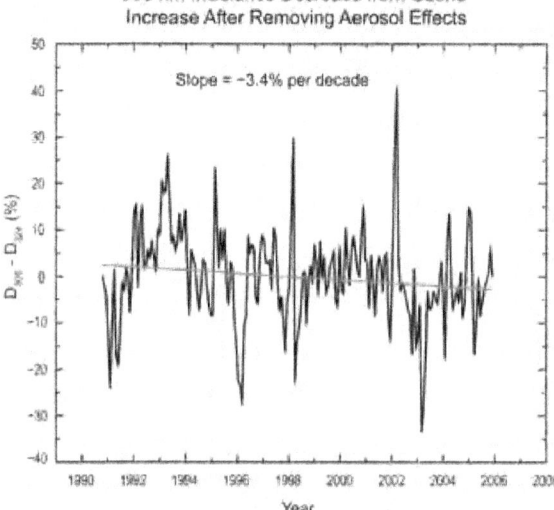

**305 nm Irradiance Decrease from Ozone
Increase After Removing Aerosol Effects**

Slope = −3.4% per decade

Figure 3.19 The difference between the 305 nm and 324 nm departures from the long-term (1990-2006) averages shown in Figure 3.18 showing the measured 3.4% per decade decrease in 305 nm irradiance caused by an ozone amount increase of 0.9% per decade.

A climatology of UV erythemal irradiance for the United States and Canada has been derived from Brewer and pyranometer data for the United States and Canada (Fioletov *et al.*, 2004). The ground-based climatology is lower by 10 to 30% than satellite estimates because of aerosol and pollution absorption that are neglected in the satellite estimates.

An excellent example of UV trend detection is from measured solar irradiances at 305 nm and 324 nm at Thessaloniki, Greece. The irradiances shown in Figure 3.18 are for cloud-free skies at a constant solar zenith angle of 63° (WMO, 2007, which are an extension of Bais and Lubin *et al.*, 2007). These data are obtained from a carefully maintained Brewer spectrometer located in an industrial area of Thessaloniki that is subjected to moderate amounts of pollution generated both locally and reaching Greece from other countries in Europe. There are also occasional dust episodes originating in northern Africa.

The radiation at 324 nm should not be significantly affected by ozone so that the cause of the upward trend at 324 nm (11.3% per decade) is almost certainly due to aerosol and pollution decreases. Decreasing amounts of aerosol and pollution that cause the upward trend at 324 nm will also affect 305 nm by

approximately the same amount. Combining the changes seen for 324 nm with those observed for 305 nm (8.1% per decade) implies that the effect of increasing ozone amount (0.9% per decade) on 305 nm irradiance is a statistically significant decrease of 3.2% (11.3% minus 8.1%) per decade. This is also shown in Figure 3.19, where the time series for 324 nm, D_{324}, was subtracted from the time series for 305 nm D_{305}. The difference, $D_{305} - D_{324}$, was fit with a linear regression having a slope of −3.4% per decade.

An easy way to check this conclusion is through the radiation amplification factor defined as part of Equation 1. The radiation amplification factor, $RAF = -\alpha\Omega \sec(\theta) = -4$ for $\Omega = 375$ DU and $\theta = 63°$, the average measured values for Thessaloniki, Greece. Based on the RAF and the observed ozone change of 0.9% per decade, the change in 305 nm UV irradiance $dF/F = RAF\ d\Omega/\Omega$ should be approximately $-4(0.9) = -3.6\%$ per decade, consistent with the measurements of −3.2% and −3.4% per decade discussed above. In addition to the smaller ozone effects, Figure 3.18 shows that a decline in air pollutants can cause increases in surface UV irradiance of 11.3% per decade in a local industrial site such as Thessaloniki.

When data from cloudy and clear days are present in the UV time series, the measured trends in UV radiation at individual stations can have sufficient variation (typically 0 to 50%, and occasionally larger caused by clouds) to make estimated long-term trends lose statistical significance. As shown in the WMO (2007) report, trend estimates for the period from 1998 through 2005 for Toronto was 1.5±5% per decade (1 standard deviation, 1σ) (WMO, 2007) during a period in which the total ozone amount was relatively constant. Even using Toronto UV radiation data going back to 1990, no statistically significant trend is observable in the extended Toronto UV data despite ozone decreases that took place during the 1990s, because of variability introduced by clouds. To

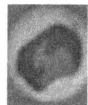

relate the estimated trends to ozone changes requires knowledge of changes in aerosol and cloud amounts, which can be obtained from a wavelength not affected by ozone.

3.3.1.4 ESTIMATING UV TRENDS: SATELLITES

The data for estimating long-term changes of surface UV irradiance can come from individual local ground-based measurements or from global estimations using satellite ozone, aerosol, and cloud data. Global estimates of surface UV irradiance, UV_{EST}, as a function of latitude and longitude have been calculated from satellite measurements of atmospheric backscattered UV and the small amount reflected from the surface. UV_{EST} data are obtained from vector radiative transfer calculations that include polarization effects, ozone absorption, cloud reflectivity and transmission, aerosol scattering and absorption, and the measured surface reflectivity climatology (Herman and Celarier, 1997). The long-term precision and stability of a satellite instrument's in-flight calibration, especially the single channel radiances used to estimate cloud transmission and reflectivity, make it very useful for estimating trends in UV_{EST}. In the absence of a widely distributed closely spaced network of well-calibrated UV spectrometers, satellite UV irradiance estimates are extremely useful, especially over ocean areas where there are no other measurements. Since ozone amount, aerosol amount, and cloud reflectivity are the measured quantities, it is straightforward to separate their respective effects on estimated UV irradiance from satellite data.

There are two ways of estimating the UV irradiance reaching the ground from satellite ozone, aerosol, and reflectivity data. First, one can enter these quantities in a detailed plane parallel radiative transfer model to compute cloud transmission, C_T, using Mie theory to approximate the cloud and aerosol properties in addition to Rayleigh scattering and ozone absorption (Krotkov et al., 1998; 2001). The second, and easier method, is to estimate the irradiance reaching the ground for a Rayleigh scattering and ozone absorbing atmosphere F_{CLEAR}, and then add the cloud and aerosol transmission as a correction factor based on the measured fractional scene R ($0 < R < 1$) and

surface reflectivity R_G, $T \approx (1 - R)/(1 - R_G)$, where $0 < T < 1$. The irradiance at the surface is then approximately

$$F_{SURFACE} = T \, F_{CLEAR} \qquad (2)$$

The two methods agree quite closely (Krotkov et al., 2001), except when there is enough multiple scattering within a cloud to give enhanced ozone absorption at wavelengths less than about 310 nm, where C_T is the better estimate. Both the C_T and the simplified method are frequently 10% higher than measured irradiance values on the ground, and sometimes 20% higher. The differences are usually caused by an underestimate in the satellite calculation of aerosol amount and aerosol absorption (Herman et al., 1999a; Krotkov et al., 1998; 2001; Kalliskota, 2000). The differences become much less when the aerosol amount is small or is known from ground-based measurements. Other sources of difference between ground-based measurements and satellite estimates of UV irradiance arise from the large satellite field of view (50×50 km^2 for TOMS and 12×24 km^2 for OMI) compared to the smaller ground-based field of view, and also from terrain height differences within a satellite field of view.

A recent comparison of measured UV erythemal irradiance from ground-based measurements and OMI satellite estimates has been made (Tanskanen et al., 2007). The comparison shows that for flat, snow-free regions with modest loadings of absorbing aerosols or trace gases, the OMI-derived daily erythemal doses have a median overestimation of 0 to 10%, and 60 to 80% of the erythemal doses are within ±20% compared to ground-based measurements.

Similar errors occur when interpolating between widely separated ground-based stations, where the aerosol, ozone, and cloud amount varies between the stations. Given the need for global coverage of UV_{EST} and the sparsely located ground-based stations, calculations of UV_{EST} from satellite-observed column ozone abundances and cloud reflectivities, which are validated by ground-based measurements, are a useful method for estimating regional, zonal average, and global UV irradiance trends.

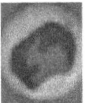

Note that year-to-year shifts in cyclic weather patterns (*e.g.*, clouds, ozone transport, *etc.*) by even a tenth of a degree in latitude and longitude (about 10 km) have a minimal effect on area-averaged satellite ozone and reflectivity measurements (and the UV estimates derived from them), but strongly affect ground-based UV measurements and their estimates of UV irradiance trends. Therefore, the surface UV changes deduced from ozone amounts and reflectivity measured by satellites, UV_{EST}, are expected to be equivalent to those from cloud-filtered ground-based observations of UV irradiance, and superior for estimating regional and global changes. Satellite measurements provide both local and global long-term coverage, which can be used to construct zonal and regional averages and long-term trends that have much less geophysical variance from clouds than corresponding ground-based measurements. The use of satellite estimates, however, presupposes ground-based measurements for validation and as a bridge between successive satellite instruments.

Satellite measures of UV_{EST} have used data from Nimbus7-TOMS (N7, 1979 to 1992), global weekly averages from multiple SBUV-2 instruments (1988 to present), global coverage from Earth-Probe TOMS (EP, 1997 to 2002), and the Aura satellite's Ozone Monitoring Instrument (OMI, 2005 to present). Other data are available from European satellites (*e.g.*, GOME).

It has been shown that cloud plus aerosol reflectivity over the United States has only changed by a small amount for the periods 1980 to 1992 (Herman *et al.*, 2001b) and for 1997 to 2007 (Herman *et al.*, 2008), where there are well-calibrated satellite reflectivity data records. Because of this, the change in UV irradiance over the United States can be estimated from just the change in satellite measured ozone amounts as shown in Figure 3.20. Fioletov *et al.* (2001) has made ground-based estimates of erythemal irradiance changes from two Brewer spectrometer stations (Montreal and Edmonton), and found that the UVB trends were similar to those expected from just changes in ozone, but with much larger uncertainty because of clouds and aerosols.

Satellite-observed long-term changes in ozone amount averaged over the United States suggest that there were significant UV changes for both erythemal irradiance and for UVB. Compared to the annual mean levels in 1980, the change in UV averaged over the United States was approximately 20% (erythemal irradiance) and approximately 40% (305 nm irradiance) early in 1993. Fortunately, these large percent changes were during the winter months when the solar zenith angles are large, so that the absolute irradiances are comparatively small. The calculated annual average irradiance increase during 1993 was about 7% and about 14% for erythemal and 305 nm irradiances, respectively. By 2007, the irradiance increase moderated to 4% and 8%, respectively, in response to a partial recovery of stratospheric ozone, which model calculations show is a direct consequence of the

Barrow, Alaska, has experienced UVB increases related to springtime ozone depletion, but these increases are a factor of ten smaller than those observed at the southern high latitudes.

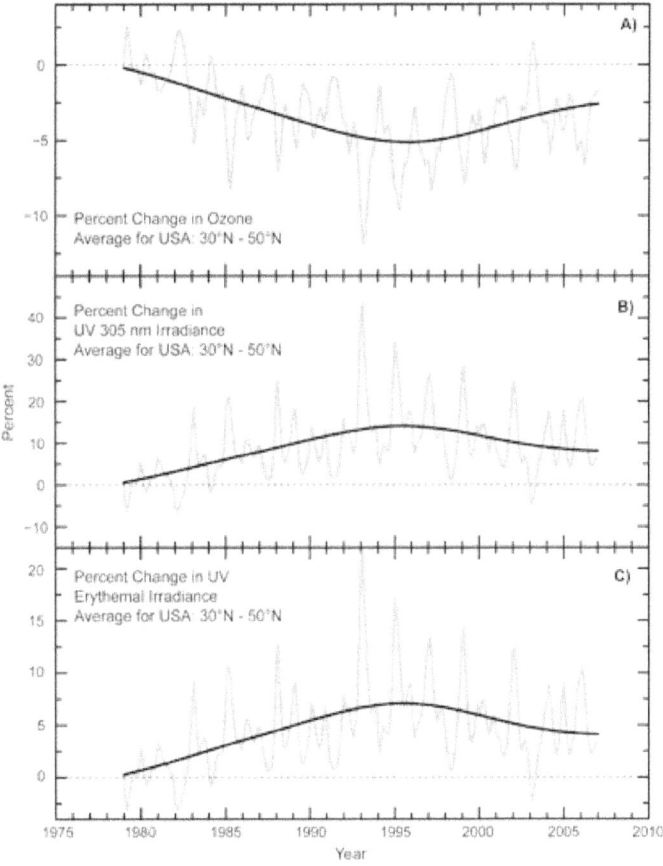

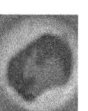

Figure 3.20 The calculated percent change in UV irradiance caused by percent changes in ozone over the continental United States. The ozone change is estimated from satellite measurements over the United States.

implementation of the Montreal Protocol and its subsequent Amendments.

3.3.2 UV in the Polar Regions

The expansion of the Antarctic polar vortex during the 1990s, both in spatial extent and temporal into early summer, has increased the frequency of elevated UVB episodes over sub-Antarctic populated areas. These episodes are no longer just small pockets of ozone-depleted stratospheric air coming from the breakup of the polar vortex, but include occasional excursions of the polar vortex edge over Ushuaia, Argentina and Punta Arenas, Chile. This occurred 44 times in the years 1997, 1998, and 2000 combined, with some episodes lasting three to four days. Surface measurements show average erythemal UV increases of about 70% over Ushuaia since 1997, and episodic total UVB increases of up to 80% over Punta Arenas (WMO, 2007 and references therein).

Diaz *et al.* (2003) show that Barrow, Alaska, has experienced UVB increases related to springtime ozone depletion in March and April, but these increases are a factor of ten smaller than those observed at the southern high latitudes. Summertime low-ozone episodes in the Arctic also affect surface UVB irradiances. These summertime events result from gas-phase chemistry involving nitrogen and hydrogen

cycles, which become very efficient during the 24-hour insolation that occurs in the Arctic summer. During summer 2000, two low-ozone episodes brought about erythemal UV increases on the order of 10 to 15%, each lasting more than five days (WMO, 2007 and references therein).

Because of the extreme Antarctic springtime ozone depletion (ozone hole) compared to all other regions, it is useful to compare (Figure 3.21) the measured amounts of UV irradiance at Palmer Station, Antarctica (64°S) with San Diego, California (32°N) and Barrow, Alaska (71°N). For seasons other than spring in Antarctica, there is a decrease in UVB irradiance caused by the increased path through the atmosphere resulting in less UVB than in San Diego. The Antarctic ozone depletion that occurs each spring causes the UVB portion of the erythemally weighted irradiance to increase dramatically to where it exceeds even the summertime values observed in San Diego at 32°N. Similar wide-area springtime low ozone amounts do not occur in the Arctic region because of the degree of meteorological wave activity in the north that leads to a weaker polar vortex and higher ozone amounts.

3.3.3 Human Exposure to UV

From the viewpoint of human exposure to UV, the maximum clear-sky UV irradiance and exposure occurs in the equatorial zone, 23.3°S to 23.3°N, following the seasonal sub-solar point, and at high mountain altitudes. In general, UV erythemal, UVA, and UVB irradiance decreases with increasing latitude outside of the equatorial zone, since the maximum daily noon solar elevation angle decreases. An exception occurs for UVB wavelengths at southern mid- to high latitudes when reduced ozone amounts from the Antarctic ozone hole remain late into the spring and are pushed away from Antarctica towards lower latitudes, which includes some populated areas. For example, UV measurements indicate equatorial irradiance levels can occur in the southern part of South America for several days.

Global images of daily-integrated UV erythemal exposure (kilojoules, kJ per m²) averaged during the months of January (Southern Hemisphere summer) and July (Northern

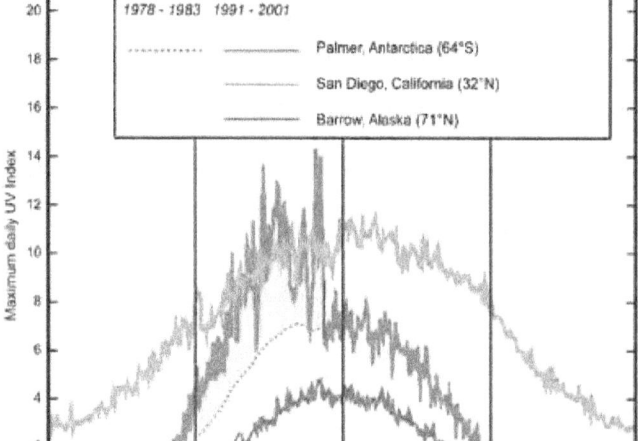

Figure 3.21 A comparison of measured erythemally weighted UV irradiance in Antarctica, the Arctic, and a midlatitude site in relative units. (Fahey, 2007)

Global Erythemal Radiation

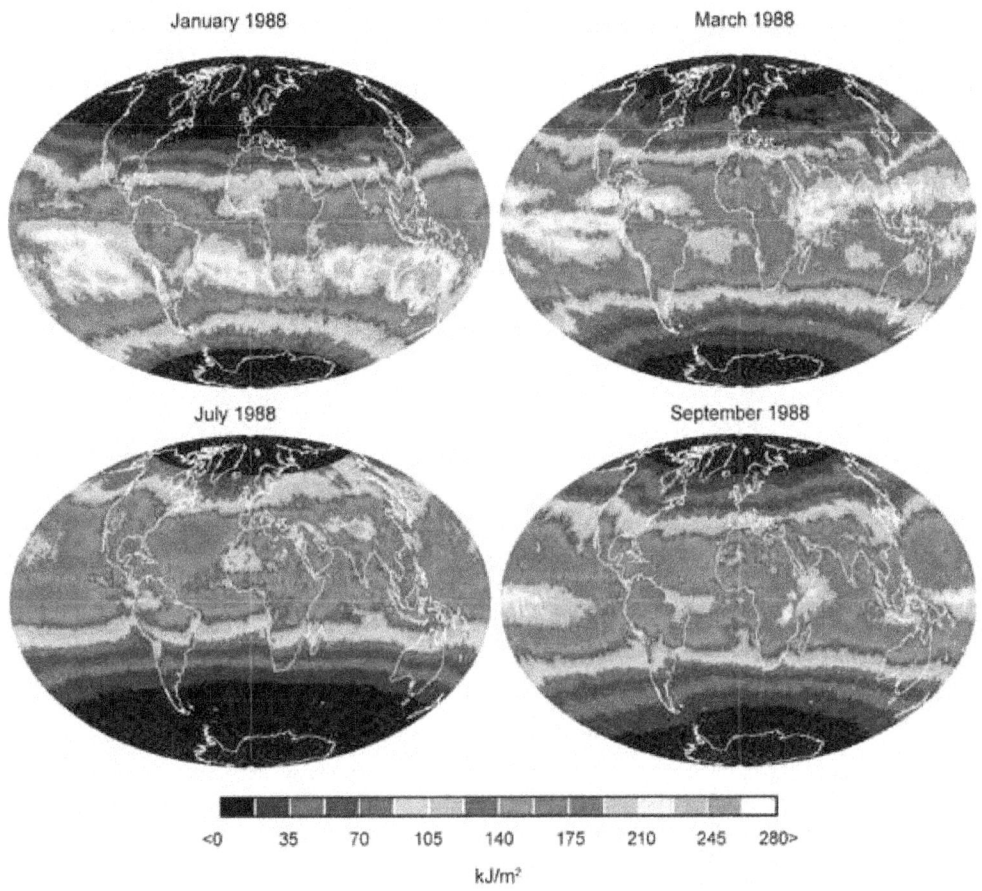

January 1988

March 1988

July 1988

September 1988

<0 35 70 105 140 175 210 245 280>

kJ/m²

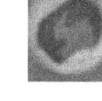

Figure 3.22 Erythemal exposure kJ per m² for the months of January, March, July, and September 1988 (from WMO, 1999) based on Nimbus-7/TOMS ozone and reflectivity data. In terms of the UV index, the numbers would be divided by 25. High UV levels are observed over Antarctica in the Southern Hemisphere late Spring and Summer (Figure 3.21). These extreme levels are not seen in the September 1988 panel because the sun is just beginning to rise over Antarctica and the 1988 ozone depletion was not extreme (Figures 3.5 and 3.7).

Hemisphere summer), and the two equinox months September and March, are shown in Figure 3.22 (based on WMO, 1999). Because of cloud cover, the high equatorial clear-sky irradiances do not translate into the highest monthly cumulative exposures. The maximum erythemal doses at the equator occur when the sun is directly overhead during March, which has lower cloud cover than during September. The difference is related to the annual cycle of the cloud cover associated with the Intertropical Convergence Zone (ITCZ), which is usually over the equator in September, but is south of the equator in March. Two extreme examples of very high UV exposures occur in the South American Andes (*e.g.*, the sparsely populated Atacama Desert in Chile at 4400 to 5600 meters altitude) during January and in the Himalayan Mountains (over 100 peaks exceeding 7000 meters in height) during July as shown in Figure 3.22. Excluding high altitude locations, the largest monthly UV exposures occur in Australia and South Africa during summer (January) because of their very low amount of day-to-day cloud cover from late spring to early autumn. Other midlatitude low altitude areas also receive high doses, *e.g.*, summertime (July) in the southwest United States and the Mediterranean countries.

Other factors contribute to the high Southern Hemisphere UV doses. There is a five million km decrease in Earth-Sun distance for the Southern Hemisphere summer solstices, as compared to the Northern Hemisphere, causing a 6.5% increase in summer solstice irradiance

The southwest United States receives high UV exposures during summertime.

107

in the Southern Hemisphere. Average summer ozone in the Southern Hemisphere (270 DU) is lower than the Northern Hemisphere (320 DU) by about 13%, which would lead to a 13% increase in 310 nm and a 26% increase in 305 nm irradiance. The exact percent increase is a function of latitude. In general, the Southern Hemisphere has less pollution aerosols, which can cause another few percent increase in UV irradiance relative to Northern Hemisphere.

In Australia and South Africa, the combination of high UV exposure and residents of European descent have lead to a major skin cancer health problem. Based on National Institutes of Health data, the same problems are present in the United States, with more skin cancer occurring at lower latitudes where the UV exposure is higher. The seriousness of the very high UV exposure problem is observed in Australia, where skin cancer rates have increased dramatically (20% for basal cell, to 788 per 100,000; over 90% for squamous cell, to 321 per 100,000 carcinomas), based on household surveys conducted in 1985, 1990, and 1995 (Staples *et al.*, 1998). This compares to the U.S. National Cancer Institute estimate of 14.5 per 100,000 for the United States. Lucas *et al.* (2006) give a comprehensive review of health problems and benefits related to UV exposure.

3.3.4 UV Summary

Measurements from ground-based instruments at different midlatitude sites around the globe show a mixture of UVB increases and decreases that depend on changes in local cloud cover, ozone, and aerosol amounts. Trends in UV in the polar regions, especially Antarctica, are dominated by changes in springtime stratospheric ozone. In the latitude range 60°S to 60°N, all three main factors governing UVB must be taken into account (for UVA, clouds and aerosols are the dominant factors). Ground-based stations located in or near urban sites have observed increases in cloud-free sky UV radiation from pollution abatement comparable to those from observed total column ozone changes.

Measurements of ozone and cloud plus aerosol reflectivity from satellites have been used to estimate the changes in UVB over the last 28 years. Based on the satellite ozone record,

the annual average clear-sky UV erythemal irradiance averaged over the continental United States increased from 1979 to the mid-1990s by about 7%. Since the mid-1990s the erythemal irradiance has decreased, so that the current level is about 4% higher than it was at the start of the record in 1979. Year-to-year and seasonal variations ranged from only a few percent to about 20% with the largest changes occurring during the winter months when UV irradiance is at an annual minimum. In the absence of the Montreal Protocol, summer maximum and annually integrated UVB doses over the United States would have been much larger with adverse consequences for public health and ecosystems (U.S. EPA, 1999).

Ground-based measurements of surface UV trends present a challenge that can be overcome with proper analysis of the data for cloud-free conditions along with simultaneous aerosol measurements. UV estimates from satellite measurements of ozone, aerosols, and cloud reflectivity are averages over large areas on the order of 25 km to 100 km, which minimizes many problems with local variability of cloud and aerosol amounts. Both ground and satellite UV estimates are critically dependent on establishing and maintaining an accurate calibration over the lifetime of an instrument and between successive instruments. Ground-based measurements are essential to provide validation of satellite calibration and as a bridge between successive satellite instruments.

While the UV irradiance maximum in 1993 was associated with the massive equatorial Mt. Pinatubo eruption in 1991, a portion of the total increase occurred before 1991 and was associated with ozone destruction from chlorine loading in the atmosphere before being limited by the Montreal Protocol. Major chlorine-driven ozone decreases and UVB increases were prevented by this and subsequent agreements that were effective for limiting releases of chloroflourocarbons (CFCs) and other chlorine-bearing compounds, with CFCs being almost completely phased out by 1995.

Through the Montreal Protocol and subsequent agreements, CFCs were almost completely phased out by 1995, preventing major chlorine-driven ozone decreases and UVB increases.

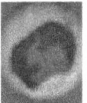

APPENDIX 3A: LAMBERT EQUIVALENT REFLECTIVITY (LER)

The Lambert Equivalent Reflectivity, R, is calculated by requiring that the measured radiance, I_{SM}, match the calculated radiance, I_S, at the observing position of the satellite (Equation A1) by solving for the single free parameter, R, in the formal solution of the radiative transfer equation

$$I_S(\Omega,\Theta,R,P_O) = \frac{RI_d(\Omega,\Theta,P_O)f(\Omega,\Theta,P_O)}{1 - RS_b(\Omega,P_O)} + I_{dO}(\Omega,\Theta,P_O) = I_{SM} \tag{A1}$$

where Ω = ozone amount from shorter wavelengths (*e.g.*, 317 nm)

       Θ = viewing geometry (solar zenith angle, satellite look angle, azimuth angle)

       R = LER at P_O $0 < R < 1$

       P_O = pressure of the reflecting surface (*e.g.*, ground or cloud)

       S_b = fraction scattered back to P_O from the atmosphere

       I_d = sum of direct and diffuse irradiance reaching P_O

       f = fraction of radiation reflected from P_O reaching the satellite

       I_{dO} = radiance scattered back from the atmosphere for R=0 and P=P_O

The quantities S_b, I_d, f, and I_{dO} are calculated from a radiative transfer solution and stored in tables. From Equation A1,

$$R = \frac{I_{SM} - I_{dO}}{I_d f + (I_{SM} - I_{dO})S_b} \tag{A2}$$

APPENDIX 3B: UV INDEX AND UNITS

Erythemal irradiance is frequently expressed in terms of the UV index = 25 milliWatts (mW) per m^2 = 2.5 microWatts (μW) per cm^2 (the units of Figure 3.14). The index is an arbitrary unit such that very high values reported by weather services have a UV index of 10. In Figure 3.14, the highest value is about 22 μW per cm^2, which is a UV index of 8.8. High altitude locations with extreme UV amounts can exceed 10 on clear days. Erythemal exposure or dose is a time-integrated quantity normally expressed in kJ per m^2.

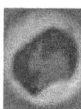

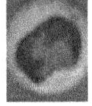

CHAPTER 4

How Do Climate Change and Stratospheric Ozone Loss Interact?

Convening Lead Author: David W. Fahey, NOAA

Lead Authors: Anne R. Douglass, NASA; V. Ramaswamy, NOAA; Anne-Marie Schmoltner, NSF

KEY ISSUES

Stratospheric ozone abundances are dependent on a balance of chemical processes that both produce and destroy ozone and dynamical processes that transport ozone throughout the stratosphere. The chemical processes depend on atmospheric temperatures and the abundances of ozone-depleting substances and other trace gases, such as water vapor and nitrogen oxides. Transport depends on heating in the atmosphere, which also depends on the distribution and abundance of trace gases, such as carbon dioxide, ozone-depleting substances, and ozone. Atmospheric temperature, transport, and trace gas amounts, for example, are all aspects of Earth's climate. As these and other climate parameters change as a result of human activities and natural variability, ozone abundances will decrease or increase in a manner that depends on a variety of factors.

This complex coupling of ozone and climate parameters is not fully defined at present and has significant uncertainties associated with known key aspects. Chemistry climate models (CCMs) of the atmosphere are in development and use by researchers aiming to reduce the uncertainty in the ozone-climate interaction and to explore other aspects of the interrelationship. Key questions related to the coupling of ozone and climate are:

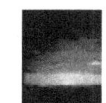

- How do ozone-depleting substances and ozone depletion contribute to the radiative forcing (RF) of climate?
- How do long-term changes in greenhouse gases affect stratospheric ozone?
- How have stratospheric temperatures changed in recent decades and what are the causes of these changes?
- Is stratospheric water vapor changing in a way that influences ozone abundances?
- How do ozone changes influence the climate of the stratosphere and troposphere?
- Will volcanic eruptions influence future ozone amounts?

KEY FINDINGS

Linking of ozone and climate change

- Ozone and climate change are linked because both ozone and ozone-depleting substances (ODSs) are greenhouse gases, which contribute to radiative forcing. The ODS contribution to global radiative forcing between 1750 and 2005 is approximately 20 percent (+0.34 W per m^2) of that from carbon dioxide, the largest human contribution. The ODS contribution is expected to decline in coming decades as ODS emissions and their atmospheric abundances continue to decline in the atmosphere.

- Each ODS contributes to ozone depletion and climate warming with a different level of effectiveness as represented, for example, by the Ozone Depletion Potentials (ODPs) and Global Warming Potentials (GWPs), respectively. For the principal ODSs, these values vary over orders of magnitude for equal mass of emissions.

- The abundance of stratospheric ozone is dependent on a balance of production and loss processes. These processes are dependent on several features of the atmosphere, namely, its chemical composition, air motions, radiation, and temperatures. Climate change will lead to changes in these features, which in turn will affect ozone abundances. Climate change has the potential to increase or decrease ozone abundances depending on the region and extent of climate change.

- Chemical ozone depletion also contributes to climate change by modifying atmospheric radiative properties. Ozone losses can also alter atmospheric temperatures and atmospheric transport. Ozone depletion can affect the climate of both the troposphere and stratosphere. The ozone depletion due to ODSs is expected to offset the direct climate forcing by ODSs. The extent of the offset is uncertain, with an estimated value of −0.05 W per m^2 with an uncertainty range of −0.15 to + 0.05 W per m^2.

Impact of climate change on ozone

- The complexity of the interactions of ozone changes with climate parameters requires development and evaluation of coupled models of Earth's atmospheric chemistry and climate processes (called chemistry climate models, CCMs). These CCMs are used to predict future ozone amounts. In addition, CCMs are needed to evaluate the sensitivity of ozone to climate parameters and the response of climate to ozone changes.

- Global average stratospheric temperatures have decreased in the observational records that begin in the 1960s. The decrease is attributed mainly to ozone depletion, increased carbon dioxide (CO_2), and changes in water vapor.

- Stratospheric temperatures influence ozone amounts through chemical and transport processes. Future increases in CO_2 will continue to contribute to global stratospheric cooling. The photochemical loss of ozone is slowed in some regions when temperatures decrease with the result that ozone recovery may be accelerated.

- Human activities are expected to increase the future abundances of greenhouse gases that influence stratospheric ozone amounts, principally, CO_2, methane (CH_4) and nitrous oxide (N_2O).

- Stratospheric water vapor has increased in recent decades but since 2001 has been decreasing in the lower stratosphere. The oxidation of methane emissions is an important contributor to increasing water vapor trends. Tropical tropopause temperatures modulate dehydration of air entering the stratosphere, and recent decreases in water vapor are well correlated with negative tropical tropopause anomalies. Future water vapor trends are uncertain because of uncertainties in projecting methane emissions and the temperatures of the tropical tropopause. Stratospheric water vapor influences stratospheric ozone through reactive hydrogen chemistry and polar stratospheric cloud formation.

- Chemistry climate model simulations predict that the atmospheric circulation between the stratosphere and troposphere will increase in a changing climate in the coming decades. Increased circulation will change stratospheric ozone amounts and increase the stratospheric flux of ozone to the troposphere.

Impact of changes in stratospheric ozone on climate change

- Depletion of stratospheric ozone since about 1980 has caused a negative radiative forcing of climate change (approximately -0.05 W per m^2, with an uncertainty that encompasses a range from -0.15 to $+0.05$ W per m^2). The forcing is a balance between a shortwave cooling of the lower stratosphere and a longwave cooling below the region of ozone depletion. In comparison, increases in ozone from pollution chemistry in the troposphere have caused a positive radiative forcing (approximately $+0.35$ W per m^2).
- Ozone depletion causes changes to the temperature and circulation of the stratosphere and troposphere. Observational analyses indicate that stratospheric ozone depletion over Antarctica has strengthened circumpolar flow throughout the troposphere over Antarctica and caused surface temperature changes. This effect has been well simulated using many different general circulation models (GCMs).

Importance of volcanic eruptions

- If volcanic eruptions that inject material into the stratosphere, referred to here as explosive eruptions, occur again in the coming decades, they will decrease stratospheric ozone levels for several years as a result of the heterogeneous reactions occurring on volcanic sulfate aerosols. These reactions increase halogen loss processes by reducing the abundance of key reactive nitrogen compounds. For a given eruption size, the resulting enhanced ozone destruction from halogens will decrease as halogen amounts in the atmosphere decrease in the coming decades.
- Explosive volcanic eruptions are expected to cause major temperature and circulation changes in the stratosphere as have occurred after past eruptions. These changes are in response to the large increases in sulfate aerosol amounts in the stratosphere following such eruptions. The increases result in only a short-term shift in stratospheric climate because natural processes remove most of the additional sulfate aerosols within two to three years after the eruption.

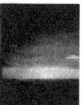

4.1 INTRODUCTION

Ozone occurs naturally in the atmosphere as a result of photochemical processes. In the stratosphere, ozone is beneficial to life on Earth because it absorbs ultraviolet radiation (UV) from the sun. Ultraviolet light absorption heats the stratosphere. Ozone is also a greenhouse gas that helps trap terrestrial infrared radiation, which leads to heating of the troposphere and stratosphere. In the natural atmosphere, ozone's warming of the planet makes it the third most important longwave greenhouse gas after water vapor and carbon dioxide (Kiehl and Trenberth, 1997). As a consequence, changes in ozone amounts have the potential to change climate parameters in the stratosphere and troposphere. Anthropogenic pollution has led to increased ozone production and abundances in the troposphere, particularly near Earth's surface. In contrast, emissions of ozone-depleting substances (ODSs) in recent decades have led to significant depletion of global stratospheric ozone, with particularly high losses in polar regions. The Montreal Protocol has been established to protect the ozone layer by reducing the global production and consumption of ODSs.

The complex interrelationship between ozone and climate change is illustrated in Figure 4.1. Multiple radiative, chemical, and dynamical processes control ozone amounts and their distribution in the troposphere and stratosphere. Production and loss cycles of ozone involve many chemical species, as well as aerosols, and are influenced by atmospheric parameters such as solar insolation and temperature. Natural processes and human activities influence ozone through changes in atmospheric composition and climate parameters. The chemical loss rate of ozone leads to an atmospheric lifetime that is relatively short compared to carbon dioxide, for example. As a result, dynamical processes such as planetary waves and the Brewer-Dobson circulation have important roles in establishing the observed non-uniform distribution of ozone in the atmosphere.

> In the natural atmosphere, ozone's warming of the planet makes it the third most important longwave greenhouse gas after water vapor and carbon dioxide.

Connections between Climate and Ozone

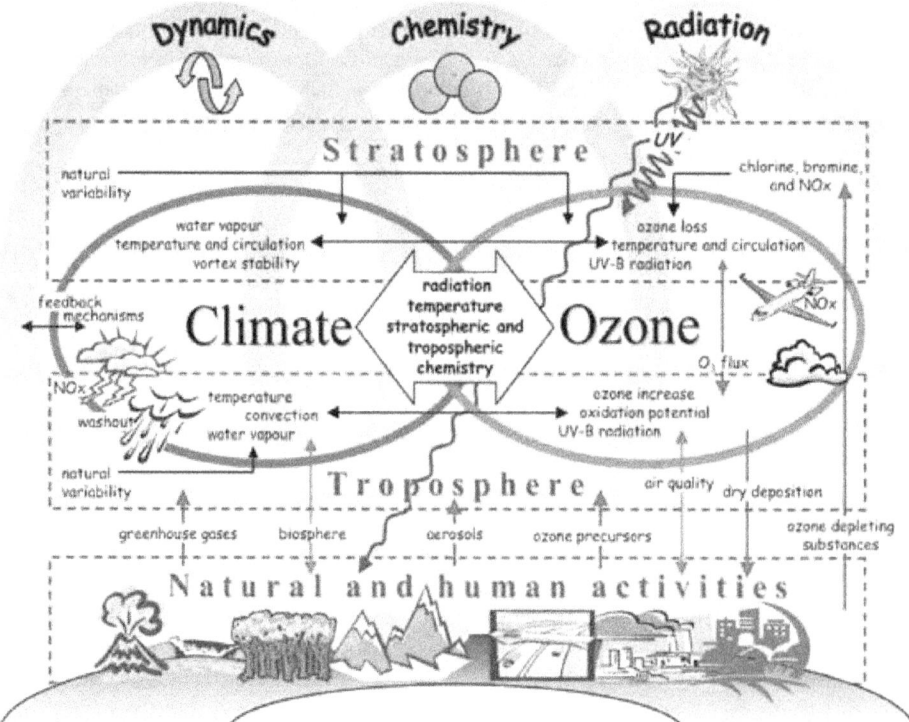

Figure 4.1 Schematic depiction of how climate change and ozone abundances are linked to each other in the stratosphere and troposphere and to natural and human activities at Earth's surface (Isaksen, 2003.)

Connections between Climate and Ozone

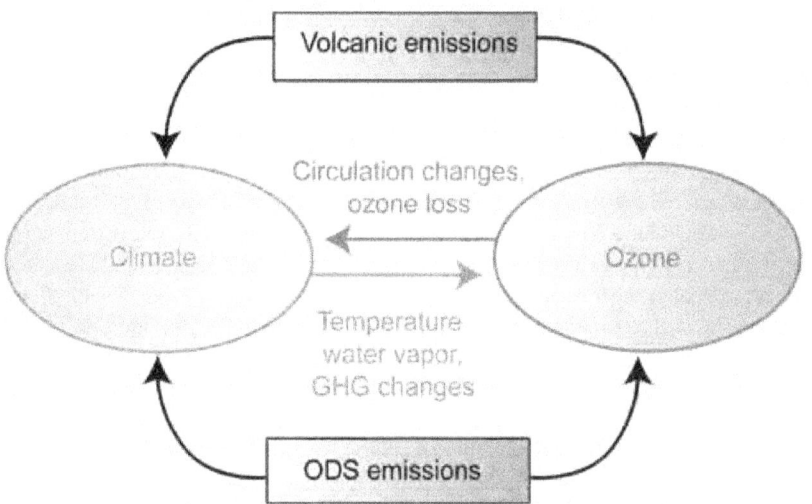

Figure 4.2 Schematic of specific processes addressed in this assessment that interconnect and influence atmospheric ozone amounts and climate parameters.

The accumulation of ozone-depleting substances in the atmosphere causes a direct radiative effect, similar to that from carbon dioxide.

The ODSs are also greenhouse gases. The radiative effect of accumulated ODS emissions is partially offset by the reduction in global ozone amounts caused by these emissions (see Section 4.2). Systematic and long-term ozone depletion can change atmospheric circulation patterns and contribute to climate change. Furthermore, changes in climate can alter ozone amounts. Changes in temperature, amounts of trace gases such as methane, nitrous oxide, and water vapor, and atmospheric circulation can all potentially lead to ozone changes in the stratosphere and troposphere. Finally, large volcanic emissions can alter both ozone and climate for temporary periods of several years.

This chapter assesses these interconnections, schematically shown in Figure 4.2, by outlining what is known about ODSs and volcanic emissions and the processes through which ozone influences climate and through which climate change will influence ozone amounts. Further detail on the coupling of climate and ozone changes can be found in recent scientific assessments (IPCC/TEAP, 2005; WMO, 2007).

4.2 RADIATIVE FORCING OF CLIMATE BY OZONE-DEPLETING SUBSTANCES AND OZONE CHANGES

Ozone and ODSs are greenhouse gases. The accumulation of ODS emissions in the atmosphere causes a direct radiative effect in a manner similar to that from emissions of CO_2, the leading anthropogenic greenhouse gas. ODS emissions also cause an indirect radiative effect, because ODSs cause the destruction of stratospheric ozone. These radiative effects are discussed in the following sections.

4.2.1 Direct Radiative Forcing by Ozone-Depleting Substances

The accumulation of ozone-depleting substances in the atmosphere contributes to the direct radiative forcing of climate. The ODS efficiencies as ozone-depleting substances (*i.e.*, ODPs) and as greenhouse gases (*i.e.*, GWPs) are contrasted in Figure 4.3 (also see Box 4.1). A large range is found in both metrics for the principal gases. The continuous measurements of ODS abundances in the atmosphere over the last two to three decades allow an accurate evaluation of their contributions to ozone depletion and climate change. Projections of emissions allow the future contribution of ODSs to ozone depletion and climate change to be estimated. ODP-weighted and GWP-weighted

ODS emissions grew substantially in recent decades but peaked in the late 1990s, as shown in Figure 4.4 (IPCC/TEAP, 2005; Velders *et al.*, 2007). The decline is in response to the provisions of the Montreal Protocol, which requires a staged phase-out of all principal ODS use in developed and developing nations. The direct radiative forcing (RF) contribution from ODSs likely would have been approximately twice as large in 2010 in the absence of the Montreal Protocol or other regulation (Figure 4.4, right panel) (Velders *et al.*, 2007).

The radiative forcing of individual ODSs varies because of differences in emissions, lifetimes, and radiative efficiencies. The RF values attributable to individual ODSs for the period 1970 to 2000 are shown in Table 4.1 along with values for CO_2 and CH_4. CFCs as a group form the largest contribution to RF amongst all ODSs. A comparison of the RF from halocarbon gases as a group with values associated with other aspects of natural and anthropogenic climate forcing is shown in Figure 4.5. The RF from halocarbon gases in 2005 is +0.34 [±0.03] W per m² (warming), which represents 13 percent of the RF from all

long-lived greenhouse gases and 21 percent of the total anthropogenic RF in 2005.

ODSs account for 94 percent (+0.32 W per m²) of the halocarbon term in Figure 4.5. The balance (6 percent) is due to the accumulation of hydrofluorocarbon (HFC) emissions, which are included in the Kyoto Protocol (UNFCCC, 1997). Emissions of HFCs, which are substitutes for ODSs in many applications, are increasing. HFCs do not deplete ozone (ODP = 0) but can have substantial GWPs (Figure 4.3).

A comprehensive evaluation of the protection of climate afforded by increases or reductions in ODS emissions must take into account two indirect effects or offsets. The case of reduced ODS emissions from actions under the Montreal Protocol were analyzed recently by Velders *et al.* (2007). As ODS emissions are reduced, global stratospheric ozone levels are restored from their depleted state. Since ozone is a greenhouse gas, ozone RF increases as ozone levels are restored, thereby offsetting the reductions in RF from ODS reductions. The second offset is the increase in emissions of HFCs, all potent greenhouse gases, which

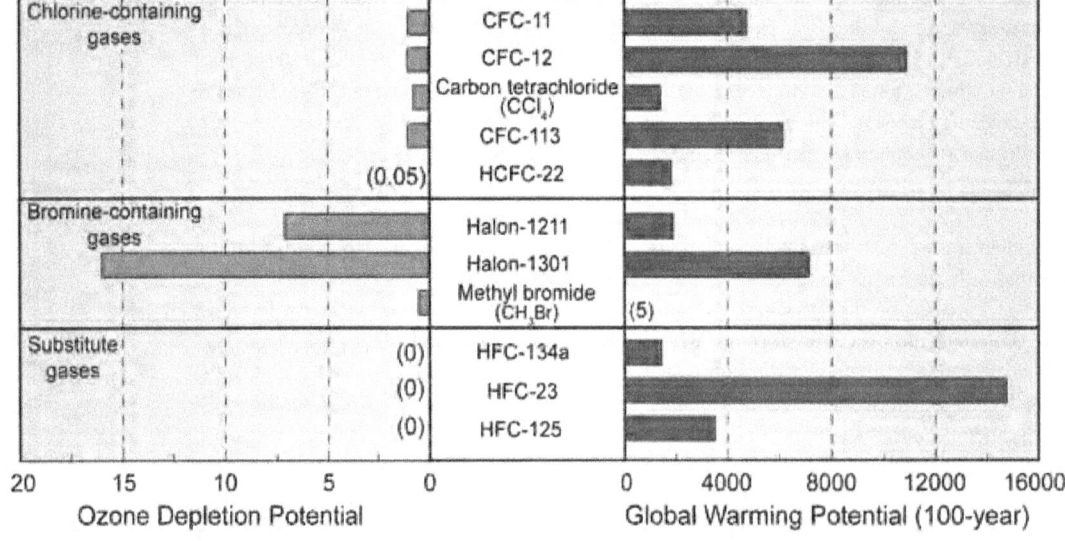

Figure 4.3 Comparison of the Ozone Depletion Potentials (ODPs) and Global Warming Potentials (GWPs) for principal ozone-depleting substances (ODSs) and hydrofluorocarbons (HFCs) (see Box 4.1). The contributions of emissions to ozone depletion and climate change increase with the ODP and GWP values, respectively. HFCs are ODS substitute gases that do not destroy ozone (*i.e.*, ODP = 0). The comparison is for emissions of equal mass. The GWPs are evaluated for a 100-year period after emission. The ODPs of CFC-11 and CFC-12, and the GWP of CO_2 are defined to have values of 1.0 (Daniel *et al.*, 1995; IPCC/TEAP, 2005; WMO, 2007).

BOX 4.1: Ozone Depletion Potential and Global Warming Potential[1]

Ozone Depletion Potential: Ozone-depleting substances are compared in their effectiveness to destroy stratospheric ozone using the "Ozone Depletion Potential" (ODP), as shown in Figure 4.3. A gas with a larger ODP has a greater potential to destroy ozone over its lifetime in the atmosphere. The ODP is calculated on a "per mass" basis for each gas relative to CFC-11, which has an ODP defined to be 1. Halon-1211 and halon-1301 have ODPs significantly larger than CFC-11 and most other emitted gases, because bromine is about 60 times more effective overall on a per-atom basis than chlorine in chemical reactions that destroy ozone in the stratosphere. The gases with small ODP values generally have short atmospheric lifetimes or fewer chlorine and bromine atoms. The production and consumption of all principal ozone-depleting substances (ODSs) in human activities are regulated under the provisions of the Montreal Protocol.

Global Warming Potential: The climate impact of a given mass of a halocarbon emitted to the atmosphere depends on both its radiative properties and its atmospheric lifetime. The two can be combined to compute the Global Warming Potential (GWP), which is a proxy for the climate effect of a gas relative to the emission of a pulse of an equal mass of CO_2. Multiplying emissions of a gas by its GWP gives the emission of that gas over a given time horizon. A value of 100 years is often chosen as a reference time horizon for intercomparisons of GWPs.

GWPs are most useful as relative measures of the climate response due to direct radiative forcing of well-mixed greenhouse gases whose atmospheric lifetimes are controlled by similar processes, which includes most of the halocarbons.

[1]Adapted from WMO (2007) and IPCC/TEAP (2005).

Worlds Avoided with the Montreal Protocol

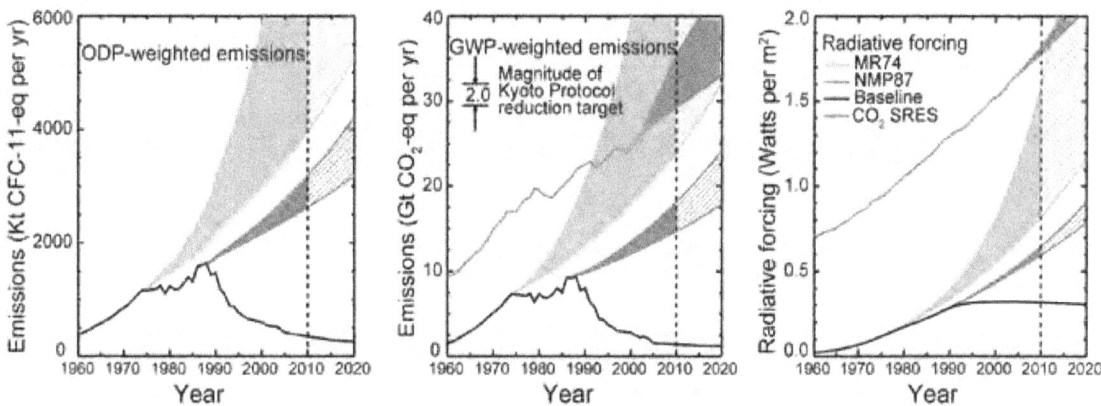

Figure 4.4 ODP-weighted emissions (left panel), GWP-weighted emissions (100-year) (middle panel), and radiative forcing (right panel) for ODS and CO_2 scenarios for 1960–2020. Four scenarios are used: (1) the baseline which represents ODS observations to date and projections for the future (black line); (2) the emissions that plausibly would have occurred in absence of the Molina and Roland warning that ODSs deplete ozone (MR74) (green shaded and striped region), (3) the emissions that plausibly would have occurred in absence of the implementation of the Montreal Protocol (NMP87) (blue shaded and striped region), and (4) the IPCC SRES scenario for CO_2 emissions beyond 2003 (red line). ODS emissions are normalized by their direct GWPs to form units of equivalent gigatons (Gt) CO_2 per year. The shaded regions reflect uncertainties in projecting ODS growth rates in the MR74 and NMP87 scenarios. The striped regions indicate larger uncertainties are associated with the scenarios past 2010. The CO_2 emissions for 1960–2003 are from global fossil fuel and cement production. All RF values represent net changes from the start of the industrial era (1750) to present (2005). The magnitude of the reduction target of the first commitment period of the Kyoto Protocol (2008 to 2012) is shown in the middle panel for reference. The target represents average emission reductions in six key greenhouse gases by Annex-1 countries (5.8 percent below 1990 values) plus the expected growth in emissions between 1990 and 2008 to 2012 (adapted from Velders *et al.*, 2007).

Table 4.1 Direct radiative forcing of CO₂, CH₄, and principal ODSs[1]

Gas	Radiative Forcing (W per m²)
CO₂	0.67
CH₄	0.13
N₂O	0.068
CFC-11	0.053
CFC-12	0.136
CFC-113	0.023
CFC-114	0.003
CFC-115	0.002
HCFC-22	0.0263
HCFC-141b	0.0018
HCFC-142b	0.0024
Halon-1211	0.0012
Halon-1301	0.0009
CCl₄	0.0029

[1] For accumulated emissions in the period 1970-2000. (Adapted from IPCC/TEAP, 2005, Table TS-1.)

is intrinsically tied to ODS reductions because HFCs are key substitute gases for ODSs. Thus, the net gain from reducing the RF contribution of ODSs must include negative offsets due to the reversal of some ozone depletion and increased abundances of other greenhouse gases. In 2010, these factors were estimated to offset about 25 percent of the RF decrease attributable to reductions of ODSs under the Montreal Protocol since 1990 (Velders *et al.*, 2007). There is a large uncertainty in this offset in radiative forcing (see below). As ODS abundances continue to decline in the atmosphere after 2010, the relative size of the ozone offset is likely to remain unchanged while the HFC offset might increase depending on growth in production and use of HFCs, which are not regulated by the Montreal Protocol.

Global Mean Radiative Forcings

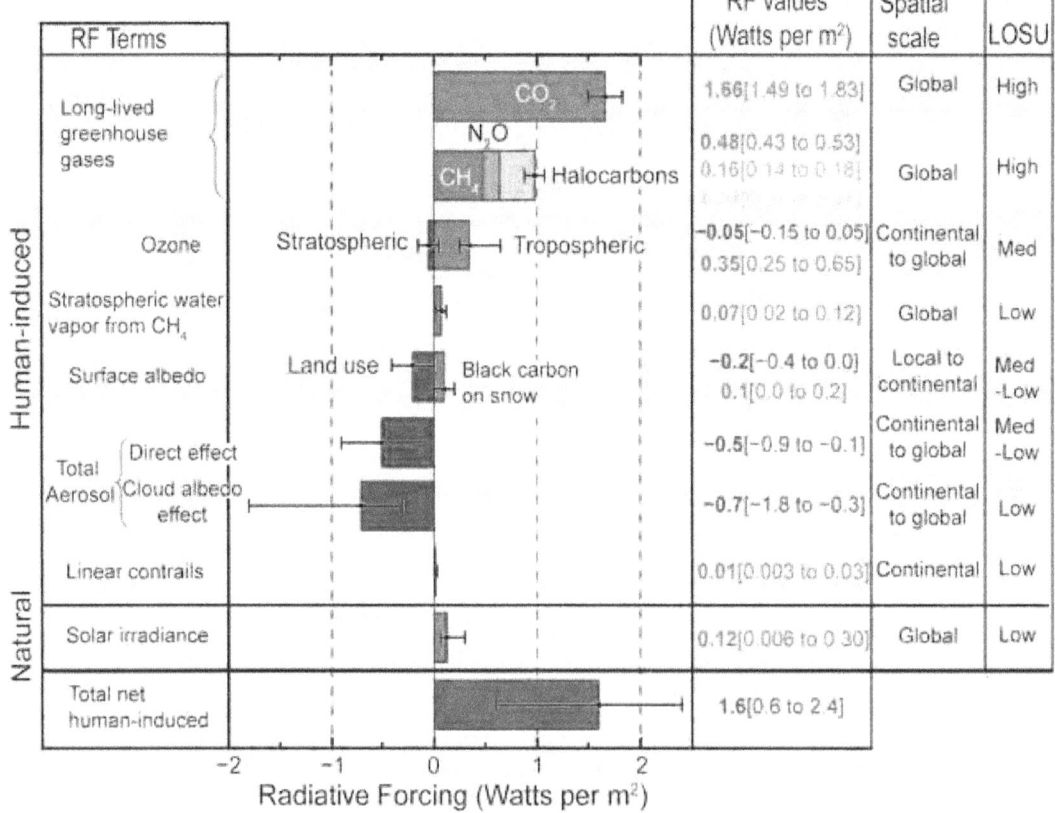

Figure 4.5 Radiative forcing values for the principal contributions to climate change from anthropogenic activities and natural processes. Each numerical value listed and indicated with a bar is a global mean value representing the best estimate of the change between preindustrial times (ca, 1750) and the present (2005). The error bars indicate the uncertainty ranges. The spatial scale and level of scientific understanding (LOSU) is also indicated for each value (adapted from IPCC, 2007, Figure SPM-2).

4.2.2 Radiative Forcing From Ozone Changes

Stratospheric ozone depletion and increases in tropospheric ozone both contribute to the radiative forcing of climate. Stratospheric ozone depletion is an indirect radiative effect of ODS emissions. The response of surface climate to ozone changes is complex, in general, because it depends on the balance between shortwave and longwave radiative effects. For example, when ozone is increased in the troposphere or lower stratosphere, surface temperatures tend to increase due to increased longwave forcing (Forster and Shine, 1997; IPCC/TEAP, 2005) (see Box 4.2). Overall, surface temperatures are most sensitive to changes in ozone concentrations near the tropopause.

Stratospheric ozone depletion has occurred primarily at extratropical latitudes with substantially larger changes in the Southern Hemisphere. Southern Hemisphere ozone values over the period from 2000 to 2003 are on average 6 percent below pre-1980 values, while Northern Hemisphere values are 3 percent lower. The net RF change from these

---

BOX 4.2: Why is the radiative forcing of stratospheric ozone negative?[1]

The radiative forcing is defined as the change in net radiative flux at the tropopause after allowing for stratospheric temperatures to readjust to radiative equilibrium (IPCC, 2001, Chapter 6). The details of this definition are crucial for stratospheric ozone, and are explained in the figure.

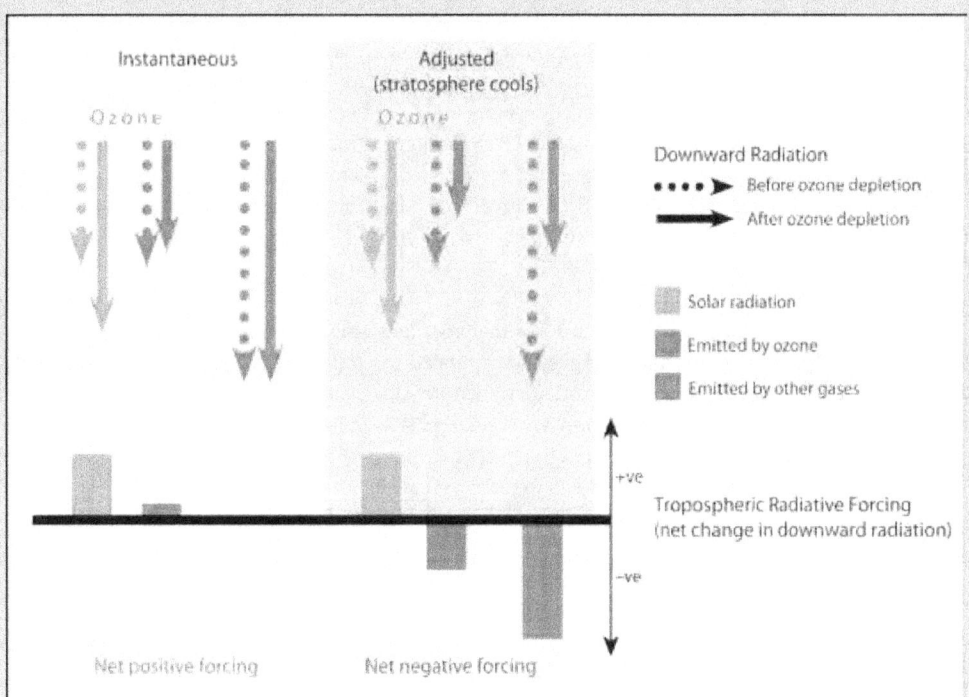

Box Figure 4.2 The instantaneous effect of stratospheric ozone depletion (left-hand side of schematic) is to increase the shortwave radiation from the Sun reaching the tropopause (because there is less ozone to absorb it), and to slightly reduce the downward longwave radiation from the stratosphere, as there is less ozone in the stratosphere to emit radiation. This gives an instantaneous net positive radiative forcing. However, in response to less absorption of both shortwave and longwave radiation in the stratosphere, the region cools, which leads to an overall reduction of thermal radiation emitted downward from the stratosphere (right-hand side of schematic). The size of this adjustment term depends on the vertical profile of ozone change and is largest for changes near the tropopause. For the observed stratospheric ozone changes the adjustment term is larger than the positive instantaneous term, thus the stratospheric ozone radiative forcing is negative.

[1]Excerpted from IPCC/TEAP (2005).

observed depletions has been assessed by the Intergovernmental Panel on Climate Change (IPCC) to be −0.05 [±0.1] W per m² (cooling) (Figure 4.5) (IPCC, 2007; Hansen *et al.*, 2005). The instantaneous response to ozone depletion in the lower stratosphere is a positive forcing because solar flux significantly increases below the tropopause and downwelling longwave radiation decreases slightly. In addition, less solar and longwave radiation is absorbed in the lower stratosphere when ozone amounts are reduced, thereby cooling the region and further reducing the downwelling longwave flux from ozone and other gases. The calculation of the net change in forcing is associated with large uncertainties because net cooling is sensitive to the location where ozone depletion occurs (see Box 4.2). When all effects and feedbacks are taken into account, stratospheric ozone depletion is estimated to cause a net reduction in RF at the tropopause and lead to a cooling effect on the atmosphere.

Stratospheric ozone depletion is included in the IPCC best estimate of the RF from global stratospheric ozone changes that have occurred between 1750 and 2005 (IPCC, 2007). The net global ozone RF estimates used in the IPCC include the possibility of a positive RF contribution due to a modeled increase in stratospheric ozone in some regions since preindustrial times, despite losses related to ODSs. The RF value is particularly sensitive to ozone changes in the tropical lower stratosphere, which are small compared to changes at high latitudes. Observational and modeling studies indicate how Northern Hemisphere ozone amounts are also influenced by changes in atmospheric dynamics, such as changes in tropopause heights (Pyle *et al.*, 2005), in addition to increased amounts of ODSs. The model results show large differences in how ozone column amounts have responded at midlatitudes to changes in ODSs and other parameters, such as circulation (Gauss *et al.*, 2006).

The total RF associated with ODS emissions is the sum of three effects: the direct warming summed over the accumulation of the contributing ODS gases (+0.34 [±0.03] W per m²); the direct warming from HFC emissions that occurred as a result of ODSs reductions;

and the indirect cooling associated with stratospheric ozone changes (−0.05 [±0.1] W per m²). The warming due to HFCs is currently small (approximately +0.02 W per m²) but is expected to grow in the future as ODS warming declines (Velders *et al.*, 2007). The uncertainty in the indirect effect is large compared to the direct effects. As previously discussed, in some cases the indirect effect as evaluated in atmospheric models includes effects not related to ODS emissions. In addition, the regional distribution of the radiative forcing from ODSs generally is different from that due to ozone changes. Thus, there are currently large uncertainties in the magnitude of global ozone cooling effects and the degree to which they offset the warming due to ODSs, as well as where and how these offsets would vary regionally.

Tropospheric ozone since preindustrial times has increased as a result of increased emissions of human pollutants, primarily nitrogen oxides, carbon monoxide, and organic compounds, including methane. Photochemical and radiative transfer models are used to calculate ozone changes and the associated RF, respectively. The changes include the net transport of ozone from the stratosphere to the troposphere, which can be altered by climate change and stratospheric ozone depletion. The tropospheric ozone RF (+0.35 W per m²) from human activities is larger than the stratospheric ozone term and associated with large uncertainties (Figure 4.5).

4.3 THE RESPONSE OF OZONE TO CLIMATE CHANGE PARAMETERS

Ozone responds to climate change parameters in a variety of ways because ozone is continually photochemically produced and destroyed in the atmosphere and thus dependent on the abundance of other gases emitted by natural processes and human activities. The complexity of the interaction of ozone with climate change parameters (Figure 4.1) requires the use of chemistry climate models (CCMs) (see Box 4.3) to diagnose the sensitivity of ozone to climate change parameters and to predict future ozone amounts in a changing climate.

> ## BOX 4.3: Models Used to Study Climate Processes[2]
>
> ***Atmospheric General Circulation Model (AGCM):*** A three-dimensional model of large-scale (spatial resolution of a few hundred km) physical, radiative, and dynamical processes in the atmosphere over years and decades. An AGCM is used to study changes in natural variability of the atmosphere and for investigations of climate effects of radiatively active trace gases (greenhouse gases) and aerosols (natural and human-made), along with their interactions and feedbacks. Usually, AGCM calculations employ prescribed concentrations of radiatively active gases, e.g., carbon dioxide (CO_2), methane (CH_4), nitrous oxide (N_2O), chlorofluorocarbons (CFCs), and ozone (O_3). Changes of water vapor (H_2O) concentrations due to the hydrological cycle are directly simulated by an AGCM. Sea surface temperatures (SSTs) are prescribed. An AGCM coupled to an ocean model, commonly referred to as an AOGCM or a climate model, is used for investigation of climate change. More recently, climate models may also include other feedback processes (e.g., carbon cycle, interaction with the biosphere).
>
> ***Chemistry Climate Model (CCM):*** An AGCM that is interactively coupled to a detailed chemistry module. In a CCM, the simulated concentrations of the radiatively active gases are used in the calculations of net heating rates. Changes in the abundance of these gases due to chemistry and advection influence heating rates and, consequently, variables describing atmospheric dynamics such as temperature and wind. This gives rise to a dynamical-chemical coupling in which the chemistry influences the dynamics (via radiative heating) and vice versa (via temperature and advection). Not all CCMs have full coupling for all chemical constituents; some radiatively active gases are specified in either the climate or chemistry modules. Ozone is always fully coupled, as it represents the overwhelmingly dominant radiative-chemical feedback in the stratosphere.
>
> [2]Excerpted from WMO (2007).

4.3.1 Calculating the Response of Ozone to Climate Change Parameters with CCMs

The approach to CCM use is schematically shown in Figure 4.6. Transport, radiation, dynamics, and chemistry and microphysics are the four principal aspects of a CCM. A CCM requires as input specific knowledge of natural process and their trends, such as in emissions, solar irradiance, and volcanic eruptions; and of human activities and their trends, primarily for emissions. These inputs define and constrain the current and future state of climate parameters. The CCM output includes a wide array of parameters and diagnostics in addition to ozone that can be compared to observations and other models.

Chemistry climate models are complex because simulating the atmosphere requires interdependence and interaction between the core aspects of the model. Important examples include the coupling between transport and radiation. Transport depends in part on atmospheric temperature gradients that are established by the distribution of radiative heating. Radiative heating is determined, in part, by long-lived greenhouse gases and ozone.

Photochemical reaction rates also depend on ambient temperatures. Thus, the photochemical balance controlling the abundances of ozone and other species depends substantially on the atmospheric abundances of greenhouse gases. Sea surface temperatures, land-sea temperature differences, and other factors influence wave propagation into the stratosphere, thereby affecting meridional transport rates. These couplings are discussed in more detail in WMO (2007) and Eyring *et al.* (2005).

The validation of CCM output for ozone and other parameters has become a focus topic because of the heightened need to project future ozone abundances with reliable uncertainty estimates (Eyring *et al.*, 2005; 2006). Reasonable agreement is found between many CCMs and global ozone trends, but for polar ozone trends the CCMs show a large spread in results. Uncertainties in CCM results reflect limitations in our understanding of how to represent atmospheric processes and their feedbacks in model simulations and, therefore, limit the precision and accuracy of our projections of future ozone amounts and the influence of climate change.

> *A chemistry climate model requires as input specific knowledge of natural processes and their trends, such as in emissions, solar irradiance, and volcanic eruptions. It also requires knowledge of human activities and their trends, primarily for emissions.*

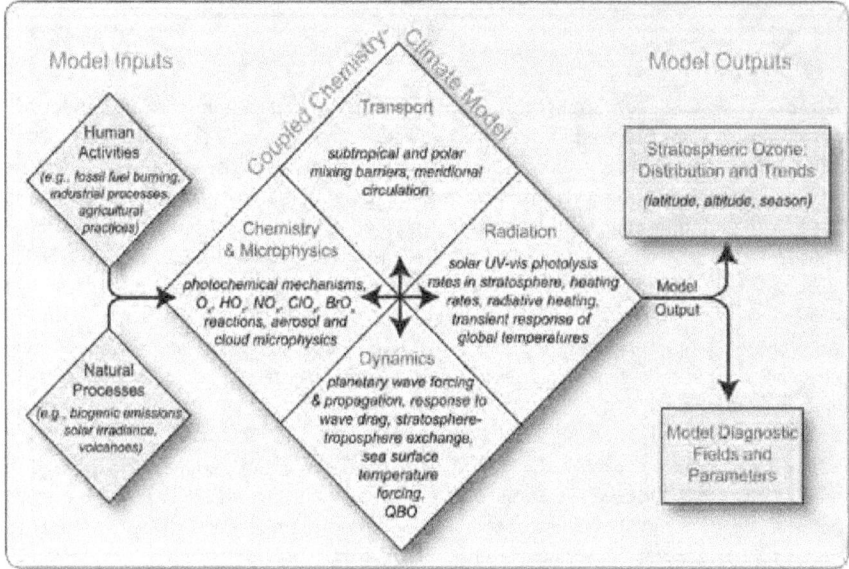

Stratospheric temperature has decreased by about 0.5 K per decade in the lower stratosphere and by about 1 to 2 K per decade in the middle stratosphere.

Figure 4.6 The processes in a Chemistry Climate Model (CCM) are represented by four basic groups: transport, dynamics, radiation, and stratospheric chemistry and microphysics. Significant interactions occur between aspects within each group. The CCM requires inputs describing human activities and natural processes. The CCM provides as output projections of future ozone abundances and their distribution, along with a large variety of other parameters and diagnostics. See Box 4.3 (Eyring *et al.*, 2005).

4.3.2 Stratospheric Temperature Changes

Stratospheric temperatures have decreased over the last three to four decades. Observations from satellites beginning in 1979 and radiosonde observations from about 1960 both reveal the cooling. The trend is about −0.5K per decade in the lower stratosphere and about −1 to −2K per decade in the middle stratosphere (about 25 to 30 km in altitude). The latitude dependence of the temperature trends is not fully consistent across the various datasets, especially in the tropics, and remains a topic of research (WMO, 2007). The time series of temperatures reveals a non-monotonic decrease in the lower stratospheric temperatures (Figure 4.7). Volcanic aerosols formed in the aftermath of a volcanic eruption that injects sulfur dioxide in to the stratosphere lead to a warming of the stratosphere for a few years following an eruption. Both the El Chichón (1982) and Mt. Pinatubo (1991) volcanic eruptions increased stratospheric aerosol amounts (Figure 4.7) (McCormick *et al.*, 1995; Pawson *et al.*, 1998). In the evolution of the global lower stratospheric temperature, a sharp increase, lasting for approximately two years, is found immediately following the El Chichón (1982) eruption and is followed by a period of quasi-steady temperatures that are

lower than the pre-eruption value. After the eruption of Mt. Pinatubo (1991), temperatures again increased sharply and were followed by a steady period in which the temperatures became lower than before this eruption. There is a slightly reduced cooling towards the end of 1990s and beginning of 2000s (Mears *et al.*, 2003).

Climate model simulations also show that the combined influences of the agents that are known to "drive" the climate system offer a reasonable quantitative explanation of the observed non-monotonic decrease of the temperatures in the global lower stratosphere (Seidel and Lanzante, 2004; Dameris *et al.*, 2005; Ramaswamy *et al.*, 2006). The global stratospheric temperature trends over the past two to three decades are attributed in modeling studies to a combination of increases in well-mixed greenhouse gases and water vapor, and decreases in ozone (Ramaswamy and Schwarzkopf, 2002; Schwarzkopf and Ramaswamy, 2002; Langematz *et al.*, 2003; Shine *et al.*, 2003; Santer *et al.*, 2006). The above studies indicate that attribution of the cooling trend is possible on the global-annual and zonal-annual scales, and for the springtime Antarctic, but smaller spatial scales and

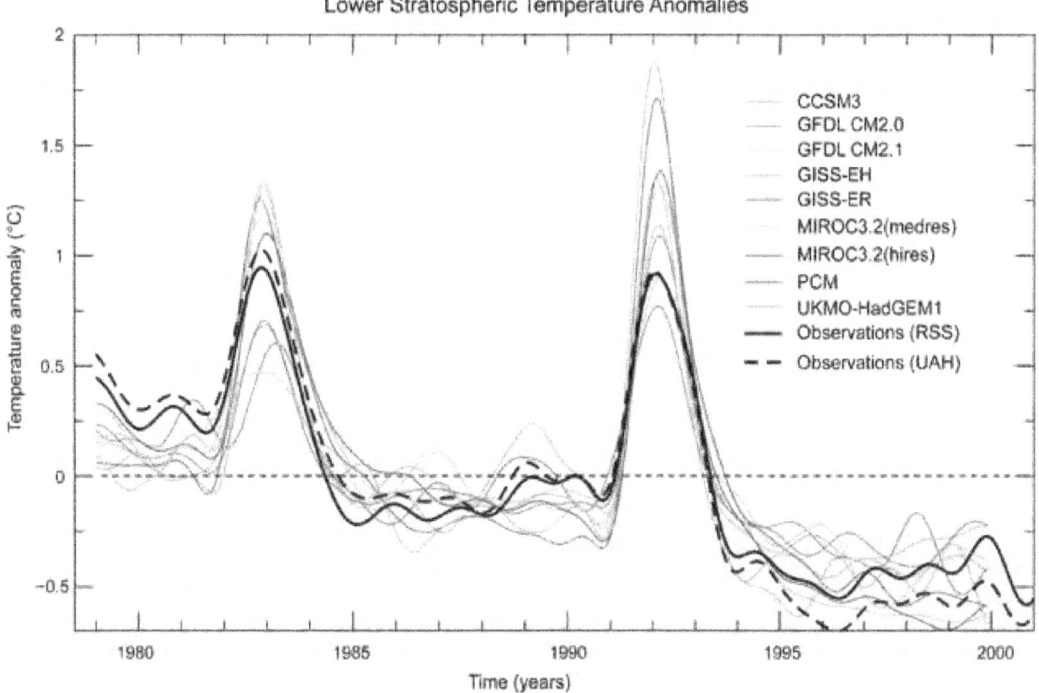

Figure 4.7 Temperature anomalies in the lower stratosphere as calculated by an ensemble of climate models. All anomalies are expressed as departures from a 1979 to 1999 reference period average. These models were chosen because they satisfy certain minimum requirements in terms of the forcings applied in the model run. All were driven by changes in well-mixed greenhouse gases (WMGHGs), sulfate aerosol direct effects, tropospheric and stratospheric ozone, volcanic aerosols, and solar irradiance. Model results are compared with observations derived from satellite datasets (RSS and UAH). Further details are discussed in Section 4.3.2 (From Santer *et al.*, 2006: Figure 5.2 and WMO, 2007: Figure 5-3).

seasonal behavior pose problems in attribution owing to the dynamical variability. The results from an ensemble of climate models are shown in Figure 4.7 for stratospheric temperature anomalies calculated as global and monthly means (Santer *et al.*, 2006). In addition to stratospheric ozone depletion, the models all include climate forcings from changes in well-mixed greenhouse gases (WMGHGs), sulfate aerosol, volcanic aerosol, and solar irradiance. The temperature anomalies are differences from a 1979 to 1999 reference period. The models, in general, account for the long-term decrease in stratospheric temperatures and the short-term increases caused by the two large volcanic eruptions.

Hansen *et al.* (2005) examined the effects on climate of a wide range of forcings, showing that different forcings produce different response patterns in the vertical temperature profile. Results from climate model simulations as outlined in CCSP (2006) show:

- Increases in greenhouse gases warm the troposphere and cool the stratosphere
- Volcanic aerosols warm the stratosphere and cool the troposphere
- Increase in solar forcing warms most of the atmosphere
- Increases in tropospheric ozone warm the troposphere
- Decreases in stratospheric ozone cool the stratosphere
- Sulfate aerosols cool the troposphere and slightly warm the stratosphere.

The projections for the twenty-first century by coupled atmosphere-ocean general-circulation models (AOGCMs) using IPCC emissions scenarios show that average global temperatures will continue to decrease in the stratosphere and increase in the troposphere (Figure 4.8). This result is primarily a consequence of increases in WMGHGs (mainly CO_2). Changes in the thermal gradients in the stratosphere and troposphere, initiated by greenhouse gas and aerosol changes, could additionally

For the twenty-first century, average global temperatures are projected to continue to decrease in the stratosphere and increase in the troposphere, primarily as a consequence of increases in well-mixed greenhouse gases.

123

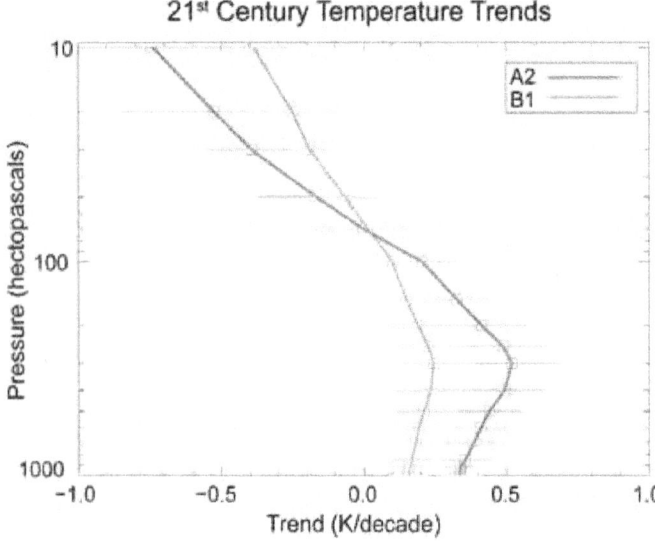

Figure 4.8 Temperature trends in the troposphere and stratosphere calculated as global and annual means for the twenty-first century using atmosphere-ocean GCMs (AOGCMs, with no ozone chemistry). Temperatures are expected to continue to increase in the troposphere and decrease in the stratosphere. The calculations were made for two IPCC emission scenarios: A2 (high) and B1 (low). The symbols indicate the average trend computed for all models, while the thin horizontal lines indicate the range. The vertical scale represents altitude from the surface (1000 hectopascal, hPa) to approximately 30 km (10 hPa). The pressure of the tropopause, which divides the stratosphere above from the troposphere below, varies between 100 and 300 hPa over most of the globe (WMO, 2007).

When temperatures decrease in the upper stratosphere, ozone loss slows in the dominant photochemical cycles.

alter stratosphere-troposphere interactions and the state of the stratosphere. Changes of water vapor in the stratosphere arising from tropospheric warming, and possible changes in convective activity and transport of water to the stratosphere, also can affect the stratospheric thermal state. Volcanoes can be expected to substantially alter the climate and chemistry of the stratosphere for a few years through the particulates produced and impacts on atmospheric circulation if they erupt more frequently and/or if they inject much more material into the stratosphere than the Mt. Pinatubo eruption.

4.3.2.1 RESPONSE OF OZONE TO STRATOSPHERIC TEMPERATURE CHANGES

With the coupling of ozone with climate parameters as outlined above, the effect of temperature changes on ozone is difficult to isolate. However, model simulations reveal some strong tendencies arising due to temperature changes. In the upper stratosphere, ozone amounts are controlled primarily by photochemical processes rather than transport,

and these processes are considered well understood. When temperatures decrease, ozone loss slows in the dominant photochemical cycles (reactive nitrogen [NO_x], reactive chlorine [ClO_x], and reactive hydrogen [HO_x]) (Figure 4.9). For example, 15-20 percent ozone increases were calculated in the upper stratosphere for a climate with doubled (CO_2) concentrations (Jonsson *et al.*, 2004). In the lower stratosphere, the destruction rate decreases for lower temperatures, but production and destruction rates are lower than in the upper stratosphere and transport plays a more important role. As a result, temperature changes have less influence on steady state ozone values in the lower stratosphere than in the upper stratosphere.

In the polar lower stratosphere, the reduction in photochemical loss with lower temperatures can be completely offset by increased activation of reactive chlorine and bromine, which increases ozone loss. Lower temperatures promote the formation of polar stratospheric clouds (PSCs), which facilitate heterogeneous reactions that form reactive halogens from reservoir gases. In the Arctic region, increased reactive halogens have the largest effect in controlling the ozone response to lower temperatures. For Northern Hemisphere winters from 1993 to present, a strong linear relationship is found between winter/early spring ozone depletion and the volume of air containing PSCs during the winter. The ODS abundances are nearly constant during this time period (Rex *et al.*, 2004). Arctic ozone depletion might increase if further reductions occur in Arctic stratospheric temperatures because temperature decreases can lead to increases in the duration and frequency of PSCs (Douglass *et al.*, 2006).

In the Antarctic lower stratosphere, winter temperatures are well below the thresholds for heterogeneous conversion of halogen reservoirs for much longer periods and for much larger fractions of the polar vortex than found in the Arctic. Antarctic ozone depletion is currently much more extensive and complete, and decreasing temperatures would have less of an effect (Tilmes *et al.*, 2006). Under current conditions, seasonal Antarctic ozone depletion is more sensitive to reductions in ODS amounts than to small decreases in temperature (see

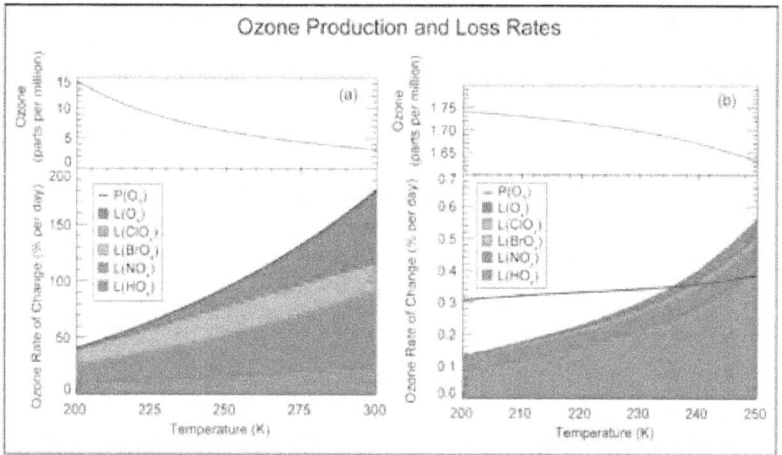

Figure 4.9 Comparison of ozone production and loss rates as a function of stratospheric temperature for 40-km altitude (left panel) and 20-km altitude (right panel) conditions at 45°N at equinox (end of March). The colored regions indicate the contribution from the principal loss cycles of ozone: odd-oxygen (O_x), reactive chlorine (ClO_x), reactive bromine (BrO_x), reactive nitrogen (NO_x), and reactive hydrogen (HO_x). The fractional contribution of each cycle varies with temperature differently in the two regions. The top trace in each panel is the ozone value at the end of 20-day runs of a chemical box model starting from climatological values for ozone, other constituents, and temperature (250K at 40 km; 215K at 20 km). At 40 km, the production rate coincides with the sum of loss rates because ozone is in photochemical balance at all temperatures shown. At 20 km, ozone production can be higher or lower than total loss depending on temperature because transport plays a more important role. The changes in ozone after 20-day runs are much smaller at 20 km than at 40 km, confirming that small temperature changes in the upper stratosphere will significantly alter ozone abundances (adapted from IPCC/TEAP, 2005).

Water vapor has increased about 5 to 10% per decade at a height of 17 to 22 km since 1980.

Chapter 3) (Newman *et al.*, 2004). As ODS abundances decrease in the coming decades, polar ozone destruction due to reactions with halogen species ultimately will decrease in both hemispheres regardless of changes in the frequency and duration of PSCs.

4.3.3 Stratospheric Water Vapor Changes

Atmospheric water vapor is the most important and abundant greenhouse gas. Change in the global distribution of water vapor is one of the important responses to the human-caused climate forcings summarized in Figure 4.5. Water vapor enters the stratosphere primarily through the tropical tropopause. The water vapor abundance is reduced in dehydration processes involving low tropical tropopause temperatures and the formation and sedimentation of ice particles. Methane, released in the troposphere and oxidized in the stratosphere, is the underlying cause of the water-vapor component of human-caused radiative forcing as summarized in Figure 4.5.

Stratospheric water vapor has been measured by a wide variety of instruments and platforms, including balloons, aircraft, and satellites. The longest time series of continuous measurements is from small-balloon observations beginning in 1980. These measurements show that water vapor has increased at all levels between 15 and 26 km altitude. At 17 to 22 km, the increase can be expressed as a trend of about 5-10 percent per decade (Figure 4.10) (Oltmans *et al.*, 2000; Rosenlof *et al.*, 2001). Other stratospheric observations up to 30 to 35 km also show increasing trends, but over shorter time periods and with a high degree of variability (SPARC, 2000; Rosenlof *et al.*, 2001). Part of the long-term increase in water vapor is attributable to increases in methane abundances due to human emissions. Methane, which has increased by about 0.55 parts per million by volume (ppmv) since the 1950s, is oxidized in the stratosphere, producing two water molecules for each molecule of methane. The methane water vapor source in the stratosphere increases radiative forcing from water vapor by an estimated 0.1 W per m² (Myhre *et al.*, 2007).

Stratospheric Water Vapor Trend

Figure 4.10 Time series of stratospheric water vapor mixing ratios (ppm) for the period 1980 to 2005 showing increases up to about 2001 followed by a decreasing trend. The measurements were made with a balloon-borne frost point hygrometer over Boulder, Colorado (40°N, 105°W). The data points are averages over 17 to 22 km altitudes. The thin line is a smoothed fit to the measurements. HALOE satellite observations for 1992-2005 are shown with the heavy line for the same altitude near Boulder (latitude 35°N to 45°N, longitude 80°W to 130°W). Preliminary revisions to the frost-point data reveal a slightly smaller trend (Scherer *et al.*, 2007). Updated from Randel *et al.* (2004) (WMO, 2007).

Increases in reactive hydrogen (HO$_X$) and the temperature threshold for polar stratospheric cloud (PSC) formation lead to increased ozone destruction in polar regions at constant ozone-depleting substance (ODS) amounts. However, the sensitivity of ozone to PSCs will decrease as ODS amounts decrease because less chlorine and bromine will be available to participate in ozone destruction reactions.

Since about 2000, the water vapor in key balloon and satellite observations in the mid- to lower stratosphere have shown significant decreases (Randel *et al.*, 2004). As a possible explanation, an analysis of the tropical tropopause temperatures for 1992 to 2005 shows that satellite water vapor amounts are consistent with interannual changes in the cold point temperatures and with the occasions of anomalously low tropopause temperatures (Randel *et al.*, 2004; 2006). Tropopause temperatures modulate the dehydration of air entering the lower stratosphere from the troposphere. In contrast, the earlier, longer water record from balloon measurements is not fully consistent with the record of tropopause temperatures (Seidel *et al.*, 2001). In general, the attribution of the causes of observed water vapor changes and trends in the stratosphere is incomplete, suggesting that projections of future amounts are uncertain.

4.3.3.1 RESPONSE OF OZONE TO STRATOSPHERIC WATER VAPOR CHANGES

Increases in stratospheric water lead to increases in reactive hydrogen (HO$_x$), which catalyze the chemical destruction of ozone (Wennberg *et al.*, 1994; Brasseur and Solomon, 1986). Ozone destruction is chemically buffered with a combination of loss cycles so that the response to increased HO$_x$ is generally not linear and varies with temperature and location in the stratosphere (Figure 4.9). Model simulations show that a 1 percent increase per year, long-term trend in water vapor would increase ozone loss due to increases in HO$_x$ and delay the recovery of the ozone layer (Dvortsov and Solomon, 2001). Increased water vapor also increases the temperature threshold for PSC formation in both polar regions because PSC are formed, in part, from water-vapor condensation. A higher threshold increases heterogeneous conversion of chlorine and extends the time period over which PSCs can form in the winter season (Stenke and Grewe, 2005). Both effects lead to increased ozone destruction in polar regions at constant ODS amounts. However, the sensitivity of ozone to PSCs will decrease as ODS amounts decrease, because less chlorine and bromine will be available to participate in ozone destruction reactions.

4.3.4 Changes in Ozone From Increases in Long-Lived Gases in the Stratosphere

The atmospheric concentrations of the three long-lived greenhouse gases, CO$_2$, CH$_4$, and N$_2$O, have increased significantly due to human activities since 1750 and are expected to continue increasing in the 21st century (IPCC, 2007). These continuing increases have consequences for ozone amounts and, hence, indirectly influence climate through the changes they produce in ozone (Portmann and Solomon, 2007). Calculations with a two-dimensional, chemical-radiative-dynamical model illustrate the sensitivity of ozone to each of these gases (Figure 4.11). Carbon dioxide increases, as previously discussed, reduce stratospheric temperatures and ozone loss rates, and consequently, increase ozone amounts in the mid- to upper stratosphere. The increased ozone in the upper stratosphere can lead to reduced ozone in the lower stratosphere because of the

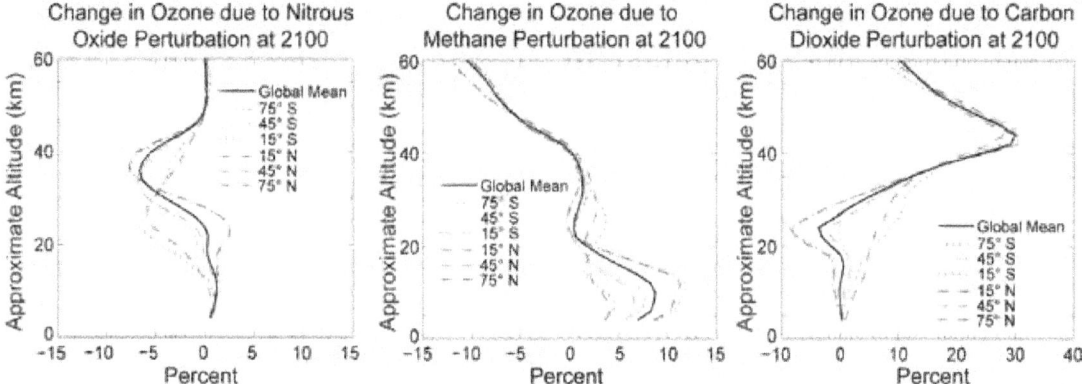

Figure 4.11 Comparison of perturbations to ozone amounts for increases in three principal greenhouse gases, N_2O, CH_4, and CO_2, as a function of altitude in the troposphere, stratosphere, and mesosphere. The changes are computed for six 30°-wide latitude bands with the NOCAR 2-D model and expressed as the percent change from 2000 to 2100. The ozone changes vary widely in magnitude and show increases as well as decreases. The changes result from a variety of chemical and radiative processes. The global mean change is shown with the black line in each panel. The halogen changes follow the WMO (2003) scenario and greenhouse gas increases follow the IPCC A2 scenario. The latter is high compared to other scenarios but was chosen to maximize the ozone response in the model (adapted from Portmann and Solomon, 2007).

reduced penetration of solar UV into the lower stratosphere. Increases in N_2O lead to increases in the NO_x catalytic loss cycle for ozone in the mid- to upper stratosphere, because N_2O decomposes to form NO_x in the stratosphere. The effect of increased NO_x is less in the lower stratosphere, because the NO_x loss cycle plays a less prominent role, competing with the HO_x and ClO_x catalytic loss cycles (Figure 4.7) (Wennberg *et al.*, 1994). Finally, the oxidation of CH_4 increases H_2O and ozone losses in the HO_x catalytic cycle in the upper stratosphere and lower mesosphere. In the troposphere, ozone is increased because oxidation of CH_4 catalyzed by NO_x produces ozone.

4.3.4.1 CHANGES IN OZONE FROM STRATOSPHERIC CIRCULATION CHANGES

The net mass exchange between the troposphere and stratosphere is associated with the large-scale Brewer-Dobson circulation (Holton *et al.*, 1995), with a net upward flux in the tropics balanced by a net downward flux in the extratropics. Model studies indicate that climate change will impact the mass exchange rates across the tropopause. For a doubled CO_2 concentration, all 14 climate-change model simulations analyzed by Butchart *et al.* (2006) showed an increase in the annual mean troposphere-to-stratosphere exchange rate, with a mean trend of about 2 percent per decade. Consequences of such an increase include

shorter lifetimes and more rapid removal from the atmosphere for long-lived gases, including CFCs, CH_4 and N_2O (Butchart and Scaife, 2001), and increased mass flux of ozone from the stratosphere to the troposphere at mid- and high latitudes. A model simulation for 2100 with projected climate change (Zeng and Pyle, 2003) shows that a strengthened Brewer-Dobson circulation would increase ozone amounts in the lower stratosphere and the flux of ozone to the troposphere. A larger flux results from increased transport downward across the tropopause and enhanced ozone amounts in the extratropical lower stratosphere. The enhanced ozone results from the strengthened circulation and decreases in ODSs and temperatures.

4.4 THE EFFECT OF OZONE CHANGES ON CLIMATE PARAMETERS

Ozone and climate change are highly coupled, as illustrated in Figures 4.1 and 4.2. The response of ozone to changes in stratospheric temperature and water vapor is discussed in Section 4.3. In this section, changes in atmospheric temperatures and circulation are described as examples of the responses in the climate system to ozone depletion.

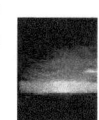

Model studies indicate that climate change will impact the mass exchange rates across the tropopause.

4.4.1 Response of Stratospheric and Tropospheric Temperatures to Ozone Depletion

Temperatures have decreased throughout the stratosphere in recent decades as described above in Section 4.3.2. Furthermore, model simulations show that a combination of increases in greenhouse gases and water vapor and decreases in ozone can account for observed temperature changes. A more detailed examination of the influence of ozone on temperature was carried out with the SKYHI GCM for ozone decreases observed in the period from 1979 to 1997 (Ramaswamy and Schwarzkopf, 2002). The results in Figure 4.12 indicate that in the lower to middle stratosphere

(100 to 5 hPa; 21 to 38 km), ozone changes create a larger decrease in temperature than increases in WMGHGs. In this case, these include CO_2, CH_4, N_2O, CFC-11, CFC-12, CFC-113, and HCFC-22. However, above about 5 hPa (about 38 km) changes in both ozone and WMGHGs contribute significantly to temperature decreases. Thus, depletion in stratospheric ozone plays a significant role throughout the stratosphere in creating a reduction in stratospheric temperatures in GCM simulations.

The response of temperature to stratospheric ozone depletion extends into the upper troposphere. The Reading Narrow Band Model was used to calculate temperature changes for observed ozone depletion with the assumption of fixed dynamical heating (Forster *et al.*, 2007). Model cooling occurs in the 70 to 30 hPa (13 to 25 km) region due to ozone depletion. Shortwave absorption and upwelling longwave radiation are both reduced and contribute comparably to the cooling in this region. The missing ozone also causes a decrease in the downwelling longwave radiation that causes reduced temperatures at altitudes below the ozone depletion region (150 to 100 hPa; 14 to 21 km). This response or coupling of temperatures in different altitude regions is found at all latitudes in the model and may be a cause of upper tropospheric temperature trends.

4.4.2 Response of Surface Temperatures to Antarctic Ozone Depletion

The largest depletion in stratospheric ozone is found over the Antarctic in late winter/early spring. Studies of Antarctic ozone depletion have revealed strong evidence for responses in the temperatures and circulation of the Antarctic troposphere (Gillet and Thompson, 2003; Thompson and Solomon, 2002). Severe ozone depletion strengthens the circumpolar winds of the Antarctic winter vortex in many model simulations. Recent observations show that strengthened circumpolar winds extend to the surface, especially in the summer months, with changes in geopotential heights (a vertical coordinate system referenced to Earth's mean sea level) serving as a proxy. A model with high vertical resolution was used to show that anomalies in geopotential height

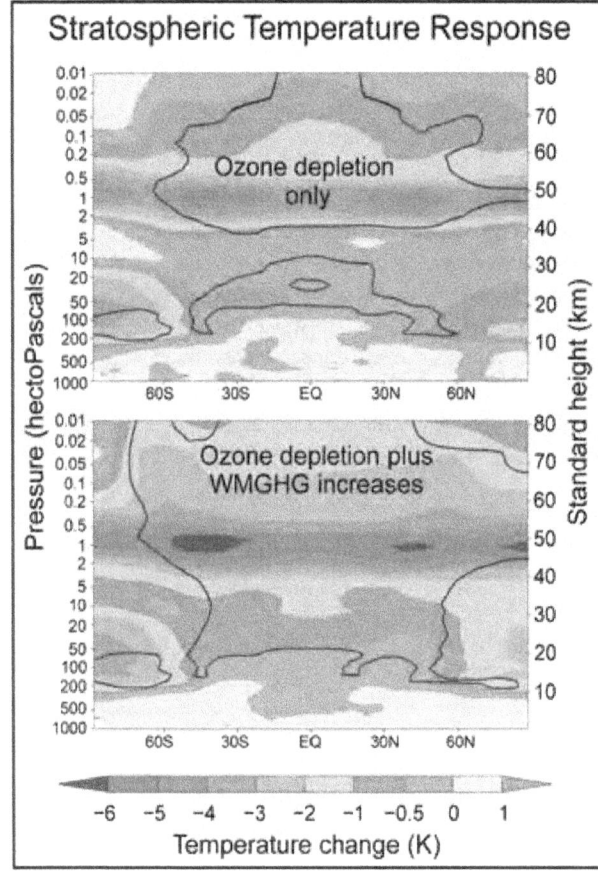

Figure 4.12 The response of stratospheric temperatures calculated with a global climate model (GCM) using observed ozone depletion and changes in well-mixed greenhouse gases (WMGHGs) between 1979 and 1997. The upper panel shows modeled temperature response to ozone changes alone and the lower panel shows the response to ozone changes plus increases in greenhouse gases over the same period. Ozone changes alone cause a significant temperature response in the period. Solid lines enclose regions of statistical significance. The vertical scale shows pressure (left) and altitude (right) (adapted from Ramaswamy and Schwarzkopf, 2002).

Response to Stratospheric Ozone Depletion

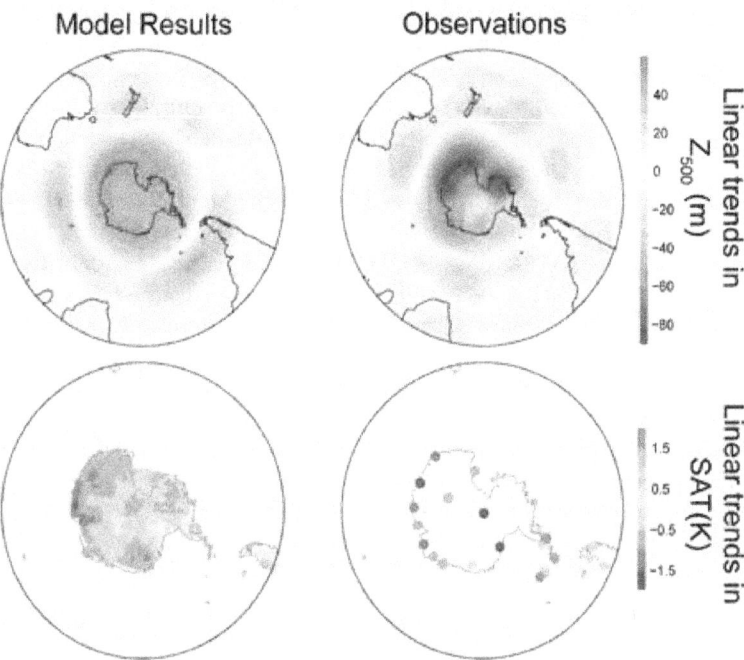

Figure 4.13 Comparison of model results (left column) with observations (right column) for changes in 500-hPa geopotential height (m) (upper row) and near-surface temperature (K) (lower row) in the Southern Hemisphere in response to stratospheric ozone depletion between 1979 and 1997. The observed and simulated patterns in geopotential height and surface temperature show strong similarities, reinforcing the conclusion that stratospheric circulation influences tropospheric circulation patterns and, hence, that intense stratospheric ozone depletion can effect changes in surface climate parameters (adapted from Thompson and Solomon, 2002).

in the troposphere could be well simulated in intensity and seasonality (Gillet and Thompson, 2003). Changes in surface circulation also lead to cooling over most of the Antarctic continent and modest warming of the Antarctic Peninsula. Figure 4.13 shows model results compared to observed changes in 500-hPa geopotential height over a 22-year period and in surface temperature over a 32-year period (1969 to 2000), both averaged over December to May. The observed and simulated patterns in geopotential height and surface temperature show strong similarities, reinforcing the conclusion that stratospheric circulation influences tropospheric circulation patterns and, hence, that intense stratospheric ozone depletion can bring about changes in surface climate parameters.

4.5 IMPORTANCE OF VOLCANOES

4.5.1 The Effect of Volcanic Aerosol on Ozone

Large volcanic eruptions are those that inject significant quantities of sulfur dioxide (SO_2) into the stratosphere. The SO_2 is subsequently oxidized to sulfuric acid, which condenses onto preexisting aerosols, causing significant increases in aerosol surface area and volume in the lower stratosphere. As a consequence, heterogeneous reactions occurring on these surfaces gain prominence in the chemical production and loss balance of ozone, leading to decreased ozone amounts (WMO, 2007: Figures 3-26). These reactions convert NO_x to a more stable form, nitric acid (HNO_3). In the lower stratosphere, reduced NO_x increases the role that reactive halogen compounds (ClO_x) play in destroying ozone. Analysis following the most recent large volcanic eruption, Mt.

The intense ozone depletion above Antarctica brought about changes in surface climate parameters of that continent.

Pinatubo in 1991, shows that ozone amounts reached record lows and that halogen reactions, aided by temperature variability, could explain the observed losses (Solomon *et al.*, 1998; Tie and Brasseur, 1995).

As ODS amounts decrease to pre-1980 levels in the coming decades, the sensitivity of ozone to depletion caused by volcanic aerosol reactions will also decrease. Global ozone levels decreased by about 2 percent following the 1991 Mt. Pinatubo eruption. Sedimentation and transport removal of volcanic aerosol occurs over a two to three year period following an eruption, so the effects are short lived compared to ODS atmospheric lifetimes, which are 45 to 100 years for principal species (*e.g.*, CFC-11 and CFC-12). Thus, expectations for the long-term recovery of ozone are not significantly affected by episodic volcanic eruptions.

The plumes of large volcanic eruptions contain significant amounts of hydrochloric acid (HCl) that are removed in the troposphere by uptake and sedimentation of the liquid aerosols formed (Tabazadeh and Turco, 1993). However, some eruptions inject non-negligible amounts of HCl into the stratosphere, adding to inorganic chlorine. For example, temporary increases in HCl column amounts of up to 40 percent were observed after the Mt. Pinatubo eruption (Coffey, 1996). Overall, the frequency of explosive volcanic eruptions has been low in the past two decades, thereby precluding significant, long term, volcanic enhancements in global stratospheric chlorine.

Volcanic aerosols have a direct radiative impact by scattering incoming solar radiation and absorbing solar infrared radiation. Absorption increases stratospheric temperatures while decreasing surface temperatures. The tropospheric cooling that results can be expected to change the tropospheric circulation, as well as the interaction between the stratosphere and the troposphere. Lower stratospheric temperatures, for example, following the eruptions of El Chichón and Mt. Pinatubo, were observed to increase by about 1K near 20 km altitude (Figure 4.7). The loss of ozone following an eruption also adds to the temperature perturbation. Lower stratospheric temperatures influence water vapor amounts through dehydration of air parcels

entering the stratosphere from the troposphere and influence ozone amounts through the sensitivity of ozone chemical reaction rates. Climate-chemistry model simulations of the temperature perturbations after the eruptions of El Chichón and Mt. Pinatubo often show larger increases than observed (Figure 4.7). The elevated temperatures are evident for several years and are followed by an overall slow cooling. The strength of the volcanic temperature response varies substantially between the different CCM and climate models (see Eyring *et al.*, 2006).

Volcanic eruptions, while not predictable, are expected to occur in the future atmosphere. A large volcanic eruption is likely to occur in the next 30 years, based on the historical record (Roscoe, 2001). Infrequent large volcanic eruptions would affect ozone with timescales as observed for previous large eruptions (Figure 4.7). A period of frequent large eruptions in the next century could enhance ozone depletion from ODSs for many years but the potential for enhancement will lessen as global ODS abundances decline in the coming decades (Figure 2.12). Whenever the stratosphere is cleansed of volcanic aerosol, ozone abundances are expected to recover fully from volcanic effects.

4.6 SUMMARY

Stratospheric ozone and climate change are linked through a variety of processes. Radiative forcing of climate occurs from depletion of stratospheric ozone, as well as increases in ozone-depleting substances. Global ozone depletion is a principal cause of decreasing temperature trends in the stratosphere and upper troposphere. Severe ozone depletion over Antarctica has changed the circulation over the continent in both the stratosphere and troposphere and altered surface temperatures. Other important components of human-caused climate change arise from emissions of long-lived greenhouse gases, such as carbon dioxide. Observed and anticipated changes in climate parameters include decreases in stratospheric temperatures and increases in stratospheric water vapor, carbon dioxide, methane, and nitrous oxide. Lower stratospheric temperatures reduce ozone loss rates in the mid- to upper

A period of frequent volcanic eruptions in the next century could enhance ozone depletion from ozone-depleting substances (ODSs) but the potential for enhancement will lessen as global ODS abundances decline.

stratosphere, thereby aiding the recovery from ozone depletion. Enhanced water vapor alters ozone destruction rates in reactive hydrogen photochemistry and can increase the frequency and extent of polar stratospheric clouds, which aid ozone destruction. These varied composition changes contribute to circulation changes in the stratosphere and between the stratosphere and troposphere that can cause significant changes in the ozone distribution. The increases in stratospheric aerosols that follow explosive volcanic eruptions create several-year changes in climate parameters in the stratosphere and troposphere and increase ozone depletion.

The complexity of the interactions between ozone and climate involving changes in atmospheric composition pose a challenge to our understanding of basic stratospheric and tropospheric processes. Tools of the complexity of chemistry climate models (CCMs) are required to combine stratospheric transport, dynamics, radiation, and chemistry and microphysics to analyze past ozone amounts and project future amounts. CCMs guided by atmospheric observations help define the sensitivity of ozone to future climate changes and reduce the uncertainties in our understanding of ozone and climate interactions. As ozone depletion slows and ozone amounts recover from ODSs in the coming decades, expected changes in climate parameters will increase in importance in influencing stratospheric ozone amounts.

4.6.1 Relevance for the United States

Human activities have led to changes in ozone abundances and climate parameters. Ozone depletion is attributed primarily to the accumulation of ozone-depleting substances and climate change is attributed to increases in long-lived greenhouse gases, changes in aerosols and clouds, and surface albedo changes (Figure 4.5). Ozone is further influenced by changes in climate parameters such as stratospheric temperatures and composition, and atmospheric circulation. Since activities in the United States have caused significant emissions of greenhouse gases and ozone-depleting substances, the changes in ozone and climate attributable to human activities are, in part, attributable to the United States.

Decisions initiated or supported by U.S. policymakers have great potential to influence ozone and climate in the future. Important decisions could be taken on the following topics or issues:

- *Increased stringency of Montreal Protocol regulations.* The Montreal Protocol regulates production and consumption of ODSs in developed and developing nations. Stratospheric ODS amounts will decline to pre-1980 values around the middle of this century based on current regulations. More stringent regulations could accelerate this decline. For example, recent unratified regulation, supported by the United States, accelerates hydrochlorofluorocarbon (HCFC) production in developed and developing nations.
- *Increased destruction or capture of ODS banks.* Banks of ODS compounds represent large sources of future ODS emissions (see Chapters 2 and 5).
- *Increased climate protection under the Montreal Protocol.* ODS compounds are also greenhouse gases. Reducing ODS production and consumption under the Montreal Protocol has led to significant reductions in ODS atmospheric abundances and their associated radiative forcing of climate. Further reductions in ODS production, as well as emissions, will further protect climate. The accelerated HCFC phase-out under the Montreal Protocol represents a large potential benefit to climate. In addition, promotion of low-GWP compounds as replacements for ODSs in widespread applications can help minimize the climate consequences of new and existing Montreal Protocol regulations.
- *Reductions in the future growth rates of carbon dioxide, methane, and nitrous oxide emissions.* Ozone and climate are strongly influenced by carbon dioxide, methane and nitrous oxide emissions. Emissions of these gases ultimately affect the photochemical production and loss of ozone in the troposphere and/or stratosphere. All are greenhouse gases that have increased significantly due to human activities.

CHAPTER 5

The Future and Recovery

Convening Lead Author: Malcolm Ko, NASA

Lead Authors: John S. Daniel, NOAA; Jay R. Herman,
NASA; Paul A. Newman, NASA; V. Ramaswamy, NOAA

KEY ISSUES

This chapter presents results on how future halogen loading is expected to affect the behavior of total column ozone and the prospect for the detection/validation of the anticipated recovery trend. In a hypothetical argument, if circulation, climate, and the background atmosphere were to remain unchanged as of the present-day, the projection of ozone could be based essentially upon future halogen loading. However, the reality is that the concentrations of other trace gases that affect ozone (e.g., methane [CH_4], nitrous oxide [N_2O], and water vapor) are changing because of changes in emissions; climate, which also affects ozone abundances, is changing as well. The model-simulated results show that the ozone increase expected between now and 2025 is largely due to the anticipated decrease in halogen loading, and ozone-depleting substances (ODSs) will remain one of the drivers of human-caused ozone depletion up until the middle of the twenty-first century, when the halogen loading is expected to approach its 1980 value. Reductions in emissions of these chemicals represent the only known currently acceptable method to reduce this depletion in this period. The effects of climate change (largely driven by increases in carbon dioxide, CO_2) and changes in other trace gases (e.g., CH_4, N_2O, and hydrofluorocarbons) will play an increasing role in the ozone behavior.

Ozone is only one of many factors that affect UV (ultraviolet radiation) at the surface. Future changes in UV will be discussed in this chapter in the context of the projected ozone change in the stratosphere. The Equivalent Effective Stratospheric Chlorine (EESC) is used to compare the relative impacts of various ODS emissions scenarios on future ozone. Included in this discussion is the radiative forcing associated with the halocarbons as well as the hydrofluorocarbons (HFCs) used as replacements for the ODSs. The contribution from the United States to the future halocarbon loading will be addressed in the context of EESC and radiative forcing.

The key issues, in the form of questions, that are addressed in this chapter include:
- What is the future behavior of ozone as predicted by numerical models?
- What is the future behavior of ultraviolet radiation at the Earth's surface?
- Are there any new findings concerning projected future emissions of ODSs?
- What is the radiative forcing associated with ODSs and HFCs emitted as replacement chemicals for the ODSs and how will it likely change in the future?
- To the extent that the emissions from a specific country can be used to estimate its contribution to global ozone depletion and radiative forcing by ODSs and their replacements, what is the United States' contribution?

Key Findings

Equivalent Effective Stratospheric Chlorine (EESC) is a useful index for comparing relative merits of different halocarbon emission scenarios in minimizing ozone depletion. Current scenarios being considered include projected emissions of long-lived source gases only and do not include emissions of very short-lived source gases.

- The time for EESC to return to the 1980 level (the EESC recovery date) and the integrated EESC values (to the EESC recovery date) provide useful metrics to compare the relative merits of various emissions scenarios.
- There have been suggestions that the mean age of air, and age-dependent release factors should be used in the calculation of EESC. This is potentially useful for calculating EESC values that are more representative of polar ozone depletion, in particular. The new recipe will change the absolute values of the metric for global ozone depletion, but should not qualitatively affect the relative benefit estimates of the different scenarios.

Two-dimensional chemistry transport models (both with and without climate feedback) and three-dimensional climate chemistry models (3-D CCMs) were used to simulate the behavior of ozone in the twenty-first century using the WMO (2003) baseline scenario.

The scenario includes changes arising from natural and anthropogenic activities other than ODS emissions. Thus, it is expected that the model-simulated ozone recovery date will differ from the EESC recovery date. The model-simulated ozone results will be discussed in the context of the EESC recovery date between 2040 and 2050 for midlatitudes, and between 2060 and 2070 for the polar regions. Analyses of simulation results indicate that:

For the model-simulated ozone content between 60°N and 60°S:

- Between now and 2020, the simulated total ozone content will increase in response to the decrease in halogen loading.
- Some 3-D CCMs predict that stratospheric cooling and changes in circulations associated with greenhouse gas emissions will enable global ozone to return to its 1980 value up to 15 years earlier than the EESC recovery date.
- Based on the assumed scenario for the greenhouse gases (which includes CH_4 and N_2O), the ozone content between 60°N and 60°S is expected to be 2% above the 1980 values by 2100. Values at midlatitudes could be as much as 5% higher.

For the model-simulated Antarctic ozone:

- The model-simulated ozone recovery date (the year when ozone returns to its 1980 value) for Antarctic ozone behavior depends on the diagnostics chosen. The minimum ozone value is projected not to start increasing until after 2010 in several models, while a decrease in ozone mass deficit in most models has occurred by 2005.
- Model simulations show that the ozone amount in the Antarctic will reach the 1980 values 10 to 20 years earlier than the 2060 to 2070 time frame.

For the model-simulated Arctic ozone:

- Ozone in the Arctic region is expected to increase. Because of large interannual variability, the simulated results do not show a smooth monotonic recovery. The dates of the minimum column ozone from different models occur between 1997 and 2015.
- Most CCMs show ozone values at 2050 larger than the 1980 values, with the recovery date between 2020 and 2040.
- Results from the majority of the models indicate that Arctic ozone depletion will not be significantly worse than what has already occurred.

With the current scenarios, anthropogenic halogens identified in the Montreal Protocol should have a minimal effect on midlatitude ozone beyond 2050. In order to predict the future trend of ozone in that time frame, one must consider projections for climate changes and changes in trace gases such as other halogens, CH_4, and N_2O.

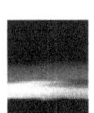

Applying ozone trend analysis techniques to future observations should enable one to confirm the time when halogen loading has minimal influence on global ozone within five to ten years after it occurs.

The future UV trend at the surface is likely to be more dominated by changes in clouds, aerosols, and tropospheric air quality than by ozone changes in the stratosphere. Equivalent Effective Stratospheric Chlorine (EESC) will still be a useful predictor for the relative effects of ODSs on future UV in terms of evaluating the different scenarios.

Future halocarbon emissions are derived using a new bottom-up approach for estimating bank sizes. The new method gives future chlorofluorocarbon (CFC) emissions that are higher than previously estimated in WMO (2003).
- Current projected concentrations for EESC in the twenty-first century are higher than reported in WMO (2003) because the most recent CFC bank estimates, which are believed to be more accurate, are larger and lead to larger emissions, and the estimated emissions due to future production of hydrochlorofluorocarbons (HCFCs) from Article 5(I) countries are also larger.
- The EESC in the baseline scenario returns to the 1980 value in the year 2049, about five years later than the date based on the WMO (2003) baseline scenario.
- Compared to the WMO (2007) baseline scenario, cessation of all future emissions will bring the EESC recovery date earlier by 15 years to 2034. Integrated EESC (from 2007 to the EESC recovery date) from ODSs already in the atmosphere as of 2007 is 58% of the integrated EESC for the baseline scenario.
- If no future ODS production is assumed, the date when EESC returns to the 1980 level is moved earlier by six years to 2043. The integrated EESC from ODSs produced after 2007 is 17% of the integrated EESC for the baseline case.

Direct radiative forcing from ODSs and HFC replacement chemicals is approximately 0.34 W per m^2 in 2005 and is expected to stay below 0.4 W per m^2 through 2100 (according to the WMO (2007) scenario for ODSs and the IPCC Special Report on Emission Scenarios (SRES) scenarios considered for HFCs). This is to be compared with forcing from CO$_2$ of 1.66 W per m^2 in the 2005 atmosphere, increasing to as high as 5 W per m^2 by 2100 for the SRES A1B scenario.
- Direct forcing from CFCs will decrease to 0.1 W per m^2 by 2100. Direct forcing from HCFCs and other anthropogenic ODSs is expected to be negligible by 2100.
- Forcing from HFCs is 0.15 W per m^2 and 0.24 W per m^2 in 2050 and 2100, respectively, for the SRES A1B scenario, whereas other scenarios indicate that it will be lower. However, current observations suggest that the present atmospheric radiative forcing of the HFCs has been larger than computed for the SRES scenarios, primarily due to higher HFC-23 concentrations.

The forcing associated with the observed ozone depletion was estimated to be about −0.05±0.10 W per m^2 in 2005, corresponding to one-sixth of the direct forcing due to ODSs.
If one assumes that all of the observed ozone depletion is due to ODSs, that would imply that the indirect effect of ODSs due to ozone depletion is one-sixth of the direct effects for the mix of ODSs present in the atmosphere at that time. Current estimates assume that the indirect forcing from ODSs will decrease to zero approximately when EESC returns to its 1980 levels, while the direct forcing (mainly from CFCs and HFCs remaining in the atmosphere) will continue. The indirect forcing remains highly uncertain and is discussed in more detail in Chapter 2.

Using available historical and projected United States and global emissions estimates, we find that emissions from the United States contribute between about 15% and 37% to global EESC due to man-made emissions at 2030. For the same year, the United States' contribution to direct radiative forcing from ODSs, HFCs, and PFCs (perfluorocarbons) is 19% to 41%.

5.1 INTRODUCTION

This chapter presents results on how future halogen loading will affect the future behavior of total column ozone and the prospect for the detection/validation of the expected recovery trend. In a hypothetical argument, if the transport circulation, the climate, and the background atmosphere were to remain unchanged as of the present-day, the projection of ozone could be based essentially upon future halogen loading. Chapter 2 discussed the concept of Equivalent Effective Stratospheric Chlorine (EESC) and how the values for midlatitude EESC and polar EESC could be used as metrics to approximate the effects of ODSs on ozone behavior. Daniel *et al.* (1995) pointed out that global stratospheric ozone losses became apparent from observations around the late 1970s to around 1980 (see also Figure 3.2 in this report). They also considered that the total ozone over Halley Bay dropped nearly linearly from 1980 to 1990. From these facts, they assumed that the ozone trend due to halogen loading was close to zero for EESC smaller than the 1980 value. If one assumes that the same holds for the future and nothing else is different, ozone is expected to recover to the 1980 value at the same time as EESC. The time for EESC to return to its 1980 value thus has been adopted as a metric to compare various control options. For this reason, the date when the future EESC value returns to its 1980 value is given some significance, and is referred to as the EESC recovery date (EESC RD).

Since policy decisions are being made based on EESC, it would be prudent to perform analyses to see how well the EESC-based predictions correlate with model simulations of ozone recovery. Unfortunately, the scenarios used in the model simulations reported in WMO (2007) include changes in trace gases other than the halogens that affect ozone directly through chemical interactions, or indirectly through climate change. Thus, the model-simulated ozone recovery date (MS ORD, *i.e.*, the date ozone returns to its 1980 value) for these simulations is expected to be different from the EESC RD.

The results of numerical simulations of the future behavior of ozone as reported in WMO (2007, Chapter 6) report are presented in Section 5.2. A discussion of how future observed ozone behavior can be used to detect and confirm the ozone recovery date is presented in Section 5.2.3. Section 5.3 discusses how future ozone may affect UV. Sections 5.4 and 5.5 focus on expected future trends of the halocarbons through 2050. The future emissions and abundances of the CFCs and HCFCs are discussed in Section 5.4. The EESC will be used to compare the relative impacts of various ODS emissions scenarios on future ozone in Section 5.5. Included in this section is a discussion of the radiative forcing associated with the ODSs as well as the HFCs used as replacements for them. The contribution from the United States to the global future halocarbon loading will be addressed in the context of EESC and radiative forcing in Section 5.6.

5.2 MODEL SIMULATIONS AND ANALYSES OF OZONE BEHAVIOR

Analyses of the over 40-year time series of global ozone data between 1964 and 2006 (see discussion in Chapter 3, Figure 3.2) indicate that it is possible to attribute the observed ozone behavior to several processes that affect ozone. These include the responses to the seasonal cycle, to the quasi-biennial oscillation (QBO), to the 11-year solar cycle, to episodic volcanic eruptions, and to halogen loading from halocarbons. In particular, most of the decreasing trend in ozone during this period can be correlated with EESC and attributed to the increase in halogen loading. It is anticipated that the decrease in halogen loading in the next 20 years will have a large influence on the decadal trend of ozone. To predict the future trend of ozone, one must identify all processes that may affect ozone, determine how the driving mechanisms may change (*i.e.*, the scenarios), and employ numerical models to simulate the ozone behavior. The projected behavior will depend on the adopted scenario. The results presented in this section also show that different models predict different results for the same scenario. This indicates that there is still disagreement on how processes are represented in the models, and one must depend on further comparison with observations to resolve these issues. Finally, the purpose for presenting the model results in this chapter is to illustrate, in general terms,

The decreasing trend in ozone during the period 1964 to 2006 can be mostly attributed to the increase in halogen loading. It is anticipated that the decrease in halogen loading in the next 20 years will have a large influence on the decadal trend of ozone.

how the expected ozone behavior differs from the parameterized behavior based on EESC. It is beyond the scope of this report to address the various outstanding issues associated with simulating ozone behavior. Such attempts would greatly benefit from studies of changes in local ozone as functions of altitude.

5.2.1 Processes and Scenarios Used in Model Simulations

It is clear that the model simulations must include the effects from changes in halogen loadings. The model simulations use prescribed surface concentrations of the halocarbons derived from projected emissions. The method for deriving the surface concentrations from emissions will be discussed in more detail in Section 5.4. The current best-estimate scenario for future halocarbon emissions (A1) is discussed in Table 8-4 and the surface concentrations are summarized in Table 8-5 of the WMO (2007, Chapter 8) report. Note that this scenario, as with all other scenarios previously considered in the WMO reports, considers only relatively long-lived (lifetime > 0.5 years) chlorine and bromine source gases. However, it has become clearer that very short-lived (VSL) ODSs also contribute to stratospheric ozone depletion, particularly short-lived bromocarbons. A more detailed discussion of the contribution of VSL compounds to stratospheric chorine and bromine loading can be found in Chapter 2 of WMO (2007). Note that the standard procedure for estimating EESC from emissions of long-lived source gases (Box 2.7 in Chapter 2 of this report) should not be applied to VSL source gases in its current form. It was estimated in WMO (2007) that VSL compounds might contribute 50 ppt (parts per trillion by mole) of stratospheric chlorine and 3 to 8 ppt of stratospheric bromine. It was unclear whether any trend in these VSL compounds should be expected in the future or has occurred in the recent past. Enhancement in convective activities associated with future climate changes may increase the Ozone Depletion Potentials (ODPs) of the VSL species.

Because the chapters in the WMO (2007) report were prepared in parallel, there was not sufficient time to use this most updated scenario in the model simulations. The model results presented in Chapter 6 of the WMO (2007) report were simulated using the scenario (Ab) as summarized in Table 4B-2 in the WMO (2003, Chapter 4) report. Using assumed values for the atmospheric lifetimes, the release factors of the halocarbons, and the transport lag from the tropopause, one can compute the date when midlatitude and polar EESC will reach its 1980 value. For scenario Ab, the midlatitude EESC RD is calculated to be 2044. However, because of the uncertainties associated with the lifetimes and the release factors, Chapter 6 of the WMO report chose to discuss the results relative to an EESC RD range between 2040 and 2050. The model-simulated ozone recovery date (MS ORD) could be earlier than the EESC RD if the net effect from other non-ODS factors (see below) causes an increase in ozone relative to the 1980 value. Finally, the MS ORD for ozone at specific latitudes is likely to be different for different latitudes.

Note that even in simulations that keep all other parameters the same except for halogen loading, the EESC RD and MS ORD is still likely to be different because the lifetimes, the release factors, and the bromine efficiency factor in the model are likely to be different from what are assumed in the EESC calculation. To resolve this issue, comparison and validation of model-simulated atmospheric lifetimes and release factors should be a priority. Until this is accomplished, direct comparison between the EESC RD with the MS ORD is not as meaningful as it could be.

Variations in natural factors such as changes in the Sun's energy output and volcanic events will continue to have impacts on the ozone abundances. Changes in solar UV between solar cycles are assumed to be small. Effects on ozone from variations within each 11-year cycle can be isolated as demonstrated in Figure 3.2. Once identified, the effect can be removed in interpreting the observed ozone changes. Thus, it is not crucial whether the solar cycle effect is included in the simulations. Effects from volcanoes are not included as there is no reliable way to predict volcanic eruptions in the future. Their effects can be removed in the analyses several years after it occurs. The philosophy here is that, like the solar cycle, the effect can be removed from the observation before they are compared with the model simulated trends.

Variations in natural factors such as changes in the Sun's energy output and volcanic events will continue to have impacts on the ozone abundances.

Chapter 4 discussed how climate change due to increased CO_2 (and other well-mixed greenhouse gases), changes in water vapor in the stratosphere, and changes in long-lived source gases (CH_4 and N_2O) could affect ozone. Climate change can affect ozone through changes in temperature and transport circulation. Cooling of the stratosphere associated with greenhouse gases is expected to slow gas-phase ozone loss reactions and increase ozone. This is particularly effective in the upper stratosphere. Water vapor in the stratosphere plays a particularly interesting role. It affects ozone concentration through the odd-hydrogen (HO_x) chemistry, as well as contributing to the cooling of the stratosphere. Its concentration can be changed due to changes in methane and changes in climate. In the scenario calculations, changes in water are not prescribed. They are calculated from the CH_4 increase and from changes associated with climate in chemistry climate models (CCMs).

The scenario for CO_2, CH_4, and N_2O used in the simulations is summarized in Table 5.1. Based on sensitivity simulations from two-dimensional (2-D) chemistry transport models (CTMs) reported in WMO (1999, Chapter 12), a 15% increase in CH_4 at 2050 from its 2000 values would have added about 0.5% in column ozone at midlatitudes. A 15% N_2O increase would have decreased ozone by about 1%. Thus, in the scenario shown, the combined effects from CH_4 and N_2O increase ozone around 2050.

The WMO reports also discussed changes in aerosol and oxides of nitrogen (NO_x) from aviation (WMO, 2003, Chapter 4), emissions from rocket launches (WMO, 2003, Chapter 4), and changes in molecular hydrogen (H_2) (WMO, 2007, Chapter 6). The effects from these processes are not included in the

WMO simulations. Emission of NO_x from subsonic airplanes increases ozone in the upper troposphere. The IPCC (1999) estimates that a 0.4% increase in column ozone at midlatitudes in the current atmosphere can be attributed to en route emissions from aircraft. Anticipated doubling to tripling of emissions by 2050 could add another 1%. Detailed projections of future emissions based on demands and technology advances are not yet available. Previous estimates suggest that the current rocket launch schedule may have caused a small (less than 1%) column decrease due to loss in the lower stratosphere. Future trends will depend on growth and mix of solid fuel and liquid fuel propellants. Estimates for changes in H_2 are based on the assumption that liquid hydrogen may become an important energy source for the economy and leakage from storage and usage may cause a dramatic increase in H_2. Not enough is known to do any reliable projections.

5.2.2 Results From Model Simulations

Three types of models were used to simulate the future behavior of ozone in WMO (2007, Chapter 6):

1. Two-dimensional chemistry-transport models (2-D CTMs) use fixed temperature and circulation. They are most useful for isolating the effects of different source gases;

2. Interactive 2-D models partially account for the changes in transport associated with climate change by calculating the residual circulation from heating rates allowing for interaction of planetary waves with the mean circulation. However, the feedback from changes in wave forcing is not simulated; and

3. Three-dimensional climate chemistry models (3-D CCMs) incorporate all the identified feedbacks and are generally

Climate change can affect ozone through changes in temperature and transport circulation. Cooling of the stratosphere associated with greenhouse gases is expected to slow gas-phase ozone loss reactions and increase ozone.

Table 5.1 Future concentrations of CO_2, CH_4, and N_2O used in the model simulations. The CO_2 values are from the Integrated Science Assessment Model (ISAM) as listed in Appendix II of IPCC (2001).

Year	2000	2010	2020	2030	2040	2050	2060	2070	2080	2090	2100
CO_2 (ppm)	369	391	420	454	491	532	572	611	649	685	717
CH_4 (ppb)	1760	1871	2026	2202	2337	2400	2386	2301	2191	2078	1974
N_2O (ppb)	316	324	331	338	344	350	356	360	365	368	372

better able to represent the key processes related to 3-D transport in the atmosphere (particularly the polar regions).

In the following discussion, both the observations and the model results will be displayed as annual mean or monthly anomalies expressed as a percentage of the pre-1980 conditions. The midlatitude EESC RD is expected to occur sometime between 2040 and 2050. In looking at the MS ORD, it has proven convenient to examine the spatial aspects of the problem in terms of the phenomena in the two polar regions (Arctic and Antarctic) and that in the tropics plus midlatitudes (approximately 60°N to 60°S). This separation accounts for the distinct stratospheric circulation patterns prevailing in the climate system, is relevant for compartmentalizing approximately the ozone chemical-dynamical interactions, and represents a convenient way to look at the "big" global picture.

5.2.2.1 TROPICS AND MIDLATITUDES

Figure 5.1 shows the simulated future behavior of column ozone from interactive 2-D models (solid lines) and non-interactive 2-D CTMs (dashed lines). The models' hind-cast predictions are compared with observations as a way to screen the 2-D CTMs. All 2-D models show that ozone amount increases with time between 2007 and 2050. The model spread among the non-interactive 2-D CTMs for northern midlatitudes is about 3% at 2050. The WMO (2007, Chapter 6) report did not discuss how changes in N_2O and CH_4 contributed to the individual model results. Based on the estimates given above and the scenario stated in Table 5.1, it would appear that CH_4 is adding about 1% while the effect of N_2O is to decrease ozone by about 0.5% in 2050. It is also evident from the figure that the sensitivities in these models differ.

Results from the interactive 2-D models show that the ozone anomaly is larger by about 2% in 2050 and 4% in 2100. The effect at midlatitudes is larger at about 3% in 2050. This is consistent with the expected ozone increase due to cooling in the stratosphere. There is no clear indication on the effect of increased upwelling in the tropics, though this could have been masked by the ozone increase in the upper stratosphere due to the cooling.

Results from 3-D CCMs are shown in Figure 5.2. Several tests were used to identify models that successfully simulate parameters important for ozone response to halogen loading (Eyring *et al.*, 2006). Models that perform better in those tests are identified using solid lines in Figure 5.2. For our purpose, we concentrate on the three models (CCSRNIES, CMAM, and WACCM) that "earned" the solid line rating and performed the REF2 simulations from 1980 to 2050. Other models performed the REF2 simulation starting in 1990 or 2000 making it difficult to compare the ozone anomaly at 2050 to the anomaly at 1980 to determine the

All two-dimensional models used to simulate the future behavior of ozone show that ozone amount increases with time between 2007 and 2050.

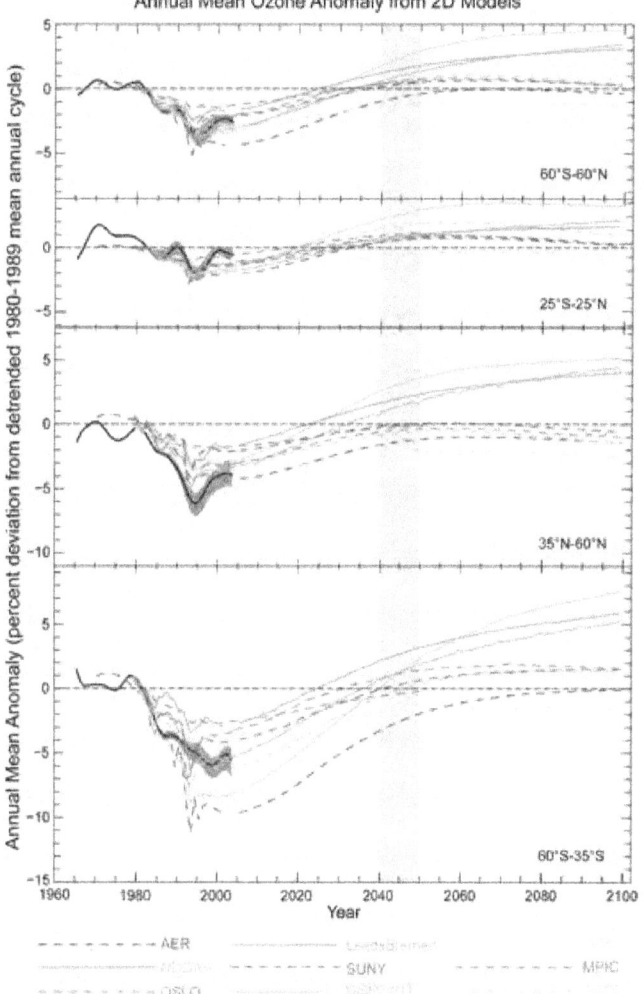

Figure 5.1 Simulated annual mean ozone anomaly from 2-D models for different latitude bands. Results from interactive models are designated by solid lines. The figure is based on Figure 6-9 in WMO (2007). See Eyring *et al.* (2006) for details on how the annual mean anomaly is computed. The black line with the grey shade represents the observed mean values and the range. The grey vertical band marks the time period when midlatitude EESC is expected to recover to the 1980 value.

Annual Mean Ozone Anomaly from 3D Models

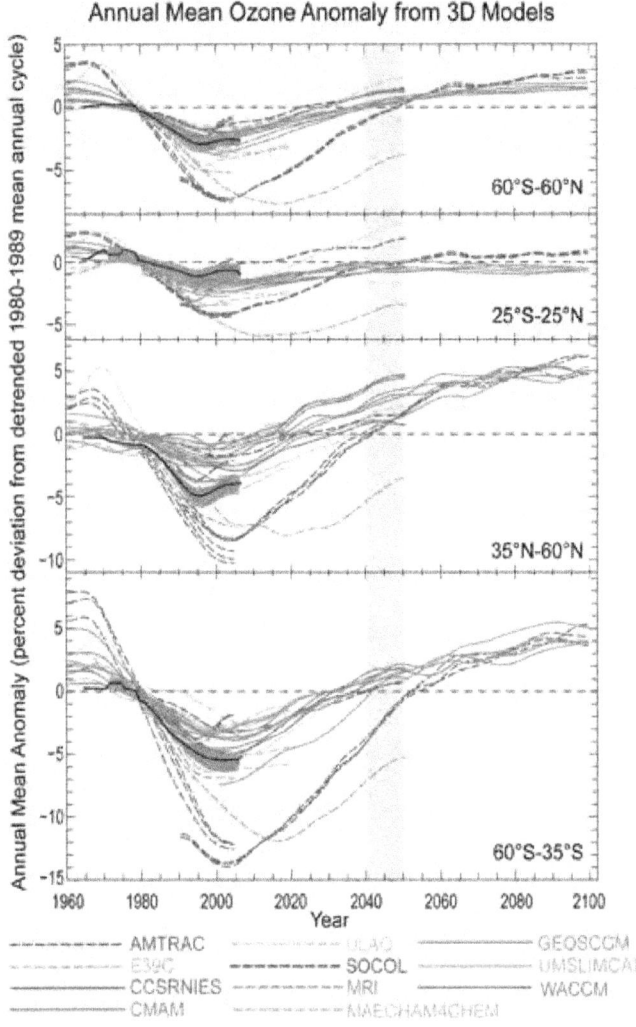

Figure 5.2 Results from 3-D CCMs. The figure is adapted from Figure 5 in Eyring *et al.* (2007) which includes additional model results computed after publication of the WMO (2007) report. See Eyring *et al.* (2006) for details on how the annual mean anomaly is computed. The solid line with the grey shade represents the observations with uncertainty. The grey vertical band marks the time period when midlatitude EESC is expected to recover to the 1980 value.

Almost all 3-D models predict that the Arctic ozone recovery date is much earlier than the Antarctic recovery date.

all between 2030 and 2040. The CMAM model presented results through 2100.

To isolate the effects of climate change, three CCMs performed a simulation where the surface concentrations of the greenhouse gases (GHGs) were kept fixed at their 1970 value (Figure 5-25 in Chapter 5, WMO, 2007). The results from WACCM show that in the absence of these GHG forcings, the MS ORD for 60°S to 60°N is around 2040 and the ozone amount in 2050 is about 1% smaller compared to the baseline scenario that includes the GHG forcings. Unfortunately, the run also kept the surface concentrations of CH_4 and N_2O fixed. Thus, the 1% effect results from both climate change and the direct chemical effects of CH_4 and N_2O.

5.2.2.2 POLAR REGION

Figure 5.3 shows the model-simulated ozone anomalies for the Arctic and the Antarctic regions. Most models show larger anomalies in the Antarctic, consistent with the fact that the temperature is colder, leading to formation of more polar stratospheric clouds (PSCs), and the vortex is more confining. Within almost all models, the Arctic ozone recovery date is much earlier than the Antarctic ozone recovery date.

The polar EESC RD is estimated to be between 2060 and 2070. We will again concentrate on the results from CCSRNIES, CMAM, and WACCM for the reason discussed in the previous section. The MS Arctic polar ORDs are 2000 for CCSRNIES, 2010 for CMAM, and 2015 for WACCM. Once the ozone anomaly reaches the 1980 value, it increases smoothly to beyond the 1960 anomaly.

MS ODR. For ozone content between 60°S and 60°N, the MS ORD are 2030 for WACCM and 2040 for CMAM and CCSRNIES. All three models show little ozone increase beyond the 1980 values in the tropics, consistent with the expectation that increase in upwelling is suppressing ozone. This is evident in the model-simulated decrease in tropical ozone below 20 hectopascals (hPa) (see Figure 6[b] in Eyring *et al.*, 2007). The MS ORDs for northern midlatitudes are 2010 for WACCM, 2020 for CMAM, and 2030 for CCSRNIES. The MS ORDs for the southern midlatitudes are

The exact time evolution of the Antarctic ozone hole is different depending on the diagnostics chosen. These include ozone amount in October, minimum ozone in September and October, ozone mass deficit, and maximum Antarctic ozone hole area between September and October. The minimum ozone value is projected not to increase until after 2010 in several models, while the decrease in ozone mass deficit in most models has occurred by 2005. If we use the ozone content poleward of 60° calculated by the three models as the metric, the MS ORDs are up to 30 years earlier. The CMAM and WACCM models produced ensemble results. The three simulations from the WACCM model produced polar ozone recovery dates between 2030 and 2040, while those from CMAM are between 2040 and 2060. The CMAM results also showed that the value for the Antarctic ozone anomaly stays closed to zero for about 20 years after the initial recovery before taking off.

5.2.3 Stages of Ozone Recovery From ODSs

The model-simulated results presented in Figure 5-25 in Chapter 5 of WMO (2007) suggest that the ozone increase between now and 2025 is largely due to the decrease in halogen loading. ODSs will remain a driver of human-caused ozone depletion up until 2040, and reductions in emissions of these chemicals represent the only known acceptable method to reduce the associated ozone depletion expected in this period. The halogen loading is expected to approach its 1980 value toward the middle of the century at midlatitudes. In the decades that follow, the effects of climate change and changes in other trace gases will determine the ozone behavior.

A good portion of Chapter 6 in the WMO (2007) report was devoted to discussion of the detection and attribution of the expected ozone recovery based on future observations. Because of interannual variability, it is not possible to identify an ozone recovery date from observation as soon as it occurs. If one were able to isolate the trend due to halogen loading from the observed ozone time series, one would expect this trend to first stop decreasing when EESC peaks (around 1997 for midlatitudes), followed by an increasing trend until the trend becomes zero again when halogen loading no longer has an effect on ozone. This is the trend-derived halogen-induced ozone recovery date (TD H-ORD). The work of Weatherhead *et al.* (2000) and Yang *et al.* (2006) clearly show that the length of observations required to detect such change depends on the quality of the data. Given the current results, it is anticipated that we should be able to confirm whether ozone is increasing due to decrease in halogen loading in the next five or six years. We are not in a position to recognize the TD H-ORD as soon as it occurs. Nonetheless, the simulations give confidence that one should be able to confirm

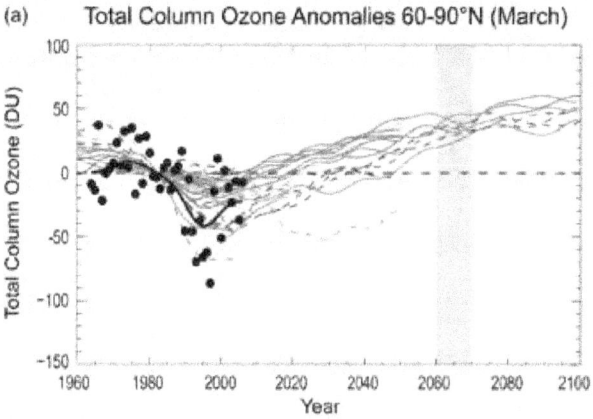

(a) Total Column Ozone Anomalies 60-90°N (March)

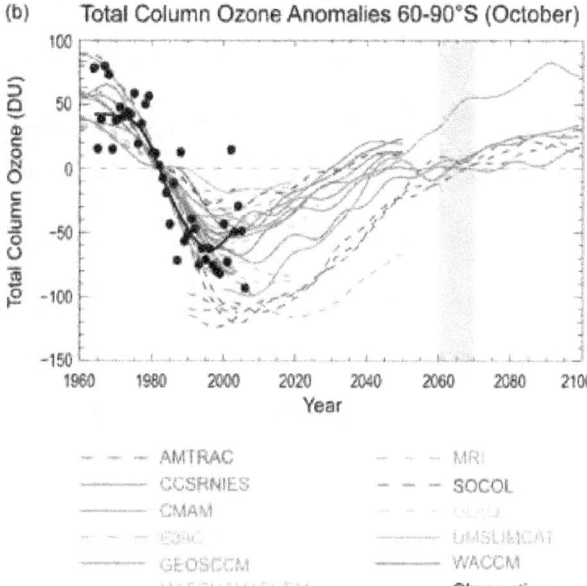

(b) Total Column Ozone Anomalies 60-90°S (October)

– – – AMTRAC	– – – MRI
——— CCSRNIES	– – – SOCOL
——— CMAM	——— ULAQ
– – – E39C	——— UMSLIMCAT
——— GEOSCCM	——— WACCM
– – – MAECHAM4CHEM	——— Observations

Figure 5.3 Zonal mean monthly ozone anomalies for the Arctic in March (panel a) and Antarctic in October (panel b). The figure is based on Figure 7 in Eyring *et al.* (2007), which was updated to include additional results after the publication of the WMO (2007) report. The observations are shown as black dots and a smooth curve representing the mean value. The grey vertical band marks the time period when polar EESC is expected to recover to the 1980 value.

the ozone recovery after the fact by waiting several years to analyze the observations and removing the interannual variability.

In the context of the rest of this chapter (Sections 5.4 and 5.5), the most important attribution issue is whether EESC is a good proxy for future ozone behavior so that one could have confidence that policy decisions based on reducing EESC would achieve the goal of similar reductions in ozone depletion. Indeed, there are concerns (*e.g.*, Hadjinicolaou *et al.*, 2005) that improper interpretation of the recent observed ozone increase after the late 1990s may give the wrong impression that the effects of halogens on ozone have been overestimated and one should relax the reduction strategy.

Chapter 6 in WMO (2007) identifies other factors that could complicate the attribution of the observed changes. These include changes in atmospheric composition of ozone-relevant compounds other than the halogens, changes in temperature and transport circulation, changes in solar cycle, and volcanic eruptions. The largest effect on the short-term (five to ten years) trend is expected to come from changes in transport circulation. The study of Yang *et al.* (2006) concluded that half of the observed increase in ozone between the late 1990s and 2005 could be attributed to changes in transport circulation in the lower stratosphere. This is not unexpected because, while EESC has stopped increasing, it essentially remained unchanged during this period. Hadjinicolaou *et al.* (2005) had a similar conclusion using a very different approach. The authors use a 3-D CTM to calculate the ozone from 1979 to 2003. The CTM uses the transport circulation from the ERA-40 analyses of the European Centre for Medium-range Weather Forecasting (ECMWF). The ozone chemistry is parameterized with the local loss frequency fixed at the 1980 conditions. The conclusion is that the ozone trends between 1994 and 2003 derived from the modeled and observed ozone agree, indicating that change in transport is the main driver in this time period. The paper also concluded that the model-calculated ozone (again with fixed loss frequency) showed a decreasing trend between 1979 and 1993, and the trend is around one-third of the trend derived from observation. More analyses (such as additional model results using full chemistry to show that the combined trend due to changes in transport and halogen loading is not significantly larger than the observed trend) are needed to support this last conclusion that changes in transport are responsible for one-third of the observed ozone trend between 1979 and 1993.

After 2050, effects from other changes (changes in CH_4, N_2O, and CO_2) will dominate. If the desire is to understand future ozone behavior beyond the effects of halogens, one must focus on the trends of the other ozone-relevant source gases and try to determine to what extent one could separate the effects on ozone in the future observations.

5.3 EXPECTED RESPONSE IN SURFACE UV

Ozone column in the atmosphere is one of many factors that affect UV at the ground. The UV community emphasizes the importance of variations in aerosol, clouds, and surface albedo on UV. The effect of ozone change on clouds through climate feedback has not been quantified at this point, but is expected to be small.

If everything else is assumed to be constant, the future UV trend will depend on the anticipated ozone change. Within the limitation that applies to EESC as a proxy for future ozone behavior, it can likewise be used as a predictor for UV. However, most UV predictions are done locally at specific latitudes, thus the relationship between EESC and typical midlatitude ozone depletion is not particularly useful due to the other factors affecting UV. In practice, model-simulated ozone changes at specific latitudes are fed into a radiative transfer model to compute the change in UV irradiance. An example of such a calculation is shown in Figure 5.4, which shows the calculated noon-time erythemal irradiance at several latitudes. Note that the recovery at the southern polar latitude occurs much later than the midlatitude values, reflecting similar behavior of midlatitude and polar ozone columns as indicated in the results from the AMTRAC model in Figure 5.2.

5.4 FUTURE SCENARIOS FOR ODSs AND THEIR REPLACEMENTS

5.4.1 Baseline Scenario
In general, the historical portion of the baseline (A1) scenario is based on the observed mixing ratio time series, while future emissions are estimated using the most current information regarding expected future demand of ODSs, future banks (the amount of a chemical that has been produced but not yet emitted or chemically altered; see Box 2.5 in Chapter 2 for a more detailed definition), and current constraints placed by the Montreal Protocol.

If all other factors remain constant, the future ultraviolet (UV) trend will depend on the anticipated ozone change.

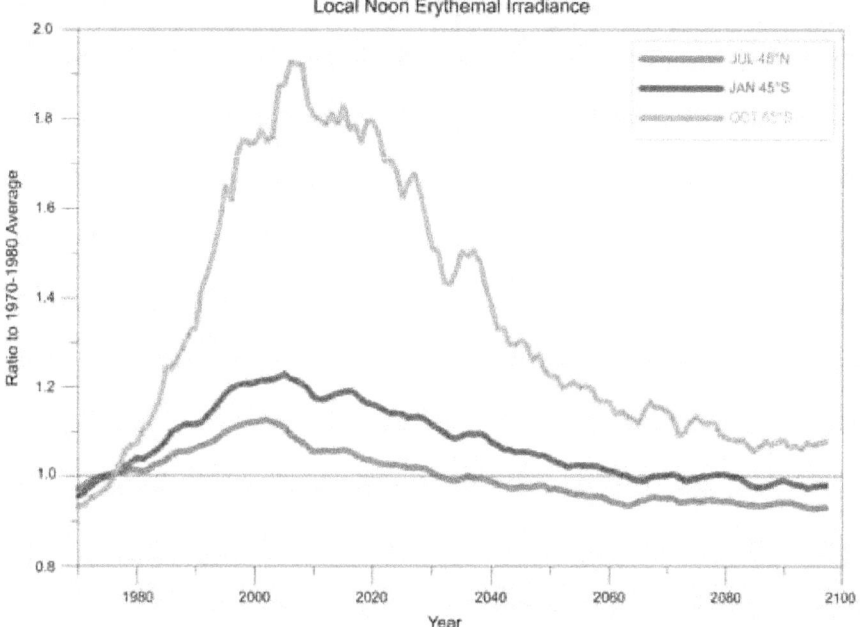

Figure 5.4 Estimated changes in erythemally weighted surface UV irradiance at local noon in response to projected changes in total column ozone as calculated by the AMTRAC CCM (see blue dashed curve in Figure 5.2) for the period 1970 to 2099, using zonal-averages in total ozone in the latitude bands 35°N-60°N, 35°S-60°S, and 60°S-90°S, and the solar zenith angle corresponding to 45°N in July, 45°S in January, and 65°S in October, respectively. At each latitude, the irradiance is expressed as the ratio to the 1970 to 1980 average. The results have been smoothed with a five-year running mean filter to remove some of the year-to-year variability in the ozone predictions in the model.

While this scenario consists of reasonable assumptions about the future, it does not represent a prediction, and future levels could be higher or lower depending on, for example, future policy actions and consumer choices. However, it represents a useful projection that is used to examine the sensitivity of ODS abundances to choices concerning future production, banks, and emissions.

The mixing ratios used to calculate the historical emissions are obtained primarily from atmospheric observations made by the NOAA Earth System Research Laboratory/ Global Monitoring Division (ESRL/GMD) (formerly Climate Monitoring and Diagnostics Laboratory, [CMDL]), the Advanced Global Atmospheric Gases Experiment (AGAGE), and the University of East Anglia (for halon-1211). South Pole firn observations are also considered for methyl chloride (CH_3Cl) and methyl bromide (CH_3Br) emissions before 1996. A box model is used to determine the emissions of the species for each year through 2005 using the observed mixing ratio time series and its

current best estimate of the steady state global lifetime. Hence, when the same box model and lifetimes are used to calculate the surface mixing ratios from the derived emissions, they produce mixing ratios in the baseline scenario that are exactly equal to the observationally based time series given in Table 8-5 of WMO (2007). The same box model and lifetimes are used to derive the mixing ratio of each species after 2005 based on projected emissions.

Projections of future demand and sizes of banks are taken from IPCC/TEAP (2005), and play a major role in the calculation of future emissions. Details relating to the production and emission projections can be found in WMO (2007). The use of future demand and bank sizes from IPCC/TEAP (2005) in WMO (2007) represents an important departure from the approach used in previous WMO reports. Previously, the evolution of the estimated bank sizes was calculated solely using the difference between estimated annual production and emission. This approach had the potential to lead to accumulating large errors in the bank

Bank estimates of ozone-depleting substances are based on inventories of equipment containing these substances, an approach called the "bottom-up" method.

sizes because the bank often represents a small difference between the two relatively large production and emission values. The IPCC/TEAP (2005) bank estimates, however, are based on inventories of equipment, an approach often referred to as a "bottom-up" method. Hence, these estimates are independent of systematic errors in production or emission. It is believed that this new approach has led to better projections of future emissions.

Comparisons between future emissions projections of WMO (2003) and WMO (2007) demonstrate that the most substantial differences arise for CFC-11, CFC-12, carbon tetrachloride (CCl_4), and the HCFCs. The increase in the CFC emissions in WMO (2007) is primarily due to larger bank estimates of the bottom-up approach than were estimated by WMO (2003). The greatest HCFC emission difference is for HCFC-22 and is due to the substantially larger estimated future consumption of this compound by Article 5(1) countries. Current and projected future carbon tetrachloride emissions are now estimated to be higher than those projected in WMO (2003) based on observed mixing ratios consistent with a smaller decrease in emissions over the last few years and the continued inability to account for all CCl_4 emissions. The resulting differences in mixing ratios are discussed in Section 5.4.3.

5.4.2 Alternate Scenarios

Alternative scenarios and test cases were examined in WMO (2007) to quantify the relative effects of making various production and/or direct emissions reductions on EESC. Three cases for different ODS groups are designed to address three issues: (1) no future emission; (2) no future production; and (3) eliminations of the 2007 bank. Results from the "no future emission" case provide the future mixing ratios due to the decay of the ODSs already in the atmosphere only. This case represents the greatest theoretically possible reduction in the future atmospheric burden of the particular compound (short of processing the air to remove the ODSs). The "no future production" case quantifies the importance of new production relative to future emissions, while the "bank elimination" quantifies the benefit of the one-time sequestration and destruction of the 2007 global bank for future

emissions. Additional cases are presented here that examine the effect of recovering and destroying the total estimated U.S. bank and the U.S. accessible bank in 2009. Estimates of these bank sizes and the technique used by the U.S. Environmental Protection Agency (EPA) to calculate these estimates are discussed in Chapter 2.

WMO (2007) also examined three alternative cases involving CH_3Br. Two cases involved removing quarantine and pre-shipment uses from 2015 onward and continuing Critical Use Exemptions at 2006 levels into the future. The third case explored the importance of the assumption that the 1992 anthropogenic emissions represented 30% of the total. Recent mixing ratio observations have suggested that this might be an overestimate with a more accurate percentage falling somewhere between 20% and 30%. These results are presented later in Section 5.5 and in Table 5.2.

A scenario based on the mitigation scenario described in IPCC/TEAP (2005) is also examined to quantify the effect of this carefully considered set of future policy options. The mitigation scenario only has a substantial effect on the bank of HCFC-22 in the scenario considered here.

After the WMO (2007) report and IPCC (2007) reports were written, the Parties to the Montreal Protocol voted to strengthen the HCFC regulations on both Article 5(1) and non-Article 5(1) countries. Approximations for the effect of this strengthening are discussed in Section 5.5.1.1.

5.4.3 Time Series of Source Gases

The mixing ratios of the current baseline scenario are compared with those of the WMO (2003) baseline scenario, and the "no future production" and "no future emission" test cases in Figure 5.5. The differences between the WMO (2007) and WMO (2003) baseline scenarios are apparent for several gases, with the differences for HCFC-22 particularly apparent. More modest, but also important are the differences for CFC-11 and CFC-12. The HCFC-22 difference is due to the increase in the expected future consumption of Article 5(1) countries of the Montreal Protocol, while the

increase in the CFCs is due to the larger bank
size estimates of IPCC/TEAP (2005). HCFC-
141b, HCFC-142b, and halon-1301 show reduced
mixing ratios in the short term compared to
WMO (2003) owing to the reduced observed
growth rates between 2001 and 2004 and the
expectation of lower future emissions.

The importance of future projected production
and bank sizes to future emissions is also
apparent for the various compounds. For
example, the "no future production" curve
for CFC-12 is only slightly different from the
baseline curve; hence the bank of CFC-12 is
expected to dominate future emissions, with
its effect represented by the difference between
the "no future emission" and "no future
production" curves. The relative importance of
future production compared to the amount in
the banks varies strongly among the ODSs, with
the future abundances of CFC-11 depending
primarily on its bank size and HCFC-22 future
abundances depending primarily on future
production. No bank is considered in the future
projections of methyl chloroform (CH_3CCl_3),
CCl_4, and CH_3Br.

5.5 CHANGES IN INTEGRATED EESC AND RADIATIVE FORCING

5.5.1 Time Series of EESC

5.5.1.1 MIDLATITUDES

The evolution of ODS mixing ratios cannot,
by themselves, be used to accurately quantify
the ozone destruction due to those ODSs. The
established relationship between stratospheric
ozone depletion and inorganic chlorine and
bromine abundances suggest that the temporal
evolution of inorganic chlorine- and bromine-
species in the stratosphere is an important
indicator of the potential damage of human
activity on the health of stratospheric ozone.
Equivalent Effective Stratospheric Chlorine
(EESC) was developed to relate this halogen
evolution to tropospheric source gases in a
simple manner (Daniel *et al.*, 1995; see also Box
2.7 in Chapter 2). This quantity sums ODSs,
accounting for a transit time to the stratosphere,
for the number of halogen atoms in the ODS,
for the greater per-atom potency of stratospheric
bromine (Br) compared to chlorine (Cl) in its
ozone destructiveness with a constant factor (α),
and also includes the varying rates with which

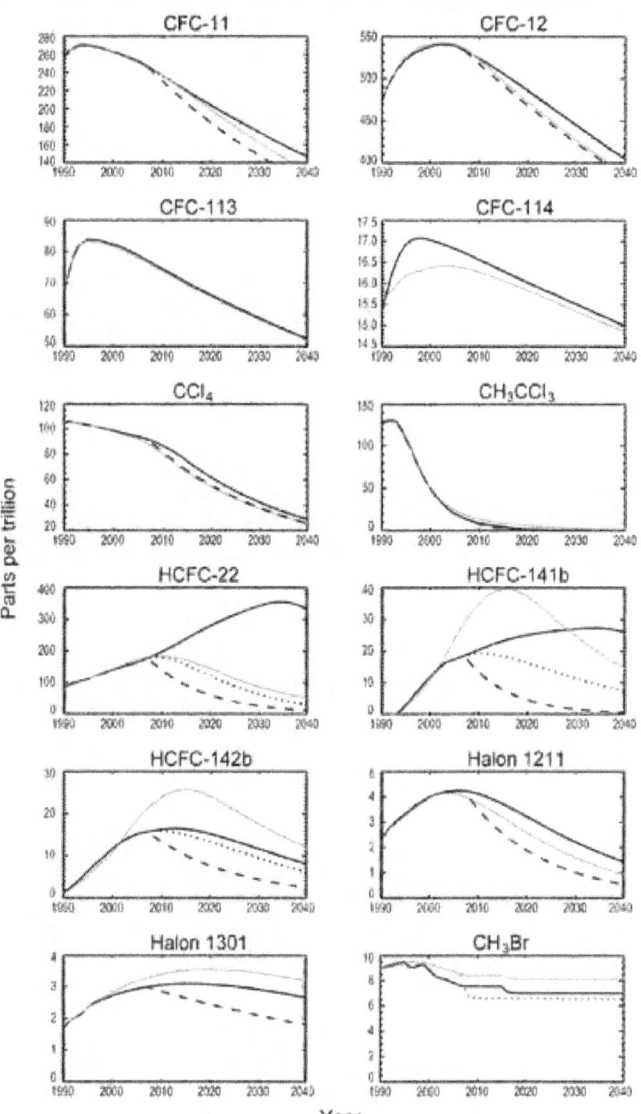

Halocarbon Abundances for Different Emission Scenarios

Parts per trillion

Year

Figure 5.5 Mixing ratio comparisons of the WMO (2007) baseline scenario (solid black) with the baseline scenario from WMO (2003) (green), the "no future emission" test case (dashed), and the "no future production" case (dotted curve). The figure is based on Figure 8-2 in WMO (2007). Note that different vertical scales are adopted for sub panels and some of the plotted values do not start from zero. For several of the gases, the solid black curve obscures the dotted or dashed curves.

Cl and Br will be released in the stratosphere
from different source gases. EESC has been
used as a proxy for the effect of human-
produced ODS abundances on future ozone
depletion (WMO, 1995, 1999, 2003, 2007).
The values for midlatitude EESC discussed
here were calculated for WMO (2007) from the
global averaged surface mixing ratios for the
ODSs, using a constant lag time of three years

and release factors given in the same report. Recent development on how to apply EESC to polar ozone and refinements in using the mean age of air will be discussed in Sections 5.5.1.2 and 5.5.2, respectively.

The relative contribution of various ODSs and ODS groups to midlatitude EESC are shown as a function of time from 2000 to 2100 on the left-hand side of Figure 5.6. The prominent role of CFCs today and into the future is apparent. The slow decline of the contribution from CFCs is primarily due to the relatively long atmospheric lifetimes of the species in this group of compounds and what is already in the atmosphere, and not to continued emission, although continued emission does play a small role. The importance of the halons and CH_3Br, all bromine-containing source gases, is also clear even though their atmospheric concentrations are substantially smaller than those of the chlorine-containing ODSs. This is because stratospheric bromine is much more effective per atom than chlorine for stratospheric ozone destruction. As stated in Chapter 2, WMO (2007) has estimated that an atom of bromine is 60 times more effective than an atom of chlorine for global stratospheric ozone destruction. The lower panels show the change in EESC due to the elimination of production and emission for CFCs, HCFCs, halons, and CH_3Br.

In the past, EESC estimates have been used to evaluate various ODS emissions scenarios primarily using two metrics. They are: (1) a comparison of the times when EESC returns to 1980 levels, the EESC RD; and (2) the relative integrated changes in EESC between 1980 or the current time and the corresponding EESC RD. Figure 5.7 demonstrates that the return of midlatitude EESC to the 1980 level is currently expected to occur around 2049 for the baseline scenario, five years later than projected in WMO (2003). This later return was primarily ascribed to higher estimated future emissions of CFC-11, CFC-12, and HCFC-22. The increase in CFC emissions is due to larger estimated current bank sizes, while the increase in HCFC-22 emissions is due to larger estimated future production. The soonest that a complete theoretical elimination of emissions could lead to a return to 1980 levels is 2034. Elimination

of all future ODS production is expected to lead to a return to 1980 EESC levels in 2043, while an elimination of the 2007 bank is expected to lead to a return in 2041.

A detailed partitioning of the effects of reductions in the various ODS groups is shown in Table 5.2. The years when EESC is expected to return to 1980 levels are also included in the table for midlatitudes and for the Antarctic vortex. The Antarctic ozone response to EESC will be discussed in more detail in Section 5.5.1.2. The table illustrates that the elimination of the future emissions of CFCs, HCFCs, and halons represents the greatest potential for reducing future ozone depletion. To accomplish this elimination for CFCs and halons, banks would have to be captured and destroyed because future emission is expected to be dominated by the release from banks. For HCFCs, future production plays a larger role in future emission than do the current banks, so emissions from both banks and future production would have to be eliminated. The technical difficulty and expense involved with capturing banks depends on the nature of the banks, while the feasibility of reducing future production will depend on replacement options for the pertinent applications. More details concerning the nature of the various ODS uses and bank types and the options available for reducing future ODS emissions can be found in other reports, including IPCC/TEAP (2005) and UNEP/TEAP (2007). It should also be recognized that these full bank recovery and zero production and emission test cases shown in Table 5.2 are meant as hypothetical cases against which more realistic scenarios can be compared. This procedure is used in Section 5.6 to evaluate the significance of the United States ODS banks.

The results for the scenario representing the IPCC/TEAP (2005) mitigation scenario are not shown in Table 5.2, but this scenario leads to an EESC response that is approximately 20% of the zero-emission case for the HCFCs, due primarily to actions that would reduce the HCFC-22 bank emission.

Future emissions of CH_3Br also have the potential to be as important as each of these three classes of compounds. The continuation

Table 5.2 Comparison of scenarios and cases[a]: the year when EESC drops below the 1980 value (i.e., EESC RD) for both midlatitude and polar vortex cases, and integrated EESC differences (midlatitude case) relative to the baseline (A1) scenario.

Scenario	Percent Difference in integrated EESC relative to baseline scenario for the midlatitude case		Year (x) when EESC is expected to drop below 1980 value	
	Midlatitude			Antarctic Vortex[b]
	$\int_{1980}^{x} EESC\ dt$	$\int_{2007}^{x} EESC\ dt$		
Scenarios				
A1: Baseline Scenario			2049	2065
Cases[a] of zero production from 2007 onwards of:				
All ODSs	−8.0	−17.1	2043	2060
CFCs only	−0.1	−0.3	2049	2065
Halons only	−0.2	−0.5	2049	2065
HCFCs only	−5.5	−11.8	2044	2062
Anthropogenic CH$_3$Br only	−2.4	−5.1	2048	2063
Cases[a] of zero emissions from 2007 onwards of:				
All ODSs	−19.4	−41.7	2034	2050
CFCs only	−5.3	−11.5	2045	2060
CH$_3$CCl$_3$ only	−0.1	−0.2	2049	2065
Halons only	−6.7	−14.4	2046	2062
HCFCs only	−7.3	−15.7	2044	2062
CCl$_4$ only	−1.3	−2.9	2049	2065
Anthropogenic CH$_3$Br only	−2.4	−5.1	2048	2064
Cases[a] of full recovery of the 2007 banks of:				
All ODS	−12.9	−27.8	2041	2057
CFCs only	−5.2	−11.3	2045	2060
Halons only	−6.7	−14.3	2046	2062
HCFCs only	−1.9	−4.1	2048	2065
CH$_3$Br sensitivity:				
Same as A1, but CH$_3$Br anthropogenic emissions set to 20% in 1992[c]	3.1	6.6	2051	2068
Same as A1, but zero QPS production from 2015 onwards	−1.5	−3.2	2048	2064
Same as A1, but Critical Use Exemptions continued at 2006 level	1.9-2.2	4.0-4.7	2050	2067

[a] Importance of ozone-depleting substances for future EESC were calculated in the hypothetical "cases" by setting production or emission of all or individual ODS groups to zero in 2007 and subsequent years or the bank of all ODS or individual ODS groups to zero in the year 2007 alone. These cases are not mutually exclusive and separate effects of elimination of production, emissions and banks are not additive.

[b] Calculated using a lag time of six years and the same release factors as in midlatitudes.

[c] In the baseline scenario this fraction was assumed to be 30% in 1992 with a corresponding emission fraction of 0.88 of production. In this alternative scenario an anthropogenic fraction was assumed to be 20% with an emission fraction of 0.56 of production. In both scenarios the total historic emission was derived from atmospheric observations and a lifetime of 0.7 years.

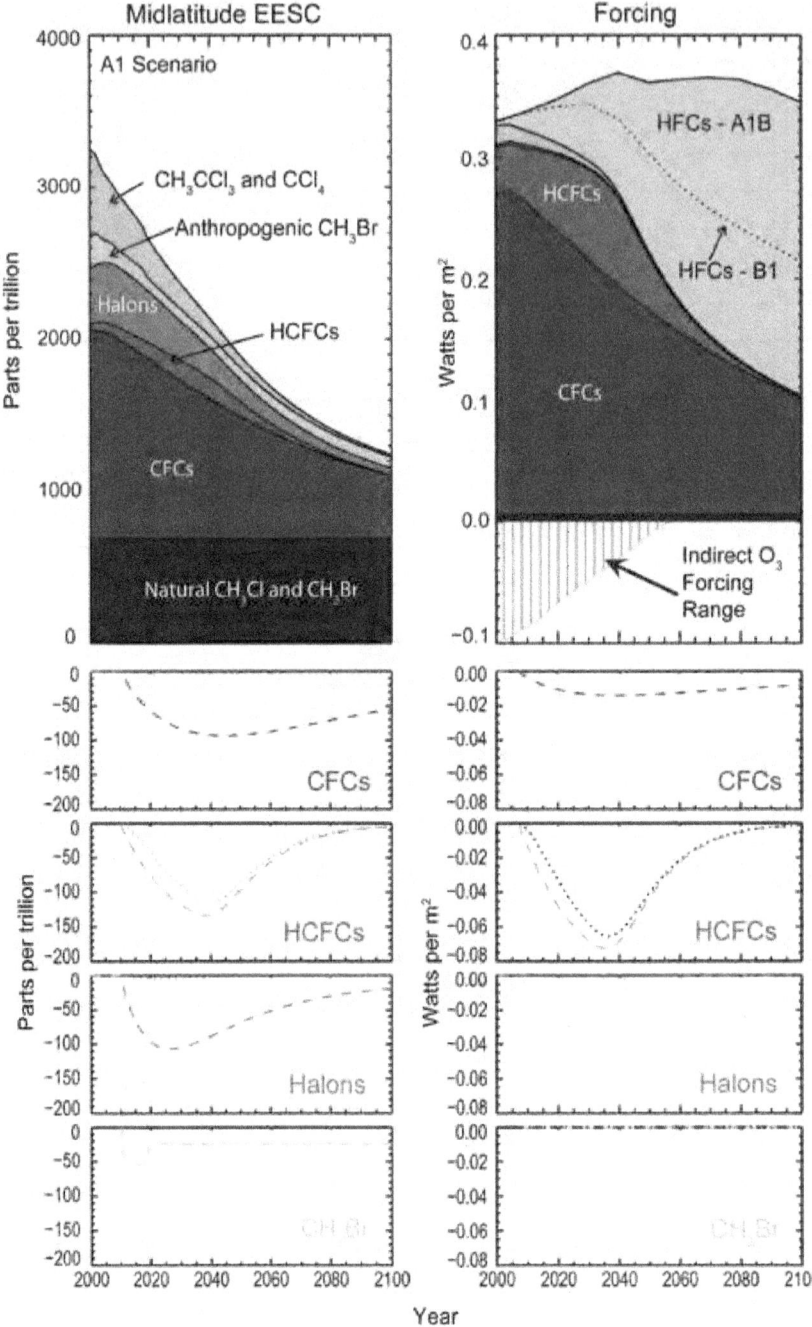

Figure 5.6 EESC and direct radiative forcing estimates from 2000 to 2100 for the A1 baseline scenario in WMO (2007) (upper panels), and expected decreases relative to the baseline scenario due to a cessation of emission (dashed curves) and production (dotted curves) in 2007 for CFCs, HCFCs, halons, and anthropogenic CH₃Br. The figure is adapted from Figure 8-5 in WMO (2007) except for the addition of the indirect forcing in the upper right hand panel. Note the difference in the vertical scales between the top panel and the bottom four panels. The "no production" curve for CFCs and halons lies almost on the zero line, indicating that future productions of these ODSs play a very small role in the baseline scenario. In contrast, the contribution from the HCFCs is mostly due to future productions. The "no emission" and "no production" curves are identical for CH₃Br because no bank was considered in its projections. The HFC forcing is shown for the B1 and A1B SRES scenarios. The indirect forcing due to ozone depletion caused by ODSs is included for comparison, but should be considered only a rough approximation owing to reasons discussed in the text of this chapter and of Chapter 2.

of the Critical Use Exemption at the 2006 level and the continuation of QPS uses both have a substantial impact on global EESC.

In late 2007, the Montreal Protocol HCFC restrictions for both production and consumption were strengthened, partly in response to the renewed awareness of the importance of HCFCs to climate forcing in addition to ozone depletion. While restrictions were tightened for both Article 5(1) and non-Article 5(1) countries, the changes for the Article 5(1) countries are much more significant for stratospheric ozone. In Figure 5.8, the former Protocol HCFC restrictions for Article 5(1) countries are compared to the newly adopted ones, as well as to the United States proposal that contributed to the strengthened restrictions. An additional curve is also shown that represents the closest scenario calculated in UNEP/TEAP (2007). For the TEAP scenario, it is estimated that integrated EESC is reduced by 2.6% and 5.6%, respectively, for the integration from 1980 to the return of EESC to 1980 levels and from 2007 to the return to 1980 levels. This is a substantial reduction even when compared to the zero emissions case for HCFCs in Table 5.2. The baseline HCFC emissions are slightly higher in UNEP/TEAP (2007) than those assumed in WMO (2007), making the effect of HCFC reductions correspondingly slightly higher. This TEAP report also examines other "practical options" that could be usefully employed to reduce future emissions of HCFCs and other ODSs. These include emission reduction measures during the use phase of applications and equipment, design and material section alternatives, end-of-life management, and early retirement of equipment. These measures were submitted by the Parties to the Montreal Protocol and organized at the 26th Open-ended Working Group Meeting of the Parties to the Montreal Protocol. The TEAP report finds that a combination of earlier HCFC phase-out described above with these additional "practical measures" leads

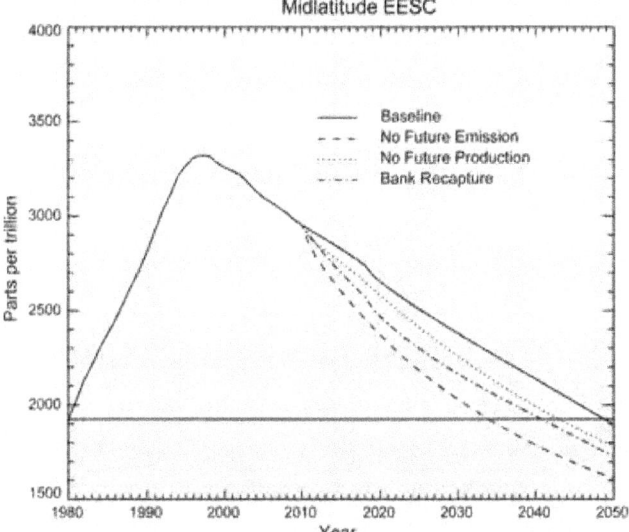

Figure 5.7 Midlatitude EESC estimates from 1980 through 2050 for the baseline scenario and the three comparative test cases considered in WMO (2007). The horizontal line represents the 1980 EESC level. This figure is redrawn from Figure 8-4 of WMO (2007).

to an integrated EESC reduction of 7.4% and 16.0% percent, respectively, for the integration from 1980 to the return of EESC to 1980 levels and from 2007 to the return to 1980 levels.

5.5.1.2 POLAR REGIONS

Compared to midlatitude EESC, Arctic EESC is less useful as a proxy for polar ozone depletion because interannual variability in meteorology has a much larger impact on the ozone response

For midlatitudes, the test case involving elimination of all future ozone-depleting substance production is expected to lead to 1980 Equivalent Effective Stratospheric Chlorine levels in 2043.

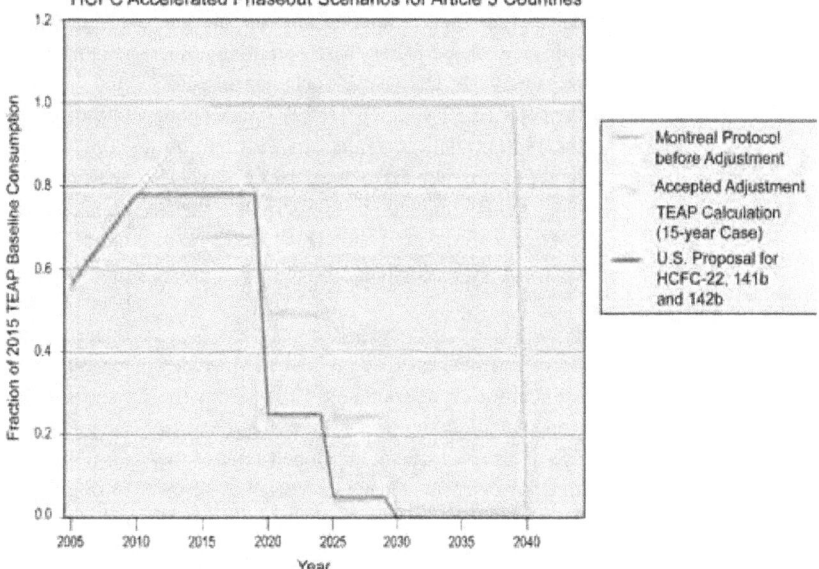

Figure 5.8 Comparison of alternate scenarios for future emissions of HCFCs.

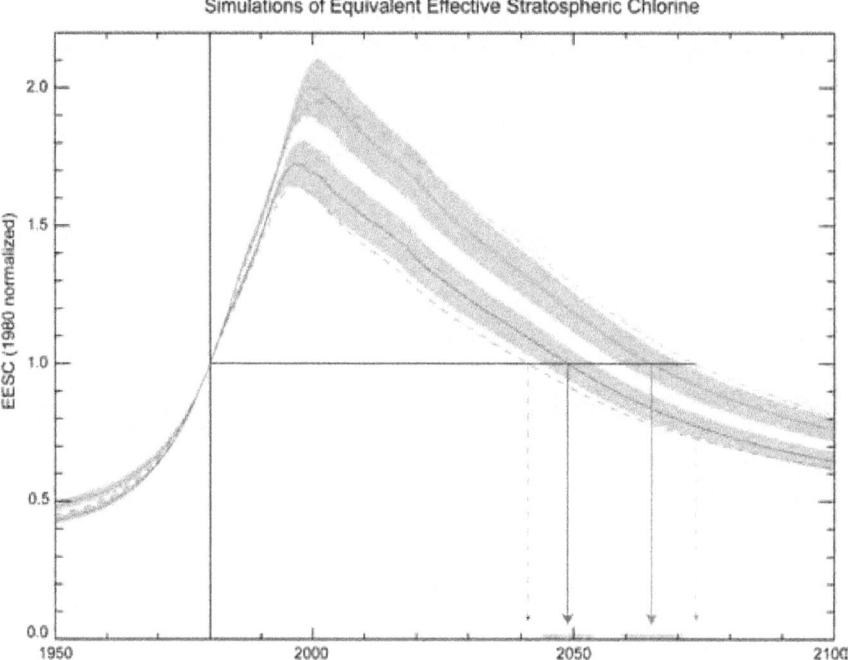

Figure 5.9 Comparison of EESC values calculated using a lag time and fixed fractional release values (solid) vs. those calculated using a mean age with an age spectrum and fractional release values parameterized as functions of mean age (dashed). The blue curves are for midlatitudes and are calculated with a mean age and lag time of three years. The curves for the polar region are in red and are calculated with a mean age and lag time of six years.

to inorganic halogen loading. In the core of the Antarctic vortex during early spring, the interannual variability is smaller, suggesting that EESC could provide a useful proxy for ozone hole recovery (Newman *et al.*, 2006). The far right column in Table 5.2 shows the results calculated using a time lag of six years and the same release factors based on midlatitude measurements. Because of the larger time-lag, the polar EESC value in 1980 is smaller than the 1980 midlatitude value. In addition, the larger lag time also makes the polar EESC value larger than the midlatitude EESC value in 2050. Therefore, the polar EESC RD is 15 to 17 years later than the midlatitude EESC RD.

5.5.2 EESC and Mean Age of Air
Previous EESC calculations have not included a distribution of transport times from the tropopause into the stratosphere (the so called age-of-air spectrum) or any dependence of the fractional chlorine release value on the age-of-air. Newman *et al.* (2006) reformulated EESC to account for both an age-of-air spectrum and age-dependent fractional release rates. Those

results were discussed in Box 8-1 of WMO (2007). In this section, we will summarize how the EESC estimates derived from Newman *et al.* (2006) differ from the results in Section 5.5.1.2.

The dashed lines of Figure 5.9 show EESC for mean ages of three years (blue) and six years (red) as estimated using the Newman *et al.* (2006) technique, while the solid lines show EESC for constant age shifts (time-lags) of three years (blue) and six years (red). The solid lines duplicate the EESC used in WMO (2007). The main difference in the recovery dates between the two methods in each case (midlatitude and polar) is a result of the differences in fractional release values. In the WMO (2007) case, the release factors are fixed values, while the release factors in the Newman *et al.* (2006) curves are mean-age dependent. Note that the Newman *et al.* (2006) release factors at midlatitudes lead to smaller EESC values relative to 1980 levels than the values used in WMO (2007) and that an earlier EESC RD is projected. In contrast, the Newman *et al.* (2006) release factors at the pole for the six-year mean age are larger than the WMO (2007) values, and result in a later EESC RD.

Newman *et al.* (2006) raised the issue of the uncertainty in predicting the EESC RD associated with the choice of mean age and release factors to represent midlatitude and polar conditions. While the use of different mean ages would change the absolute timing of the recovery for the baseline case and other test cases, we are reasonably confident that it would not change the conclusion about the relative effects of different test cases.

5.5.3 Time Series of Radiative Forcing

To adhere to the requirements of the Montreal Protocol, several courses of action have been adopted, including not-in-kind replacements of ODSs and changes in operations that reduce emissions. Applications that previously used CFCs are now performed with CFC replacements, HCFCs and HFCs, with HFCs likely to play a larger role in the future as HCFCs are phased out by the Montreal Protocol. Because HFCs contain no chlorine, bromine, or iodine, they do not destroy stratospheric ozone and therefore are not considered in WMO (2007). Furthermore, because future HFC emissions and production are not regulated by the Montreal Protocol, as are HCFCs, future projections of HFC concentrations are generally much more uncertain than those of ODSs and are heavily dependent on future economic growth assumptions and policy decisions. The forcing from HFCs will be included here as part of the discussion. However, it should be pointed out that the replacement strategy may also involve changes in other greenhouse gas emissions associated with the life cycle analyses (IPCC/TEAP, 2005). The change in forcing associated with those greenhouse gases is not included in the discussion.

Once the mixing ratio time series has been determined, it is a simple matter to estimate the future direct radiative forcing due to the various compound groups from the radiative efficiencies of each ODS (WMO, 2007). This forcing time series, calculated by multiplying the mixing ratio time series by the radiative forcing efficiencies of the particular ODSs, is shown on the right-hand side of Figure 5.6.

Direct radiative forcing from ODSs and HFC replacement chemicals is approximately 0.34 W per m^2 in 2005 and is expected to stay below 0.4 W per m^2 through 2100 (according to the WMO (2007) scenario for ODSs and SRES A1B and B1 scenarios for HFCs). This is to be compared with forcing from CO_2 of 1.66 W per m^2 in the 2005 atmosphere, increasing to as high as 5 W per m^2 by 2100 for the SRES A1B scenario. The continued importance of the CFCs, along with the importance of the HCFCs, are perhaps the most striking features of this figure. The direct forcing contributions of the halons and CH_3Br are small because of their small atmospheric mixing ratios. The effect of the bromocarbons on ozone depletion is enhanced because of the higher per-atom efficiency of bromine compared to chlorine in destroying ozone; such a chemically-caused enhancement does not apply to radiative forcing.

The potential reduction in direct forcing due to the elimination of future production and emission is shown for CFCs, HCFCs, halons, and CH_3Br in the lower four panels of Figure 5.6. Overall, direct forcing from CFCs is expected to decrease to 0.1 W per m^2 by 2100. Direct forcing from HCFCs and other ODSs are expected to be negligible by 2100. It is evident that elimination of future HCFC emission has the largest effect on radiative forcing of the ODSs among the test cases considered here. This peak forcing reduction of almost 0.07 W per m^2 in 2040 represents slightly less than 5% of the CO_2 radiative forcing in 2000 and less than half of the N_2O radiative forcing in 2000. In comparison, elimination of future CFC emissions will bring about a reduction of 0.015 W per m^2 in 2040.

The forcing of the HFCs, generally used as replacements for the ODSs, are included in the figure as the orange shaded region for the SRES (Nakićenović *et al.*, 2000) A1B scenario. The line within the orange region represents the alternate forcing of the HFCs in the B1 SRES scenario. Atmospheric observations suggest that the 2000 forcing due to the HFCs is slightly larger than that estimated by the SRES scenarios, primarily due to the higher abundances of HFC-23 observed. Projected forcing from HFCs is 0.15 W per m^2 and 0.24 W per m^2 in 2050 and 2100, respectively, for

To adhere to the requirements of the Montreal Protocol, several courses of action have been adopted, including not-in-kind replacements of ozone-depleting substances and changes in operations that reduce emissions.

the SRES A1B scenario, while other scenarios indicate that it will be lower.

The direct forcing of the ODSs provides only a partial story of their total radiative forcing. Their destruction of ozone leads to an additional, indirect forcing that is thought to be significant but remains highly uncertain. An estimate of the indirect forcing from the ODSs is shown in Figure 5.6 as the red hatched region. It represents an uncertainty range of ±100% around the best estimate for the period 1979-1998 (*i.e.*, -0.05 ± 0.05 W per m^2), taken from IPCC (2007). While the ozone forcing itself is highly uncertain, additional uncertainties exist that are not quantified here associated with, for example, the simplifying assumption of a linear relationship between ozone depletion and EESC above the 1980 threshold and even the amount of the observed ozone trend that is due to ODSs. The central forcing value (-0.05 W per m^2) suggests that ozone depletion offsets about one-sixth of the total direct ODS forcing around 2000. This figure also shows that the indirect forcing will gradually diminish in the coming decades, and is expected to return to near zero around 2050 as ozone becomes insensitive to the level of ODSs. This results from the assumption that the EESC value in 1980 represents a threshold, below which ozone depletion does not respond to changing EESC levels. While such a picture is likely imperfect, global ozone data do suggest that the response of ozone to EESC may have changed at this EESC level. Finally, there have been studies suggesting that ozone radiative forcing may lead to a substantially different temperature response than does the same radiative forcing change from CO$_2$ (Joshi *et al.*, 2003 and references therein). Such a situation would imply that direct and indirect radiative forcing comparisons, as shown in Figure 5.6, could be misleading in estimating climate response. Additional details concerning the indirect forcing of the ODSs due to ozone depletion can be found in Chapter 2, particularly in Box 2.2. Accurate quantification of this forcing and of indirect GWPs currently represents a scientific gap due to the significance of the associated uncertainties.

In addition to direct forcing by ozone-depleting substances, the destruction of ozone by ozone-depleting substances also leads to an indirect forcing that is thought to be significant but remains highly uncertain.

5.6 UNITED STATES CONTRIBUTIONS TO EESC AND TO RADIATIVE FORCING BY ODSs AND THEIR REPLACEMENTS

Because of the long-lived nature of the ODSs, EESC and the radiative forcing arising from emissions of these compounds should be thought of as global quantities. This allows the contribution to midlatitude EESC and global radiative forcing to be apportioned to individual countries if their emissions are accurately known. As discussed in Chapter 2 in this Report, ODS production and consumption amounts for the United States are reported to the United Nations Environment Programme (UNEP), as required by the Montreal Protocol (UNEP, 2007). Data are also compiled by the Alternative Fluorocarbons Environmental Acceptability Study (AFEAS, 2007) for certain ODSs and for HFC-134a, although the amount reported to AFEAS has represented a smaller fraction of global production in the last decade or so when compared with the UNEP data. A discussion of the comparison of the data from these two compilations with observed atmospheric mixing ratio observations is provided in Chapter 2. Also in response to a requirement associated with the United States being a signatory to the Montreal Protocol, the U.S. EPA uses a vintaging model to estimate annual emissions from the estimated production and consumption values after 1985. Chapter 2 discusses the results from the U.S. EPA's vintaging model through the past and the assumptions made to estimate United States emissions prior to 1984. Here, we use these assumptions along with the U.S. EPA's projections to estimate future levels of source gases attributable to the United States and their contributions to both EESC and radiative forcing.

5.6.1 Contribution to EESC

It is useful to note that the EESC in 2030 will be 2400 ppt, with about one-third from the natural CH$_3$Cl and CH$_3$Br. For the remaining two-thirds attributed to manmade emissions, about 15% is due to emissions prior to 1975, and about 20% is due to emissions between 1975 and 1984. The contributions from United

Table 5.3 Comparison of global and United States bank elimination projections in terms of integrated EESC and EESC RD. The global test cases are taken from WMO (2007) and consider elimination of the global bank in 2007. U.S. cases assume elimination of the full U.S. bank, or the accessible U.S. bank in 2009.

Scenario	Percent Difference in integrated EESC relative to baseline scenario for the midlatitude case		Year (x) when EESC is expected to drop below 1980 value	
	Midlatitude			Antarctic Vortex
	$\int_{1980}^{x} EESC\, dt$	$\int_{2007}^{x} EESC\, dt$		
Scenarios				
A1: Baseline Scenario			2048.9	2065
Cases of full recovery of the 2007 banks[b] of:				
B0: All ODS (global)	−12.9	−27.8	2040.8	2056.7
CFCs (global)	−5.2	−11.3	2045.1	2060.4
CFC-11 (U.S., accessible)	−0.0[a]	−0.0[a]	2048.1	2064.1
CFC-12 (U.S., accessible)	−0.0[a]	−0.1	2048.9	2065.0
CFC-11 (U.S., total)	−1.1	−2.3	2048.1	2064.1
CFC-12 (U.S., total)	−0.3	−0.7	2048.7	2064.8
Halons (global)	−6.7	−14.3	2045.7	2062.0
halon-1211 (U.S., accessible)	−0.1	−0.2	2048.9	2065.0
halon-1301 (U.S., accessible)	−0.3	−0.6	2048.7	2064.8
halon-1211 (U.S., total)	−0.1	−0.2	2048.9	2065.0
halon-1301 (U.S., total)	−0.3	−0.6	2048.7	2064.8
HCFCs (global)	−1.9	−4.1	2048.4	2064.8
HCFC-22 (U.S., accessible)	−0.3	−0.6	2048.8	2065.0
HCFC-22 (U.S., total)	−0.3	−0.7	2048.8	2065.0
HCFC-141b (U.S., total)[c]	−0.4	−0.8	2048.8	2065.0
HCFC-142b (U.S., total)[c]	−0.1	−0.1	2048.9	2065.0

[a] Values reported as −0.0 are smaller than 0.05% in magnitude.
[b] Note that the U.S. numbers are for 2009 banks.
[c] Accessible bank values for HCFC-141b and HCFC-142b are not provided because the U.S. EPA estimates a zero accessible bank size for these compounds.

States emissions to the loading in 2030 due to manmade emissions are 4.5 to 9% from United States pre-1975 emissions, 2 to 9% from United States emissions between 1975 and 1984, and 9 to 19% from United States emissions after 1985. Summing the contributions, we estimate that in 2030 the midlatitude anthropogenic EESC amount resulting from United States emissions represents about 15 to 37% of the EESC amount resulting from all global anthropogenic emissions.

5.6.2 Contribution to Radiative Forcing by ODSs and Their Replacements

The United States' contribution to global direct radiative forcing from ODSs, HFCs, and PFCs is expected to be 19 to 41% by 2030. The lower end of this range remains roughly constant until 2030, while the upper end gradually declines from about 47% in 2010. As was done for the previous EESC contribution results, these forcing estimates are calculated only considering anthropogenic emissions. Future United States perfluorocarbon (PFC) and sulfur hexafluoride (SF_6) emissions, which are projected to contribute very little to future

The United States' contribution to global direct radiative forcing from ODSs, HFCs, and PFCs is expected to be 19 to 41% by the year 2030.

radiative forcing (less than 10^{-5} W per m^2 through 2030), are estimated from the U.S. EPA vintaging model, while the future global abundances of these compounds are taken from the A1B and B1 SRES scenarios (Nakićenović *et al.*, 2000). If global PFCs and SF6 were not considered in the radiative forcing calculation, United States emissions are projected to contribute about 20 to 43% of global radiative forcing by ODSs and their replacements in 2030.

5.6.3 Options for United States ODS Banks

The accessible and total bank size estimates for United States equipment and applications in 2005 are compared to global bank estimates from WMO (2007) in Table 5.3. Additional reduction cases are shown in Table 5.3 that quantify the importance of the recovery and destruction of United States total banks and United States accessible banks. The U.S. EPA has defined "accessible" banks as the quantity of ODSs that is contained in equipment (*i.e.*, fire protection equipment and refrigeration/air conditioning systems). Furthermore, it is assumed that the amount of ODS recoverable from this equipment is equal to the full equipment charge minus the average annual loss rate from leakage and servicing. It is possible that some of the non-accessible banks could be recovered and destroyed with appropriate policy measures, market-based incentives, and/or certain technological advances. Table 5.3 shows that the halon-1301 and HCFC-22 United States accessible banks are the most substantial in terms of contributing to potential future integrated EESC reductions. If the total United States banks are considered, CFC-11, HCFC-141b, CFC-12, HCFC-22, and halon-1301 banks are most important. The calculations for the United States halon banks do not include stockpiles, and so should be considered underestimates of the full potential benefit of recovering and destroying these banks.

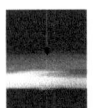

CHAPTER 6

Implications for the United States

Convening Lead Authors: A.R. Ravishankara, NOAA;
Michael J. Kurylo, NASA

Lead Authors: John S. Daniel, NOAA; David W. Fahey, NOAA;
Jay R. Herman, NASA; Stephen A. Montzka, NOAA; Malcolm Ko,
NASA; Paul A. Newman, NASA; Richard Stolarski, NASA

6.1 INTRODUCTION

The depletion of the stratospheric ozone layer due to human-produced ozone-depleting substances (ODSs) is a global phenomenon. Emissions of ODSs from around the globe contribute to depletion of the ozone layer throughout much of the stratosphere. ODSs emitted from different locations are well mixed within most of the lower atmosphere before they reach the stratosphere, where they contribute to chemical ozone depletion. Consequently, ozone depletion above a specific location is caused collectively by ODS emissions from different locations around the globe.

The observed pattern of the ozone depletion is not uniform around the globe; depletion above one region may differ from that above another region. However, this is not because of variations in emissions among regions, but because chemical and dynamical processes in the stratosphere cause regional variations in ozone and ozone loss rates. The extent of ozone depletion over a given region also varies with season and its overall magnitude changes with time. Consequently, the increase in ultraviolet (UV) light at the Earth's surface due to the depletion of the stratospheric ozone layer also varies with region and time.

Because of these factors, a simple connection cannot be drawn between emissions of ODSs from a country or a region and the depletion of stratospheric ozone above that country or region. For example, there is substantial ozone depletion each austral spring over Antarctica, a continent with essentially no emissions of ozone-depleting substances. In contrast, the decrease in stratospheric ozone at northern midlatitudes, where the dominant emissions of ODSs occur, is significantly less.

The decades of release of ODSs, the associated decreases in stratospheric ozone and increases in surface UV radiation, along with the influence of ozone depletion and of ODSs on climate, have important implications for the United States. These implications can be viewed by examining three areas: impacts, accountability, and potential management options. Each area will be summarized in the following sections. The discussion of impacts will highlight past, present, and future changes in stratospheric ozone, surface UV radiation, and globally averaged radiative forcing. The section on accountability will address the United States' contributions to the emissions of ODSs and the associated contributions to Equivalent Effective Stratospheric Chlorine (EESC) and radiative forcing of the Earth's climate. In the section on potential management options, we will attempt to provide a scientific perspective of ODS issues that can be used for supporting future management decisions. These issues will be discussed generically without addressing any specific option.

Ozone depletion above a specific location is caused by the emissions of ozone-depleting substances from around the globe.

6.2 IMPACTS

The changes in stratospheric ozone and surface UV radiation vary considerably among regions of the United States, which stretches over a wide range of latitudes in the Northern Hemisphere. Arctic ozone losses impact Alaska most prominently, while subtropical ozone changes affect Hawaii as well as Guam, Puerto Rico, and other United States territories.

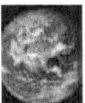

Ozone above the continental United States and other northern latitudes decreased to a minimum in 1993 and has increased since that time.

6.2.1 Changes in Ozone over the United States

The changes in total column ozone amounts for various regions around the globe have been derived from observations made by both satellite and ground-based instruments. The ozone trends reported here are derived primarily from the satellite data.

- Ozone decreases above the continental United States have essentially followed those occurring over the northern midlatitudes: a decrease to a minimum in about 1993, followed by an increase since that time. The minimum total column ozone amounts over the continental United States reached in 1993 were about 5 to 8 percent below those present prior to 1980. The decreases roughly followed the increases in atmospheric abundance of ozone-depleting substances, which reached its maximum in roughly 1995. The ozone minimum occurred earlier than the maximum in ozone-depleting substances in large part due to the atmospheric changes brought about by the eruption of Mt. Pinatubo in 1991 (as discussed in Chapter 3). Column ozone increases since 1993 have brought the ozone deficit back to about 2 to 5 percent below the pre-1980 amounts. Model calculations suggest that these midlatitude ozone changes may have a significant contribution from the mixing of lower stratospheric ozone-depleted air from the northern polar latitudes during the spring period.
- Ozone depletion over northern high latitudes, such as over northern Alaska, is strongly influenced by Arctic springtime total ozone values, which have been significantly lower than those observed in the 1980s. However, these Arctic springtime

ozone depletions are highly variable from year to year.

- There has been no significant ozone depletion at the lower latitudes of the tropics and subtropics around the globe. Hence, column ozone over the parts of the United States in these regions has been essentially unchanged from their 1980 values.

6.2.2 Changes in UV over the United States

Changes in UV levels over the United States have been obtained from ground-based and satellite-based measurements. Surface UV levels are strongly affected by clouds, atmospheric fine particles, and air pollution, in addition to stratospheric ozone depletion, making it difficult to attribute changes in UV to long-term changes in stratospheric ozone alone. This difficulty is particularly acute since stratospheric ozone depletion during the past three decades has been rather small (less than 10 percent), with the exception of the high latitudes. In a world without the Montreal Protocol, stratospheric ozone changes would have been much larger than have actually occurred, and the associated UV increases would have been so large as to stand out from other variability and be easily measured over a wide range of latitudes. In addition, ground-based records are of limited use in relating UV increases to ozone decreases that occurred during the 1980s and 1990s since many stations did not initiate measurements until the late 1990s, when ozone had already reached its minimum. A reliable way to derive the current changes in UV associated with ozone depletion is to use satellite measurements of atmospheric backscattered UV and the small amount reflected from the surface.

- Direct surface-based observations do not show significant UV trends for the United States over the past three decades because effects of clouds and atmospheric fine particles have likely masked the increase in UV due to ozone depletion.
- Estimates of UV based on satellite measurements of column ozone and reflectivity of the surface suggest that the clear-sky erythemal irradiance (a weighted combination of UVA and UVB wavelength ranges based on skin sensitivity) over the

continental United States increased from 1979 to the mid-1990s by about 7 percent. Currently, this irradiance is about 4 percent higher than it was at the start of the record in 1979. Year-to-year seasonal variations ranged from only a few percent to about 20 percent.

- Barrow, Alaska, has experienced UVB increases in March and April related to springtime ozone depletion. While these increases are larger than those observed at midlatitudes in the mid-1990s, they are roughly ten times smaller than those observed at the southern high latitudes due to the Antarctic ozone hole.

6.2.3 Changes in Radiative Forcing

Globally averaged radiative forcing is a good metric for the relative contributions to climate change. Positive (negative) values for radiative forcing lead to warming (cooling). It is a reasonably good assumption that the global-average impacts from long-lived greenhouse gases scale with the magnitude of the globally averaged forcing. Many ODSs are themselves greenhouse gases, and hence ODSs contribute to radiative forcing.

The combined direct radiative forcing from ODSs and substitutes, including hydrofluorocarbons (HFCs), is still increasing, but at a slower rate than in the 1980s because the use of many ODSs (particularly chlorofluorocarbons [CFCs]) has been curtailed by the Montreal Protocol. This continued increasing trend in radiative forcing arises from continued increases in the atmospheric mixing ratios of hydrochlorofluorocarbons (HCFCs) and HFCs, which are being used as substitutes for CFCs in various applications.

- The total contribution of human-produced ODSs and substitutes to direct radiative forcing is approximately 0.34 W per m^2 (representing the change between pre-industrial times, *ca.* 1750, and the present), which is about 15 percent as large as the contribution from other greenhouse gases (1.66 W per m^2 from carbon dioxide [CO_2] plus 0.6 W per m^2 from methane [CH_4] and nitrous oxide [N_2O] together). The bulk of the direct forcing from halocarbons in 2005 was from CFCs (about 80 percent);

other contributors include 10 percent from HCFCs, 7 percent from other ODSs, and 3 percent from HFCs. Projections of these contributions to 2100 under the SRES A1B emission scenario can be found in Chapter 5.

Changes in atmospheric ozone abundances contribute to climate change by modifying atmospheric radiative properties and atmospheric temperatures. Changes in stratospheric ozone are often considered to be indirect climate forcings of ODSs, though other processes have also influenced changes in stratospheric ozone over time.

- Depletion of stratospheric ozone since about 1980 is estimated to have caused a *negative* radiative forcing (cooling) on climate (of approximately −0.05 W per m^2), corresponding in absolute magnitude to about 15 percent of the total direct positive forcing by ODSs alone. However, the uncertainties on this forcing are large enough (±0.1 W per m^2, *i.e.*, a range of −0.15 to +0.05 W per m^2) to suggest that it could even be a *positive* radiative forcing (warming). Twentieth century increases in tropospheric ozone from pollution chemistry have caused a *positive* radiative forcing (of approximately +0.35 W per m^2). ODS emissions and abundances have very little influence over ozone abundances in the lower atmosphere.

6.2.4 Future Ozone and UV Changes over the United States

As stated earlier, changes in ozone over the United States should be similar to the changes occurring over similar latitudes around the globe. Ozone-depleting substances addressed by the Montreal Protocol and its amendments should have a declining effect on stratospheric ozone between now and 2050, and a small effect on stratospheric ozone beyond 2050, assuming compliance with the Montreal Protocol and if all other factors remain roughly the same. In order to predict the future trend of ozone in that time frame, one must consider projections for climate changes and changes in trace gases such as other halogens, CH_4, and N_2O (in addition to any changes in solar UV irradiance).

The total contribution of human-produced ozone-depleting substances and substitutes to direct radiative forcing is about 15% as large as the contribution from other greenhouse gases.

- Based on the prescribed surface concentrations of halocarbons used in the WMO (2007) baseline scenario (the scenario that was consistent with the Montreal Protocol provisions as of 2006), atmospheric halogen loading is estimated to recover to its 1980 value between 2040 and 2050 for midlatitudes, and between 2060 and 2070 for the polar regions.
- Between now and 2020, the simulated total ozone content between 60°N and 60°S will increase in response to this decrease in halogen loading.
- Three-dimensional climate chemistry models (3-D CCMs) predict that stratospheric cooling and changes in circulation associated with greenhouse gas emissions may enable global ozone to return to its 1980 value up to 15 years earlier than the expected halogen recovery date. Based on the assumed scenario for the greenhouse gases (which include CH_4 and N_2O), the ozone content is expected to be 2 to 5 percent above the 1980 values by 2100.
- Because of large interannual variability, the dates of the minimum in Arctic ozone from different models occur between 1997 and 2015. Most CCMs show Arctic ozone values at 2050 larger than the 1980 values, with recovery between 2020 and 2040. Results from the majority of the models indicate that future Arctic ozone depletion will not be significantly worse than what has already occurred.

The future trend in surface UV is likely to be controlled more by changes in cloud cover, atmospheric fine particles, and tropospheric air quality than by changes in stratospheric ozone. Nevertheless, Equivalent Effective Stratospheric Chlorine (EESC) will still be a useful predictor for the relative effects of ODSs on future UV in terms of evaluating different scenarios.

6.2.5 Future Changes in Radiative Forcing

The radiative forcing from CO_2 increases to +5 W per m^2 by 2100 for the IPCC Special Report on Emission Scenarios (SRES) A1B scenario (Nakićenović, 2000), a scenario involving rapid economic growth and balanced energy sources. Forcing from halocarbons and their substitutes

will decline in the future, assuming continued compliance with the Montreal Protocol, and is summarized below.

- Direct forcing from CFCs, which constitute a significant fraction of total ODS forcing in today's atmosphere, is expected to decrease from the current value of about +0.26 W per m^2 to about +0.1 W per m^2 by 2100. Direct forcing from HCFCs and other ODSs is expected to be negligible by 2100.
- The indirect forcing of ozone depletion is expected to approach zero when EESC returns to its 1980 levels, while the direct forcing due mainly to CFCs remaining in the atmosphere and the CFC substitutes that do not contain either chlorine or bromine (*e.g.*, HFCs) will continue.
- In the SRES A1B scenario, forcing from HFCs, which do not deplete stratospheric ozone, is predicted to increase to +0.15 W per m^2 and +0.24 W per m^2 by 2050 and 2100, respectively, while other scenarios result in smaller forcings from these chemicals. However, current observations suggest that the present atmospheric radiative forcing of the HFCs has been larger than computed for the SRES scenarios, primarily due to higher HFC-23 concentrations. Therefore, additional uncertainty perhaps should be attached to the SRES HFC projections.
- Changes in ozone due to changes in other trace gases (CH_4 and N_2O) and to changes in climate will also contribute to future forcing. For example, increases in atmospheric circulation due to climate change could increase the flux of ozone from the stratosphere to the troposphere, resulting in an additional positive radiative forcing.

6.3 ACCOUNTABILITY

As stated earlier, the amount of stratospheric ozone depletion at any location is, in large part, a result of long-lived ODSs emitted from all over the globe. In addition, ozone depletion in a given location is not simply and linearly related to ODS amounts in the atmosphere. Accordingly, there is no simple relationship between a country's contribution to global ODS emissions with the local ozone depletion occurring in

By the end of this century, direct forcing from CFCs is expected to decrease to roughly 40% of its current value.

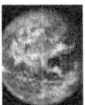

that region. To extend this association to local changes in UV radiation at the ground is further complicated by the dependence of UV on many local factors. Acknowledging this complexity, we can estimate the United States' contribution to the global emissions of ODSs to derive the United States' contribution to the atmospheric abundances of ODSs. We can then obtain a first approximation of the United States' contribution to ozone depletion at any location, and thereby estimate the United States' contribution to that portion of UV increase at that same location. In a similar manner, we can estimate the United States' contribution to changes in radiative forcing due to the emissions of ODSs.

This chapter uses several metrics to estimate the United States' ODS contributions. Ozone Depletion Potential (ODP) and Global Warming Potential (GWP) weighting is used to sum annual emissions to represent the United States' contribution in that particular year to future ozone depletion and radiative forcing. The same could be used for ODS "banks" since they represent potential future emissions. The atmospheric burden of a specific halocarbon calculated from historical United States emissions, when compared with the observed total burden, provides a measure of the United States' contribution to that halocarbon. The individual abundances can be combined using the formalism of Equivalent Effective Chlorine (EECl) and EESC to be used as a measure of ozone depletion contribution. Finally, radiative forcing calculated from the individual halocarbon abundances provides a measure of the United States' contribution to climate forcing.

6.3.1 Contribution of the United States to the Global Abundance of ODSs

It is difficult to accurately quantify the United States' contribution to the current atmospheric loading of ozone-depleting substances because of uncertainties associated with United States emission data prior to 1985. However, estimates of the United States' contributions to global consumption and emissions of ODSs for recent periods can be derived, respectively, from information compiled by the United Nations Environment Programme (UNEP) or from estimates made by the U.S. Environmental Protection Agency (EPA). It should be noted

that consumption of ODSs can have different emission patterns (in location and time) depending on the particular end use of the ODS. The U.S. EPA vintaging model calculates ODS emissions based on a variety of factors associated with the use or product application of the ODS.

Production / Consumption

- Global production and consumption of ODSs have declined substantially since the late 1980s in response to the Montreal Protocol, its amendments and adjustments, and United States policy decisions. By 2005, annual global ODP-weighted production and consumption had declined 95 percent from the peak values of the late 1980s. By 2005, annual ODP-weighted production and consumption in the United States had declined by 97 to 98 percent based on UNEP data.
- During 1986 to 1994 the United States accounted for 25($\pm$2) percent of the total annual global production and consumption of ODSs reported by UNEP when weighted by ODPs. From 2001 to 2005, this fraction has been 10($\pm$2) percent. This decline has been maintained despite a slower decline in U.S. CH_3Br consumption reported to UNEP than in other nations owing to increased U.S. Critical Use Exemptions, CUEs (critical uses that are exempted from the Montreal Protocol). Increased CUEs caused the annual U.S. contribution to global CH_3Br consumption to increase from 23($\pm$4) percent between 2000 to 2003, to 36($\pm$1) percent during 2004 to 2005.

Burdens and EECl

- Taking into account the uncertainties in United States emissions estimates for past years, atmospheric chlorine from United States emissions accounted for 17 to 42 percent of global chlorine from regulated ODSs and substitute chemicals in 2005. Atmospheric chlorine from United States and global emissions has declined since the mid-1990s.
- The U.S. EPA vintaging model suggests that in 2005 the United States accounted for approximately 17 to 35 percent of the global atmospheric bromine burden arising from industrially produced CH_3Br and ha-

From 1986 to 1994, the United States accounted for about 25% of total production and consumption of ozone-depleting substances. From 2001 to 2005, this fraction had declined to 10%.

lons, similar to that calculated for the peak year, 1998. Changes in total tropospheric bromine from the United States emissions of ODSs regulated by the Montreal Protocol mimicked global trends until 2002. Further, the vintaging model suggests that bromine emissions from the United States began increasing in 2002, due primarily to increased emissions of CH_3Br.

- The decrease in tropospheric EECl since 1994 has been about 20 percent of what is needed to return EECl to 1980 values (*i.e.*, before substantial ozone depletion was observed). Atmospheric EECl calculated from United States emissions declined between 1994 and 2004; however, it declined much more slowly from 2004 to 2005. The United States accounted for 15 to 36 percent of EECl from industrially produced chemicals measured in the troposphere in 2005.

Banks and Future Emissions

United States emissions of some ODSs in the future, like those from other developed nations, will be determined to a large extent by the size of the U.S. ODS "banks," *i.e.*, those ODSs that are already produced but not yet released to the atmosphere due to old devices, structures, and stockpiles that exist in the United States. The magnitude of halocarbon banks has been derived using a new bottom-up approach. This new method leads to larger CFC banks and yields potential future CFC emissions that are higher than previously estimated (WMO, 2003). The U.S. EPA has divided total banks into "accessible" and "non-accessible" categories, with accessible banks consisting of ODSs in refrigeration and air conditioning equipment and fire fighting equipment.

- If released in a single year, the ODS banks in the United States in 2005 would have been equivalent (in terms of their contribution to stratospheric ozone depletion) to 7 to 16 times the actual United States emissions of ODSs during that year.
- The U.S. EPA estimates that United States banks account for approximately 28 percent of global banks, whether they are accessible or not, of all ODSs (ODP-weighted). CFCs accounted for the largest fraction of the 2005 banks in the United

States as well as throughout the globe.

- Approximately one-quarter of United States banks in 2005 were classified as being accessible (consisting of 210 ODP-kilotons, Kt) and these accessible banks were comprised predominantly of halons (roughly 140 ODP-Kt), CFCs (~38 ODP-Kt), and HCFCs (~31 ODP-Kt). CFCs accounted for 18 percent of the accessible banks of ODSs as defined currently by the U.S. EPA.
- Banks play an important role in current HCFC emission rates. Future emissions of HCFCs will also be determined by the magnitude of any additional HCFC production.

6.3.2 Contribution of the United States to Climate Change via Emission of Ozone-Depleting Substances and the Resulting Ozone Changes

The increased abundances of ODSs, as well as the associated depletion of stratospheric ozone, contribute to the radiative forcing of climate. Since activities in the United States have caused significant emissions of ozone-depleting substances and other greenhouse gases, the changes in climate attributable to human activities are, in part, attributable to the United States.

- Globally, the direct radiative forcing from ODSs and substitutes was approximately +0.34 W per m^2, roughly 20 percent of that from CO_2 in 2005. When indirect forcing associated with stratospheric ozone depletion is included, the *net* forcing from ODSs and substitutes is between +0.18 and +0.38 W per m^2. These values were estimated using the 100-year direct GWPs.
- The United States' contribution to this direct forcing amounted to between +0.068 and +0.16 W per m^2, or between roughly 20 and 50 percent of the global direct radiative forcing from these chemicals. This contribution has been fairly constant over the past decade. When net GWPs (*i.e.*, that includes direct and indirect forcings) are considered, the range for the U.S. contribution is +0.04 to +0.18 W per m^2.
- Considering ODSs acting as climate forcing agents and using a 100-year direct GWP weighting, the U.S. EPA estimates

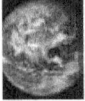

United States banks account for approximately 28 percent of total global banks of ozone-depleting substances.

that United States banks in 2005 account for about 32 percent of global banks. The range for net GWP weighting is (31-60) percent. These banks, if released to the atmosphere, would represent the equivalent of 6.2 gigatons (Gt) CO_2 emissions. Approximately one-quarter of United States banks in 2005 were classified as being accessible and they were comprised of HCFCs (0.9 to 1.1 Gt CO_2-equivalents), HFCs (0.4 Gt CO_2-equivalents), and CFCs (0.2-0.4 Gt CO_2-equivalents).

6.4 POTENTIAL MANAGEMENT OPTIONS

To provide quantitative information for assessing the societal benefit of potential future regulatory action, the future levels of ozone-depleting substances can be estimated for a variety of scenarios based on the findings noted above. These include scenarios to assess the influence of currently unrestricted uses, such as methyl bromide in quarantine and pre-shipment (QPS) applications, and unrestricted emissions from most banks and stockpiles. Equivalent Effective Stratospheric Chlorine (EESC) is a useful index for comparing the merits of different emissions scenarios. While changes in EESC do not relate in a simple way to stratospheric ozone levels that vary with location and time (due to the non-linearities that were mentioned earlier), it is clear that EESC changes are representative of the relative ozone depletion impacts. Based on projected EESC values and our understanding of atmospheric chemical and dynamical processes, we conclude the following:

- Amounts of atmospheric ozone-depleting substances, measured in terms of EESC, will be comparable to pre-1980 levels around 2050 if future emissions evolve in a manner consistent with current Montreal Protocol regulations. It is anticipated that, given the proven connection between ozone-depleting substances and stratospheric ozone loss, global ozone will also return to the pre-1980 levels roughly around the same time, assuming no other climate or atmospheric composition changes. However, as stated earlier, factors such as climate change and changes in other trace gases are predicted to accelerate global ozone recovery to pre-1980 values.

- The ozone-depleting substances (ODSs), measured in terms of EESC, in the Antarctic ozone-hole region will return to pre-1980 levels around 2060 to 2070. Thus, the Antarctic ozone hole will essentially disappear around this date assuming full compliance with the Montreal Protocol and its amendments and barring major influences by climate change and other factors.

The date at which the atmospheric abundances of ODSs return to their 1980 levels is 2049 for a baseline scenario (a scenario that is consistent with Montreal Protocol provisions as of 2006). This return can be accelerated under several scenarios.

- The hypothetical cessation of all future emissions of ozone-depleting substances (such as hydrochlorofluorocarbons (HCFCs), and chlorofluorocarbons (CFCs) from banks) starting in 2007 would hasten the decline of ozone-depleting substances to their 1980 level by roughly 15 years (to 2034).

- Under the scenario where no future production is assumed but emissions still arise from ODS banks, the EESC recovery date (i.e., to the 1980 level) is moved up by roughly six years (to 2043).

- Under the scenario where all ODS banks were recovered and destroyed in 2007, but future production is allowed to continue as in the baseline case, the EESC recovery date is moved up by eight years (to 2041).

- The significance of various United States ODS banks has been evaluated in terms of their effect on integrated EESC and compared with the significance of the global banks. Of accessible ODS banks in the U.S., halon-1301 and HCFC-22 represent the greatest potential contribution to integrated EESC above 1980 levels. If total United States banks are considered, not just accessible ones, CFC-11 banks are the most significant potential contributors to integrated EESC. Banks deemed inaccessible may still be recovered with appropriate policy measures, market-based incentives and/or certain technological advances.

Global ozone will return to its pre-1980 levels around 2050. The Antarctic ozone hole is expected to disappear around 2060 to 2070.

There are some uses of methyl bromide for which production and consumption are not limited or restricted under the currently amended Montreal Protocol.

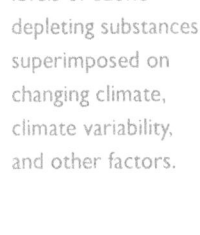

- Global consumption of methyl bromide (CH_3Br) for all fumigation-related uses declined by a factor of two from 1997 to 2005 despite substantial consumption in applications not regulated by the Montreal Protocol. Nearly half (43 percent) of the global, industrially derived emissions of CH_3Br during 2005 arose from QPS consumption not regulated by the Montreal Protocol.
- United States consumption of CH_3Br for all fumigation uses declined 40 percent from 1997 to 2005 despite enhanced Critical Use Exemptions (CUEs) and QPS consumption since 2001. Enhanced CUEs caused the annual United States contribution to global CH_3Br consumption for regulated uses to increase from 23($\pm$4) percent during 2000 to 2003 to 36($\pm$1) percent during 2004 to 2005. In the United States during 2001 to 2006, consumption of CH_3Br for fumigation not regulated by the protocol (QPS use) was, on average, 57($\pm$20) percent of the amounts used and reported to UNEP in restricted applications and had increased by 13 percent per year, on average, from 2001 to 2005.

The expected increase in stratospheric ozone over the coming decades will decrease surface UV. However, the future UV trend at the surface is likely to be more dominated by changes in cloud cover, atmospheric fine particle abundances, and tropospheric air quality than by changes in ODS abundances projected in accordance with the provisions of the Montreal Protocol.

Little further reduction in radiative forcing from ODSs can be achieved by 2100 beyond that predicted under the current provisions of the Montreal Protocol. Emissions reductions, however, could lower radiative forcing in the coming decades. Reductions in HFC emissions could also have a modest effect in this area.

- Action could be taken to limit the release of CFCs and HCFCs from banks and thus reduce their future emissions beyond what the current Montreal Protocol is expected to accomplish. If the entire estimated global CFC and HCFC banks had been recovered and destroyed in 2007, the direct radiative forcing is expected to be reduced by about 0.015 W per m^2 and 0.07 W per m^2, respectively, in 2040, compared with the radiative forcing calculated assuming future emissions consistent with the Montreal Protocol regulations. However, a complete assessment of any benefits of such action would also need to include consideration of indirect influences associated with ozone depletion changes.

6.4.1 The World Avoided

The dramatic decrease in emissions of ODSs since the late 1980s, called for by the Montreal Protocol, has been achieved in the United States through a combination of regulations restricting the use, handling, and labeling of specific compounds, a robust program to evaluate alternative compounds, voluntary industry initiatives, and outreach and education programs that have raised awareness of the threats caused by ODSs and the need to control them. Various emissions scenarios have been used to compare the ozone and UV levels of today with what might have occurred in the absence of the Montreal Protocol and the associated actions, as a way to assess the effectiveness and value of the Protocol to the United States and the world.

- Without the implementation of the Montreal Protocol, EESC levels around 2010 likely would have been more than 50 percent larger than currently expected. The abundances during the remainder of the twenty-first century would have depended on any subsequent policy actions taken. These increases in ODSs would have caused a corresponding substantially greater global ozone depletion. The Antarctic ozone hole would have persisted longer and may have been even larger than that currently observed.

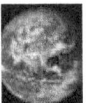

- The contributions of the United States to the ozone depletion via emission of ODSs to date have been significant. However, the United States has also contributed significantly to achieving the expected recovery of the ozone layer and associated surface UV changes, and reductions in direct climate forcing caused by ODSs by the phase-out of CFCs.

- The decline since the late 1980s of the United States' emissions of ODSs, considered on a CO_2-equivalent basis, corresponds to a climate benefit whose magnitude is large compared with the Kyoto Protocol's targets during its first commitment period. This benefit includes an offset of the -0.05 W per m^2 estimated in the Fourth Assessment Report of Working Group I of the Intergovernmental Panel on Climate Change (IPCC, 2007) due to the destruction of stratospheric ozone by ODSs (as discussed in Chapter 4).

The coming decades will be a period of changing atmospheric ODS levels superimposed on changing climate, climate variability, and other factors. Box 6.1 outlines the key gaps in scientific understanding that can be identified at this time and that could help inform future decisions regarding the continued recovery of the ozone layer back to a state that is not affected by ozone-depleting substances.

BOX 6.1: Gaps in Our Understanding and Continued Information Needs

In this document we have synthesized and assessed what is known about the depletion of the stratospheric ozone layer by ozone-depleting substances, the associated changes in surface UV radiation, and our expectations for the recovery of the ozone layer to pre-1980 values. We have described aspects of the interrelationship between stratospheric ozone depletion and climate change, such as the contribution of ozone-depleting substances to climate forcing, the impact of climate change on stratospheric ozone, and the effects of ozone depletion on climate. We have also outlined the importance of understanding the ozone-climate interrelationship, including variability of climate, in making accurate projections of future ozone as it recovers to pre-1980 values.

Evolving societal and decision-making imperatives arising from the continued global commitment to shepherd the ozone layer back to "good health" will drive future research on the stratospheric ozone issue. For example, the Parties to the Montreal Protocol recently made adjustments to the phase-out schedule to phase-out hydrochloro-fluorocarbons (HCFCs) earlier than scheduled; this adjustment agreement takes effect in mid-2008 and will be implemented over the coming few years as scheduled in the agreement. Questions still remain about topics such as the control of chlorofluorocarbon (CFC) bank emissions and the use of methyl bromide for exempted and unregulated purposes. Accurate predictions of the consequences of near-term decision options will require the United States and international scientific communities to acquire new observational data, to develop an improved understanding of the physical and chemical processes involved in ozone depletion and ozone-climate interactions, and to incorporate this understanding in global models used to project the future state of the ozone layer. Further, it will also require some reporting and documentation on production and use of ODSs and their substitutes.

At present, the scientific and regulatory communities are in the "*accountability*" phase of the ozone layer issue, because science-based regulation to protect ozone has been in place for nearly two decades. Decision makers are increasingly interested in having answers to the bottom-line questions: *Are our actions having the desired and expected effect? Is the ozone layer recovering? Are there other actions that would hasten ozone layer recovery?* As outlined in this document, scientists have addressed and/or partially answered many of these questions. However, gaps remain in our knowledge and information base and in our ability to answer these questions with sufficient clarity and accuracy for policy decisions.

BOX 6.1: Gaps in Our Understanding and Continued Information Needs *cont'd*

It has now become clear that it is critical to understand the linkages between stratospheric ozone depletion and climate change, because climate variations and change will alter the ozone recovery path and even the ozone abundance and distribution after ozone-depleting substances have returned to natural levels. This is owing to the fact that as the atmosphere moves toward a pre-1980 ODS abundance, other atmospheric conditions will not revert similarly toward their pre-1980 state. Understanding the implications of these different evolutions will be a key focus area of atmospheric ozone research.

The climate protection afforded by the Montreal Protocol regulation of ODSs has been significant over the last two decades. This protection derives from the fact that the principal ODSs are effective greenhouse gases. The findings of the IPCC Fourth Assessment Report, released in 2007, enhanced the global focus on climate protection and also increased interest in questions that lie at the nexus of these two global environmental issues. Decision support information demands an evaluation of the implications for ozone and climate protection for scenarios of future regulation under the Montreal Protocol. Decision makers need to know in detail how the ozone layer and the climate system are connected and what aspects of this linkage are likely to be most important in this evolving Earth system.

Based on our synthesis and assessment of the current state of knowledge and the above set of broad research imperatives, we have identified some key knowledge gaps. For simplicity they are listed in four different categories and are equally important. We believe that this description of the knowledge gaps will aid United States and international agencies in establishing research priorities and directions and in establishing reporting requirements.

Atmospheric Observations

Ozone observations: Precise and accurate ozone observations in the troposphere and stratosphere anchor our understanding of the present and future ozone layer. Furthermore, ozone observations must be geographically comprehensive and have extended duration. The recovery of the ozone layer is likely to manifest itself differently at different altitudes and regions in the stratosphere. Changes in the ozone abundance profiles in turn impact climate change. Therefore, more precise continued, uninterrupted, observations of temporal changes in distributions of column ozone as well as local ozone abundances over altitude and latitude are essential to identifying the path to recovery and to better predicting the future state of the ozone layer.

Observations and derivation of surface UV and associated factors: Predicting the surface UV changes, especially those due to stratospheric ozone changes, requires not only the measurements of the UV radiation but also of many associated factors. These include ozone (noted above), atmospheric fine particle abundances and properties, surface albedo, and transmission and reflection of radiation by clouds. Continuity in time, accuracy in value, and global coverage are necessary. Currently, most of the conclusions about the surface UV trends come from calculations that use observed ozone distributions. Facilitation and enhancement of this ability will better fill the need on recording and predicting UV changes.

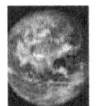

ODS observations and EESC: The accuracy of our current predictions for the recovery of the ozone layer depends directly on the accuracy of our predictions of the return of ozone-depleting substances and EESC to their pre-ozone-depletion values. We lack adequate knowledge for many of the factors that influence how these quantities return to pre-1980 values. For example, we may not have sufficiently accurate values for the atmospheric lifetimes of many ODSs, especially in a changing climate. Another concern includes the accurate quantification and the eventual release of ODS from banks and stockpiles, an emission that is likely large enough to delay the ozone recovery by many years. Also, uncertainties in bank emissions can hinder the identification of potential ODS emissions in violation of Montreal Protocol regulations. Verification of these bank emissions and emissions from other unregulated activities, such as methyl bromide from QPS use, will require more extensive atmospheric monitoring on global and regional scales.

BOX 6.1: Gaps in Our Understanding and Continued Information Needs *cont'd*

In addition, we cannot precisely quantify how much and in what chemical form the short-lived chemicals (especially those containing halogens and many of natural origin) are transported to the stratosphere, transformed to reactive compounds, and contribute to EESC and ozone depletion. Detailed knowledge of these factors will become more critical as overall ODS emissions and abundances decrease in the future, thereby increasing the relative contributions of the short-lived substances and bank and fugitive emissions to ozone depletion. Accuracy of emission information needs to be established via verification of emissions on global and regional scales.

Process Understanding

There are many specific inputs required to account for the past and predict the future ozone levels as well as climate change. They include accurate rates of chemical and photochemical processes, timescales and rates of dynamical processes including variations, and identification and quantification of many microphysical processes involved in formation, persistence, and characteristics of polar stratospheric clouds and stratospheric sulfate aerosols. The rates of many of these known processes are not sufficiently accurate and there may be some unrecognized processes that are not quantified. Examples include the recently highlighted uncertainties in the photolysis rate of dichlorine peroxide (Cl_2O_2), a molecule that plays a critical role in polar ozone depletion, and uncertainties in the destruction rates and pathways of existing ODSs (especially HCFCs) and of not yet released, but planned, substitutes for ODSs. Therefore, a continuing effort to understand and determine rates and mechanisms of such processes is essential.

Global Models

We are in the early stages of developing climate models and Earth System models that include the known interconnecting processes that link climate and ozone. Projections from three-dimensional chemistry climate models (that did not include explicit land surface and ocean interactions) were used extensively for the first time only during the WMO/UNEP ozone assessment of 2007. Such models will be essential for future evaluations. These models are highly complex because they include all known important chemical, physical, and dynamical processes that influence ozone and other atmospheric constituents. The identification and parameterization of contributing processes and the completion and validation of these maturing climate models together represent important improvements in our ability to project future ozone abundances. The models have demonstrated skill in predicting observed ozone changes and attributing the cause of global ozone decreases to ODS emissions. However, additional improvements are needed due to the demand for more precise and accurate projections of future ozone abundances that include the relevant climate feedbacks. For example, climate benefits related to reductions in ODS emissions are influenced by our understanding of the climate forcing associated with changes in stratospheric ozone, but this influence currently has large uncertainties. The ozone-climate models need to be sufficiently accurate to identify regulatory options that would optimize the dual ozone-climate benefit.

Reporting and Documentation

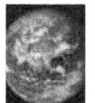

The accumulation of global emissions of ODSs and other greenhouse gases has led to ozone depletion and climate change. The United States' contributions in both emission categories have been significant. The ability to quantify the United States' contributions is limited by gaps in our knowledge of country-specific emissions. Detail is lacking for historical emissions for ODSs as well as for other greenhouse gases. Efforts to fill these historical gaps will improve the statements of attribution and benefit concerning potential future United States actions. Efforts to avoid similar gaps in the future will add credibility to and confidence in documenting United States accountability for ozone depletion and climate change and in providing guidance for United States national regulations or United States participation in new international policy discussions.

APPENDIX A

Twenty Questions and Answers About the Ozone Layer:
2006 Update

Coordinating Lead Author: David W. Fahey, NOAA

Appendix A is reprinted here with permission from:
Fahey, D.W. (Lead Author), *Twenty Questions and Answers About the Ozone Layer: 2006 Update*,
50 pp., World Meteorological Organization, Geneva, Switzerland, 2007. [Reprinted from
Scientific Assessment of Ozone Depletion: 2006, Global Ozone Research and Monitoring
Project—Report no. 50, 572 pp., World Meteorological Organization, Geneva, Switzerland,
2007.]

Repaginated for this report

Twenty Questions and Answers About the Ozone Layer: 2006 Update

Contents

INTRODUCTION

Ozone is a very small part of our atmosphere, but its presence is nevertheless vital to human well-being.

Most ozone resides in the upper part of the atmosphere. This region, called the stratosphere, is more than 10 kilometers (6 miles) above Earth's surface. There, about 90% of atmospheric ozone is contained in the "ozone layer," which shields us from harmful ultraviolet radiation from the Sun.

However, it was discovered in the mid-1970s that some human-produced chemicals could destroy ozone and deplete the ozone layer. The resulting increase in ultraviolet radiation at Earth's surface may increase the incidences of skin cancer and eye cataracts.

Following the discovery of this environmental issue, researchers focused on a better understanding of this threat to the ozone layer. Monitoring stations showed that the abundances of the ozone-depleting chemicals were steadily increasing in the atmosphere. These trends were linked to growing production and use of chemicals like chlorofluorocarbons (CFCs) for refrigeration and air conditioning, foam blowing, and industrial cleaning. Measurements in the laboratory and the atmosphere characterized the chemical reactions that were involved in ozone destruction. Computer models employing this information could predict how much ozone depletion was occurring and how much more could occur in the future.

Observations of the ozone layer showed that depletion was indeed occurring. The most severe and most surprising ozone loss was discovered to be recurring in springtime over Antarctica. The loss in this region is commonly called the "ozone hole" because the ozone depletion is so large and localized. A thinning of the ozone layer also has been observed over other regions of the globe, such as the Arctic and northern middle latitudes.

The work of many scientists throughout the world has provided a basis for building a broad and solid scientific understanding of the ozone depletion process. With this understanding, we know that ozone depletion is occurring and why. And, most important, we know that if ozone-depleting gases were to continue to accumulate in the atmosphere, the result would be more depletion of the ozone layer.

In response to the prospect of increasing ozone depletion, the governments of the world crafted the 1987 United Nations Montreal Protocol as a global means to address this global issue. As a result of the broad compliance with the Protocol and its Amendments and Adjustments and, of great significance, industry's development of "ozone-friendly" substitutes for the now-controlled chemicals, the total global accumulation of ozone-depleting gases has slowed and begun to decrease. This has reduced the risk of further ozone depletion. Now, with continued compliance, we expect recovery of the ozone layer by the late 21st century. The International Day for the Preservation of the Ozone Layer, 16 September, is now celebrated on the day the Montreal Protocol was agreed upon.

This is a story of notable achievements: discovery, understanding, decisions, actions, and verification. It is a story written by many: scientists, technologists, economists, legal experts, and policymakers. And, dialogue has been a key ingredient.

To help foster continued interaction, this component of the *Scientific Assessment of Ozone Depletion: 2006* presents 20 questions and answers about the often-complex science of ozone depletion. The answers are updates of those first presented in the previous ozone Assessment, *Scientific Assessment of Ozone Depletion: 2002*. The questions address the nature of atmospheric ozone, the chemicals that cause ozone depletion, how global and polar ozone depletion occur, and what could lie ahead for the ozone layer. A brief answer to each question is first given in italics; an expanded answer then follows. The answers are based on the information presented in the 2006 and earlier Assessment reports. These reports and the answers provided here were all prepared and reviewed by a large international group of scientists. [1]

---

[1] The update of this component of the Assessment was discussed by the 77 scientists who attended the Panel Review Meeting for the 2006 Ozone Assessment (Les Diablerets, Switzerland, 19-23 June 2006). In addition, subsequent contributions, reviews, or comments were provided by the following individuals: S.A. Montzka (special recognition), R.J. Salawitch (special recognition), D.L. Albritton, S.O. Andersen, P.J. Aucamp, M.P. Baldwin, A.F. Bias, G. Bodeker, J.F. Bornman, G.O. Braathen, J.P. Burrows, M.-L. Chanin, C. Clerbaux, M. Dameris, J.S. Daniel, S.B. Diaz, E.G. Dutton, C.A. Ennis, V. Eyring, V.E. Fioletov, N.P. Gillet, N.R.P. Harris, M.K.W. Ko, L. Kuijpers, G.L. Manney, R.L. McKenzie, R. Müller, E.R. Nash, P.A. Newman, T. Peter, A.R. Ravishankara, A. Robock, M.L. Santee, U. Schmidt, G. Seckmeyer, T.G. Shepherd, R.S. Stolarski, W.T. Sturges, J.C. van der Leun, G.J.M. Velders, D.W. Waugh, C.S. Zerefos.

I. OZONE IN OUR ATMOSPHERE

Q1: What is ozone and where is it in the atmosphere?

Ozone is a gas that is naturally present in our atmosphere. Each ozone molecule contains three atoms of oxygen and is denoted chemically as O₃. Ozone is found primarily in two regions of the atmosphere. About 10% of atmospheric ozone is in the troposphere, the region closest to Earth (from the surface to about 10-16 kilometers (6-10 miles)). The remaining ozone (90%) resides in the stratosphere, primarily between the top of the troposphere and about 50 kilometers (31 miles) altitude. The large amount of ozone in the stratosphere is often referred to as the "ozone layer."

Ozone is a gas that is naturally present in our atmosphere. Ozone has the chemical formula O_3 because an ozone molecule contains three oxygen atoms (see Figure Q1-1). Ozone was discovered in laboratory experiments in the mid-1800s. Ozone's presence in the atmosphere was later discovered using chemical and optical measurement methods. The word ozone is derived from the Greek word ὄζειν (*ozein* in Latin), meaning "to smell." Ozone has a pungent odor that allows it to be detected even at very low amounts. Ozone will rapidly react with many chemical compounds and is explosive in concentrated amounts. Electrical discharges are generally used to produce ozone for industrial processes such as air and water purification and bleaching textiles and food products.

Ozone location. Most ozone (about 90%) is found in the stratosphere, a region that begins about 10-16 kilometers (6-10 miles) above Earth's surface and extends up to about 50 kilometers (31 miles) altitude (see Figure Q1-2). Most ozone resides in the lower stratosphere in what is commonly known as the "ozone layer." The remaining ozone, about 10%, is found in the troposphere, which is the lowest region of the atmosphere, between Earth's surface and the stratosphere.

Ozone abundance. Ozone molecules have a relatively low abundance in the atmosphere. In the stratosphere near the peak of the ozone layer, there are up to 12,000 ozone molecules for every *billion* air molecules (1 billion = 1,000 million). Most air molecules are either oxygen (O_2) or nitrogen (N_2) molecules. In the troposphere near Earth's surface, ozone is even less abundant, with a typical range of 20 to 100 ozone molecules for each billion air molecules. The highest surface values are a result of ozone formed in air polluted by human activities.

As an illustration of the low relative abundance of ozone in our atmosphere, one can imagine bringing all the ozone molecules in the troposphere and stratosphere down to Earth's surface and uniformly distributing these molecules into a gas layer over the globe. The resulting layer of pure ozone would have a thickness of less than one-half centimeter (about one-quarter inch).

Ozone and Oxygen

Oxygen Atom (O)	Oxygen Molecule (O₂)	Ozone Molecule (O₃)

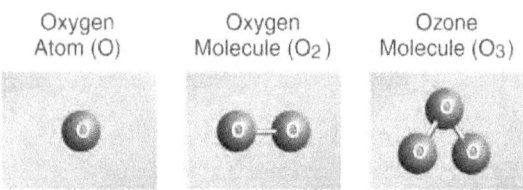

Figure Q1-1. Ozone and oxygen. A molecule of ozone (O_3) contains three oxygen (O) atoms bound together. Oxygen molecules (O_2), which constitute 21% of Earth's atmosphere, contain two oxygen atoms bound together.

Ozone in the Atmosphere

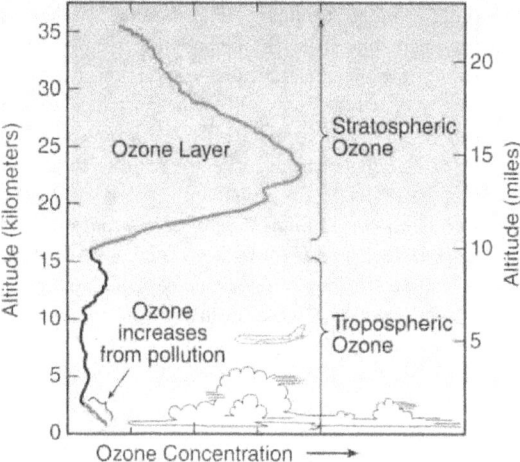

Figure Q1-2. Atmospheric ozone. Ozone is present throughout the lower atmosphere (troposphere and stratosphere). Most ozone resides in the stratospheric "ozone layer" above Earth's surface. Increases in ozone occur near the surface as a result of pollution from human activities.

Q2: How is ozone formed in the atmosphere?

Ozone is formed throughout the atmosphere in multistep chemical processes that require sunlight. In the stratosphere, the process begins with an oxygen molecule (O_2) being broken apart by ultraviolet radiation from the Sun. In the lower atmosphere (troposphere), ozone is formed in a different set of chemical reactions involving hydrocarbons and nitrogen-containing gases.

Stratospheric ozone. Stratospheric ozone is naturally formed in chemical reactions involving ultraviolet sunlight and oxygen molecules, which make up 21% of the atmosphere. In the first step, sunlight breaks apart one oxygen molecule (O_2) to produce two oxygen atoms (2 O) (see Figure Q2-1). In the second step, each atom combines with an oxygen molecule to produce an ozone molecule (O_3). These reactions occur continually wherever ultraviolet sunlight is present in the stratosphere. As a result, the greatest ozone production occurs in the tropical stratosphere.

The production of stratospheric ozone is balanced by its destruction in chemical reactions. Ozone reacts continually with a wide variety of natural and human-produced chemicals in the stratosphere. In each reaction, an ozone molecule is lost and other chemical compounds are produced. Important reactive gases that destroy ozone are those containing chlorine and bromine (see Q8).

Some stratospheric ozone is transported down into the troposphere and can influence ozone amounts at Earth's surface, particularly in remote, unpolluted regions of the globe.

Tropospheric ozone. Near Earth's surface, ozone is produced in chemical reactions involving naturally occurring gases and gases from pollution sources. Production reactions primarily involve hydrocarbon and nitrogen oxide gases and require sunlight. Fossil fuel combustion is a primary pollution source for tropospheric ozone production. The surface production of ozone does not significantly contribute to the abundance of stratospheric ozone. The amount of surface ozone is too small in comparison, and the transport of surface air to the stratosphere is not effective enough. As in the stratosphere, ozone in the troposphere is destroyed in naturally occurring chemical reactions and in reactions involving human-produced chemicals. Tropospheric ozone can also be destroyed when ozone reacts with a variety of surfaces, such as those of soils and plants.

Balance of chemical processes. Ozone abundances in the stratosphere and troposphere are determined by the *balance* between chemical processes that produce ozone and processes that destroy ozone. The balance is determined by the amounts of reacting gases and how the rate or effectiveness of the various reactions varies with sunlight intensity, location in the atmosphere, temperature, and other factors. As atmospheric conditions change to favor ozone-producing reactions in a certain location, ozone abundances will increase. Similarly, if conditions change to favor reactions that destroy ozone, abundances will decrease. The balance of production and loss reactions combined with atmospheric air motions determines the global distribution of ozone on time scales of days to many months. Global ozone has decreased in the last decades because the amounts of reactive gases containing chlorine and bromine have increased in the stratosphere (see Q13).

Stratospheric Ozone Production

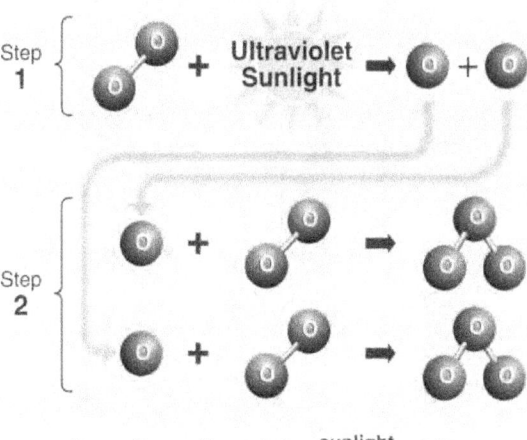

Overall reaction: $3 O_2 \xrightarrow{\text{sunlight}} 2 O_3$

Figure Q2-1. Stratospheric ozone production. Ozone is naturally produced in the stratosphere in a two-step process. In the first step, ultraviolet sunlight breaks apart an oxygen molecule to form two separate oxygen atoms. In the second step, each atom then undergoes a binding collision with another oxygen molecule to form an ozone molecule. In the overall process, three oxygen molecules plus sunlight react to form two ozone molecules.

Q3: Why do we care about atmospheric ozone?

Ozone in the stratosphere absorbs some of the Sun's biologically harmful ultraviolet radiation. Because of this beneficial role, stratospheric ozone is considered "good" ozone. In contrast, excess ozone at Earth's surface that is formed from pollutants is considered "bad" ozone because it can be harmful to humans, plants, and animals. The ozone that occurs naturally near the surface and in the lower atmosphere is also beneficial because ozone helps remove pollutants from the atmosphere.

Natural ozone. In the absence of human activities on Earth's surface, ozone would still be present near the surface and throughout the troposphere and stratosphere because ozone is a natural component of the clean atmosphere. All ozone molecules are chemically identical, with each containing three oxygen atoms. However, ozone in the stratosphere (good ozone) has very different environmental consequences for humans and other life forms than excess ozone in the troposphere near Earth's surface (bad ozone). Natural ozone in the troposphere is also considered "good" because it initiates the chemical removal of many pollutants, such as carbon monoxide and nitrogen oxides, as well as greenhouse gases such as methane.

Good ozone. Stratospheric ozone is considered good for humans and other life forms because it absorbs ultraviolet (UV)-B radiation from the Sun (see Figure Q3-1). If not absorbed, UV-B would reach Earth's surface in amounts that are harmful to a variety of life forms. In humans, increased exposure to UV-B increases the risk of skin cancer (see Q17), cataracts, and a suppressed immune system. UV-B exposure before adulthood and cumulative exposure are both important factors in the risk. Excessive UV-B exposure also can damage terrestrial plant life, single-cell organisms, and aquatic ecosystems. Other UV radiation, UV-A, which is not absorbed significantly by ozone, causes premature aging of the skin.

The absorption of UV-B radiation by ozone is a source of heat in the stratosphere. This helps to maintain the stratosphere as a stable region of the atmosphere, with temperatures increasing with altitude. As a result, ozone plays a key role in controlling the temperature structure of Earth's atmosphere.

Protecting good ozone. In the mid-1970s, it was discovered that halogen source gases released in human activities could cause stratospheric ozone depletion (see Q6). Ozone depletion increases harmful UV-B amounts at Earth's surface. Global efforts have been undertaken to protect the ozone layer through regulation of ozone-depleting gases (see Q15 and Q16).

Bad ozone. Excess ozone formed near Earth's surface in reactions caused by the presence of human-made pollutant gases is considered bad ozone. Increased ozone amounts are harmful to humans, plants, and other living systems because ozone reacts strongly to destroy or alter many other molecules. Excessive ozone exposure reduces crop yields and forest growth. In humans, exposure to

high levels of ozone can reduce lung capacity; cause chest pains, throat irritation, and coughing; and worsen pre-existing health conditions related to the heart and lungs. In addition, increases in tropospheric ozone lead to a warming of Earth's surface (see Q18). The negative effects of increasing tropospheric ozone contrast sharply with the positive effects of stratospheric ozone as an absorber of harmful UV-B radiation from the Sun.

Reducing bad ozone. Reducing the emission of pollutants can reduce bad ozone in the air surrounding humans, plants, and animals. Major sources of pollutants include large cities where fossil fuel consumption and industrial activities are greatest. Many programs around the globe have already been successful in reducing the emission of pollutants that cause excess ozone production near Earth's surface.

UV Protection by the Ozone Layer

Figure Q3-1. UV-B protection by the ozone layer. The ozone layer resides in the stratosphere and surrounds the entire Earth. UV-B radiation (280- to 315-nanometer (nm) wavelength) from the Sun is partially absorbed in this layer. As a result, the amount of UV-B reaching Earth's surface is greatly reduced. UV-A (315- to 400-nm wavelength) and other solar radiation are not strongly absorbed by the ozone layer. Human exposure to UV-B increases the risk of skin cancer, cataracts, and a suppressed immune system. UV-B exposure can also damage terrestrial plant life, single-cell organisms, and aquatic ecosystems.

Q4: Is total ozone uniform over the globe?

No, the total amount of ozone above the surface of Earth varies with location on time scales that range from daily to seasonal and longer. The variations are caused by stratospheric winds and the chemical production and destruction of ozone. Total ozone is generally lowest at the equator and highest near the poles because of the seasonal wind patterns in the stratosphere.

Total ozone. Total ozone at any location on the globe is found by measuring all the ozone in the atmosphere directly above that location. Total ozone includes that present in the stratospheric ozone layer and that present throughout the troposphere (see Figure Q1-2). The contribution from the troposphere is generally only about 10% of total ozone. Total ozone values are often reported in *Dobson units,* denoted "DU." Typical values vary between 200 and 500 DU over the globe (see Figure Q4-1). A total ozone value of 500 DU, for example, is equivalent to a layer of pure ozone gas on Earth's surface having a thickness of only 0.5 centimeters (0.2 inches).

Global distribution. Total ozone varies strongly with latitude over the globe, with the largest values occurring at middle and high latitudes (see Figure Q4-1). This is a result of winds that circulate air in the stratosphere, moving tropical air rich in ozone toward the poles. Since about 1980, regions of low total ozone have occurred at polar latitudes in winter and spring as a result of the chemical destruction of ozone by chlorine and bromine gases (see Q11 and Q12). The smallest values of total ozone (other than in the Antarctic in spring) occur in the tropics in all seasons, in part because the troposphere extends to a higher altitude in the tropics, and consequently, the thickness of the ozone layer is smallest there.

Natural variations. The variations of total ozone with latitude and longitude come about for two reasons. First, natural air motions mix air between regions of the stratosphere that have high ozone values and those that have low ozone values. Air motions also increase the vertical thickness of the ozone layer near the poles, which increases the value of total ozone in those regions. Tropospheric weather systems can temporarily reduce the thickness of the stratospheric ozone layer in a region, lowering total ozone at the same time. Second, variations occur as a result of changes in the balance of chemical production and loss processes as air moves to different locations over the globe. Reductions in ultraviolet radiation from the sun in its 11-year cycle, for example, reduce the production of ozone.

Scientists have a good understanding of how chemistry and air motions work together to cause the observed large-scale features in total ozone, such as those seen in Figure Q4-1. Ozone changes are carefully monitored by a large group of investigators using satellite, airborne, and ground-based instruments. The analysis of these observations helps scientists to estimate the contribution of human activities to ozone depletion.

Global Satellite Maps of Total Ozone

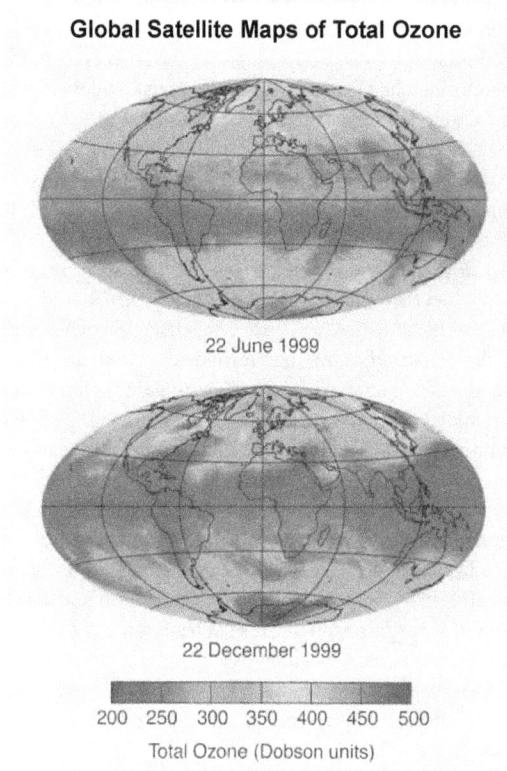

22 June 1999

22 December 1999

| 200 | 250 | 300 | 350 | 400 | 450 | 500 |

Total Ozone (Dobson units)

Figure Q4-1. Total ozone. A total ozone value is obtained by measuring all the ozone that resides in the atmosphere over a given location on Earth's surface. Total ozone values shown here are reported in "Dobson units" as measured by a satellite instrument from space. Total ozone varies with latitude, longitude, and season, with the largest values at high latitudes and the lowest values in tropical regions. Total ozone at most locations varies with time on a daily to seasonal basis as ozone-rich air is moved about the globe by stratospheric winds. Low total ozone values over Antarctica in the 22 December image represent the remainder of the "ozone hole" from the 1999 Antarctic winter/spring season (see Q11).

Q5: How is ozone measured in the atmosphere?

The amount of ozone in the atmosphere is measured by instruments on the ground and carried aloft on balloons, aircraft, and satellites. Some measurements involve drawing air into an instrument that contains a system for detecting ozone. Other measurements are based on ozone's unique absorption of light in the atmosphere. In that case, sunlight or laser light is carefully measured after passing through a portion of the atmosphere containing ozone.

The abundance of ozone in the atmosphere is measured by a variety of techniques (see Figure Q5-1). The techniques make use of ozone's unique optical and chemical properties. There are two principal categories of measurement techniques: *local* and *remote*. Ozone measurements by these techniques have been essential in monitoring changes in the ozone layer and in developing our understanding of the processes that control ozone abundances.

Local measurements. Local measurements of atmospheric ozone abundance are those that require air to be drawn directly into an instrument. Once inside an instrument, ozone can be measured by its absorption of ultraviolet (UV) light or by the electrical current produced in an ozone chemical reaction. The latter approach is used in the construction of "ozonesondes," which are lightweight, ozone-measuring modules suitable for launching on small balloons. The balloons ascend far enough in the atmosphere to measure ozone in the stratospheric ozone layer. Ozonesondes are launched regularly at many locations around the world. Local ozone-measuring instruments using optical or chemical detection schemes are also used routinely on board research aircraft to measure the distribution of ozone in the troposphere and lower stratosphere. High-altitude research aircraft can reach the ozone layer at most locations over the globe and can reach farthest into the layer at high latitudes in polar regions. Ozone measurements are also being made on some commercial aircraft.

Remote measurements. Remote measurements of ozone abundance are obtained by detecting the presence of ozone at large distances away from the instrument. Most remote measurements of ozone rely on its unique absorption of UV radiation. Sources of UV radiation that can be used are the Sun and lasers. For example, satellites use the absorption of UV sunlight by the atmosphere or the absorption of sunlight scattered from the surface of Earth to measure ozone over nearly the entire globe on a daily basis. A network of ground-based detectors measures ozone by the amount of the Sun's UV light that reaches Earth's surface. Other instruments measure ozone using its absorption of infrared or visible radiation or its emission of microwave or infrared radiation. Total ozone amounts and the altitude distribution of ozone can be obtained with remote measurement techniques. Lasers are routinely deployed at ground sites or on board aircraft to detect ozone over a distance of many kilometers along the laser light path.

Measuring Ozone in the Atmosphere

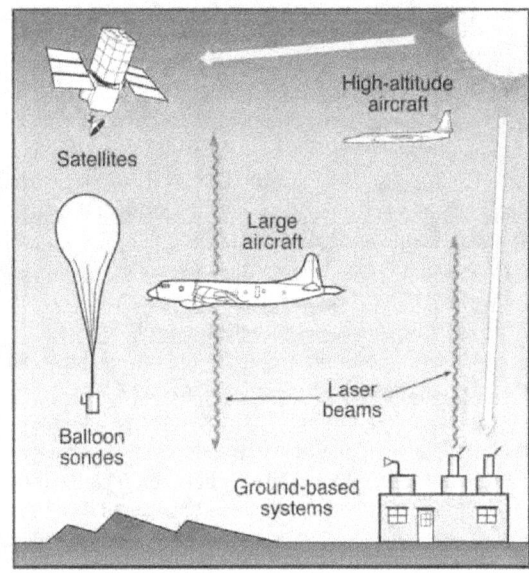

Figure Q5-1. Ozone measurements. Ozone is measured throughout the atmosphere with instruments on the ground and on board aircraft, high-altitude balloons, and satellites. Some instruments measure ozone locally in sampled air and others measure ozone remotely some distance away from the instrument. Instruments use optical techniques, with the Sun and lasers as light sources, or use chemical reactions that are unique to ozone. Measurements at many locations over the globe are made regularly to monitor total ozone amounts.

Global Ozone Dobson Network

The first instrument for routine monitoring of total ozone was developed by Gordon M. B. Dobson in the 1920s. The instrument, now called a Dobson spectrophotometer, measures the intensity of sunlight at two ultraviolet wavelengths: one that is strongly absorbed by ozone and one that is weakly absorbed. The difference in light intensity at the two wavelengths is used to provide a measurement of total ozone above the location of the instrument.

A global network of ground-based, total-ozone observing stations was established in 1957 as part of the International Geophysical Year. Today, there are about 100 sites distributed throughout the world (from South Pole, Antarctica (90°S), to Ellesmere Island, Canada (83°N)), many of which routinely measure total ozone with Dobson instruments. The accuracy of these observations is maintained by regular calibrations and intercomparisons. Data from the network have been essential for understanding the effects of chloro-fluorocarbons (CFCs) and other ozone-depleting gases on the global ozone layer, starting before the launch of space-based ozone-measuring instruments and continuing to the present day. Because of their stability and accuracy, the Dobson instruments are now routinely used to help calibrate space-based observations of total ozone.

Pioneering scientists have traditionally been honored by having units of measure named after them. Accordingly, the unit of measure for total ozone is called the "Dobson unit" (see Q4).

II. THE OZONE DEPLETION PROCESS

Q6: What are the principal steps in stratospheric ozone depletion caused by human activities?

The initial step in the depletion of stratospheric ozone by human activities is the emission, at Earth's surface, of ozone-depleting gases containing chlorine and bromine. Most of these gases accumulate in the lower atmosphere because they are unreactive and do not dissolve readily in rain or snow. Eventually, these emitted source gases are transported to the stratosphere, where they are converted to more reactive gases containing chlorine and bromine. These more reactive gases then participate in reactions that destroy ozone. Finally, when air returns to the lower atmosphere, these reactive chlorine and bromine gases are removed from Earth's atmosphere by rain and snow.

6Emission, accumulation, and transport. The principal steps in stratospheric ozone depletion caused by human activities are shown in Figure Q6-1. The process begins with the *emission*, at Earth's surface, of source gases containing the halogens chlorine and bromine (see Q7). The halogen source gases include manufactured chemicals released to the atmosphere by a variety of human activities. Chlorofluorocarbons (CFCs) are an important example of chlorine-containing gases. Emitted source gases *accumulate* in the lower atmosphere (troposphere) and are eventually *transported* to the stratosphere. The accumulation occurs because most source gases are unreactive in the lower atmosphere. However, small amounts of these gases dissolve or are taken up in ocean waters.

Some emissions of halogen gases come from natural sources (see Q7). These emissions also accumulate in the troposphere and are transported to the stratosphere.

Conversion, reaction, and removal. Halogen source gases do not react directly with ozone. Once in the stratosphere, halogen source gases are chemically converted to reactive halogen gases by ultraviolet radiation from the Sun (see Q8). The rate of conversion is related to the atmospheric lifetime of a gas (see Q7). Source gases with lifetimes greater than a few years may circulate between the troposphere and stratosphere multiple times before full conversion occurs.

The reactive gases formed in the eventual conversion of the halogen source gases react chemically to destroy ozone in the stratosphere (see Q9). The average depletion

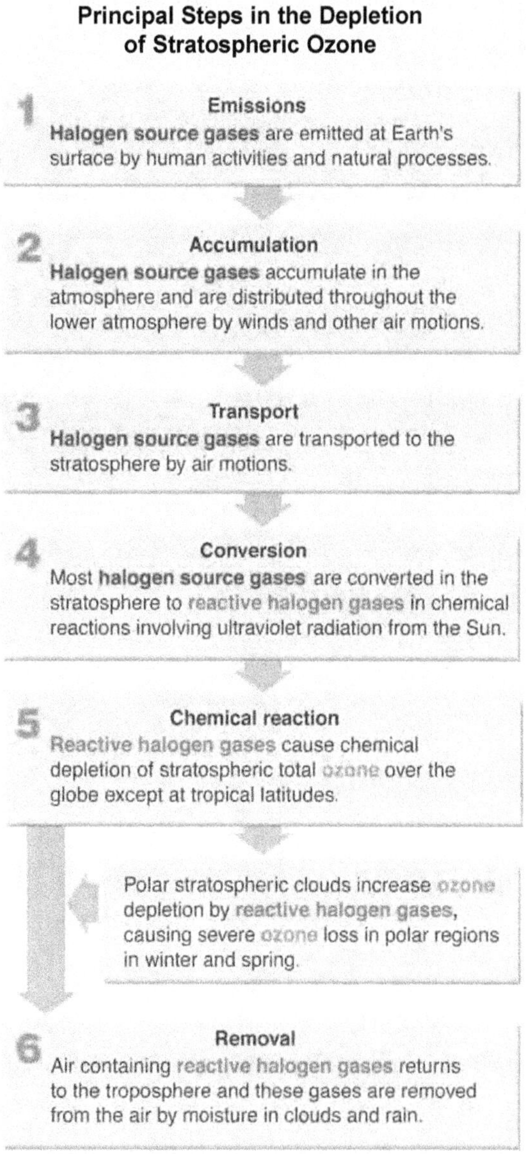

Figure Q6-1. Principal steps in stratospheric ozone depletion. The stratospheric ozone depletion process begins with the emission of halogen source gases at Earth's surface and ends when reactive halogen gases are removed by rain and snow in the troposphere and deposited on Earth's surface. In the stratosphere, the reactive halogen gases, namely chlorine monoxide (ClO) and bromine monoxide (BrO), destroy ozone.

Principal Steps in the Depletion of Stratospheric Ozone

1 Emissions
Halogen source gases are emitted at Earth's surface by human activities and natural processes.

2 Accumulation
Halogen source gases accumulate in the atmosphere and are distributed throughout the lower atmosphere by winds and other air motions.

3 Transport
Halogen source gases are transported to the stratosphere by air motions.

4 Conversion
Most **halogen source gases** are converted in the stratosphere to reactive halogen gases in chemical reactions involving ultraviolet radiation from the Sun.

5 Chemical reaction
Reactive halogen gases cause chemical depletion of stratospheric total ozone over the globe except at tropical latitudes.

Polar stratospheric clouds increase ozone depletion by reactive halogen gases, causing severe ozone loss in polar regions in winter and spring.

6 Removal
Air containing reactive halogen gases returns to the troposphere and these gases are removed from the air by moisture in clouds and rain.

of total ozone attributed to reactive gases is smallest in the tropics and largest at high latitudes (see Q13). In polar regions, the presence of polar stratospheric clouds greatly increases the abundance of the most reactive halogen gases (see Q10). This results in substantial ozone destruction in polar regions in winter and spring (see Q11 and Q12).

After a few years, air in the stratosphere returns to the troposphere, bringing along reactive halogen gases. These gases are then removed from the atmosphere by rain and other precipitation and deposited on Earth's surface. This removal brings to an end the destruction of ozone by chlorine and bromine atoms that were first released to the atmosphere as components of halogen source gas molecules.

Tropospheric conversion. Halogen source gases with short lifetimes (see Q7) undergo significant chemical conversion in the troposphere, producing reactive halogen gases and other compounds. Source gas molecules that are not converted accumulate in the troposphere and are transported to the stratosphere. Because of removal by precipitation, only small portions of the reactive halogen gases produced in the troposphere are also transported to the stratosphere. Important examples of gases that undergo some tropospheric removal are the HCFCs, which are used as substitute gases for other halogen source gases (see Q15 and Q16), bromoform, and gases containing iodine (see Q7).

Understanding Stratospheric Ozone Depletion

Scientists learn about ozone destruction through a combination of laboratory studies, computer models, and stratospheric observations. In *laboratory studies* scientists are able to discover and evaluate individual chemical reactions that also occur in the stratosphere. Chemical reactions between two gases follow well-defined physical rules. Some of these reactions occur on the surfaces of particles formed in the stratosphere. Reactions have been studied that involve a wide variety of molecules containing chlorine, bromine, fluorine, and iodine and other atmospheric constituents such as oxygen, nitrogen, and hydrogen. These studies show that there exist several reactions involving chlorine and bromine that can directly or indirectly cause ozone destruction in the atmosphere.

With *computer models*, scientists can examine the overall effect of a large group of known reactions under the chemical and physical conditions found in the stratosphere. These models include winds, air temperatures, and the daily and seasonal changes in sunlight. With such analyses, scientists have shown that chlorine and bromine can react in catalytic cycles in which one chlorine or bromine atom can destroy many ozone molecules. Scientists use model results to compare with past observations as a test of our understanding of the atmosphere and to evaluate the importance of new reactions found in the laboratory. Computer models also enable scientists to explore the future by changing atmospheric conditions and other model parameters.

Scientists have conducted *observations* to find out which gases are present in various regions of the stratosphere and at what concentrations. They have monitored the change in these abundances over time periods spanning a daily cycle to decades. Observations have shown that halogen source gases and reactive halogen gases are present in the stratosphere at expected amounts. Ozone and chlorine monoxide (ClO), for example, have been observed extensively with a variety of instruments. Instruments on the ground and on board satellites, balloons, and aircraft detect ozone and ClO at a distance (remotely) using optical and microwave signals. High-altitude aircraft and balloon instruments detect both gases locally in the stratosphere (see Q5). For example, these observations show that ClO is present at elevated amounts in the Antarctic and Arctic stratospheres in the late winter/early spring season, when the most severe ozone depletion occurs (see Q8).

Q7: What emissions from human activities lead to ozone depletion?

Certain industrial processes and consumer products result in the emission of "halogen source gases" to the atmosphere. These gases bring chlorine and bromine to the stratosphere, which cause depletion of the ozone layer. For example, chlorofluorocarbons (CFCs), once used in almost all refrigeration and air conditioning systems, eventually reach the stratosphere, where they are broken apart to release ozone-depleting chlorine atoms. Other examples of human-produced ozone-depleting gases are the "halons," which are used in fire extinguishers and contain ozone-depleting bromine atoms. The production and consumption of all principal halogen source gases by human activities are regulated worldwide under the Montreal Protocol.

Principal human-produced chlorine and bromine gases. Human activities cause the emission of *halogen source gases* that contain chlorine and bromine atoms. These emissions into the atmosphere ultimately lead to stratospheric ozone depletion. The source gases that contain only carbon, chlorine, and fluorine are called "chlorofluorocarbons," usually abbreviated as CFCs. CFCs, along with carbon tetrachloride (CCl_4) and methyl chloroform (CH_3CCl_3), historically have been the most important chlorine-containing gases that are emitted by human activities and destroy stratospheric ozone (see Figure Q7-1). These and other chlorine-containing gases have been used in many applications, including refrigeration, air conditioning, foam blowing, aerosol propellants, and cleaning of metals and electronic components. These activities have typically caused the emission of halogen-containing gases to the atmosphere.

Another category of halogen source gases contains bromine. The most important of these are the "halons" and methyl bromide (CH_3Br). Halons are halogenated hydrocarbon gases originally developed to extinguish fires. Halons are widely used to protect large computers, military hardware, and commercial aircraft engines. Because of these uses, halons are often directly released into the atmosphere. Halon-1211 and halon-1301 are the most abundant halons emitted by human activities (see Figure Q7-1). Methyl bromide, used primarily as an agricultural fumigant, is also a significant source of bromine to the atmosphere.

Human emissions of the principal chlorine- and bromine-containing gases have increased substantially since the middle of the 20th century (see Q16). The result has been global ozone depletion, with the greatest losses occurring in polar regions (see Q11 to Q13).

Other human sources of chlorine and bromine. Other chlorine- and bromine-containing gases are released regularly in human activities. Common examples are the use of chlorine gases to disinfect swimming pools and wastewater, fossil fuel burning, and various industrial processes. These activities do not contribute significantly to stratospheric amounts of chlorine and bromine because either the global source is small or the emitted gases are short-lived (very reactive or highly soluble) and, therefore, are removed from the atmosphere before they reach the stratosphere.

Natural sources of chlorine and bromine. There are a few halogen source gases present in the stratosphere that have large natural sources. These include methyl chloride (CH_3Cl) and methyl bromide (CH_3Br), both of which are emitted by oceanic and terrestrial ecosystems. Natural sources of these two gases contribute about 17% of the chlorine currently in the stratosphere and about 30% of the bromine (see Figure Q7-1). Very short-lived source gases containing bromine, such as bromoform ($CHBr_3$), are also released to the atmosphere primarily from the oceans. Only a small fraction of these emissions reaches the stratosphere, because these gases are rapidly removed in the lower atmosphere. The contribution of these very short-lived gases to stratospheric bromine is estimated to be about 24%, but this has a large uncertainty. The contribution to stratospheric chlorine of short-lived chlorinated gases from natural and human sources is much smaller (< 3%) and is included in the "Other gases" category in Figure Q7-1. Changes in the natural sources of chlorine and bromine since the middle of the 20th century are not the cause of observed ozone depletion.

Lifetimes and emissions. After emission, halogen source gases are either naturally removed from the atmosphere or undergo chemical conversion. The time to remove or convert about 60% of a gas is often called its atmospheric "lifetime." Lifetimes vary from less than 1 year to 100 years for the principal chlorine- and bromine-containing gases (see Table Q7-1). Gases with the shortest lifetimes (e.g., the HCFCs, methyl bromide, methyl chloride, and the very short-lived gases) are substantially destroyed in the troposphere, and therefore only a fraction of each emitted gas contributes to ozone depletion in the stratosphere.

The amount of a halogen source gas present in the atmosphere depends on the lifetime of the gas and the

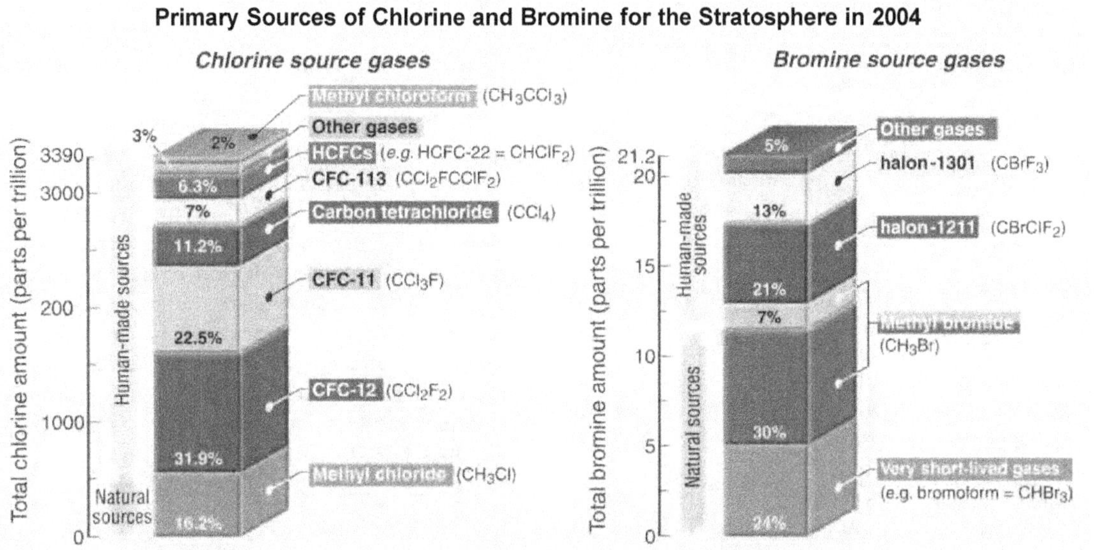

Primary Sources of Chlorine and Bromine for the Stratosphere in 2004

Figure Q7-1. Stratospheric source gases. A variety of gases transport chlorine and bromine into the stratosphere. These gases, called halogen source gases, are emitted from natural sources and by human activities. These partitioned columns show how the principal chlorine and bromine source gases contribute to the respective total amounts of chlorine and bromine as measured in 2004. Note the large difference in the vertical scales: total chlorine in the stratosphere is 160 times more abundant than total bromine. For chlorine, human activities account for most that reaches the stratosphere. The CFCs are the most abundant of the chlorine-containing gases released in human activities. Methyl chloride is the most important natural source of chlorine. HCFCs, which are substitute gases for CFCs and also are regulated under the Montreal Protocol, are a small but growing fraction of chlorine-containing gases. The "Other gases" category includes minor CFCs and short-lived gases. For bromine that reaches the stratosphere, halons and methyl bromide are the largest sources. Both gases are released in human activities. Methyl bromide has an additional natural source. Natural sources are a larger fraction of total bromine than of total chlorine. (The unit "parts per trillion" is used here as a measure of the relative abundance of a gas in air: 1 part per trillion indicates the presence of one molecule of a gas per trillion other air molecules.)

amount emitted to the atmosphere. Emissions vary greatly for the principal source gases, as indicated in Table Q7-1. Emissions of most gases regulated by the Montreal Protocol have decreased since 1990, and emissions from all regulated gases are expected to decrease in the coming decades (see Q16).

Ozone Depletion Potential. The halogen source gases in Figure Q7-1 are also known as "ozone-depleting substances" because they are converted in the stratosphere to reactive gases containing chlorine and bromine (see Q8). Some of these reactive gases participate in reactions that destroy ozone (see Q9). Ozone-depleting substances are compared in their effectiveness to destroy stratospheric ozone using the "Ozone Depletion Potential" (ODP), as listed in Table Q7-1 (see Q18). A gas with a larger ODP has a greater potential to destroy ozone over its lifetime in the atmosphere. The ODP is calculated on a "per mass" basis for each gas relative to CFC-11, which

has an ODP defined to be 1. Halon-1211 and halon-1301 have ODPs significantly larger than CFC-11 and most other emitted gases, because bromine is much more effective overall (about 60 times) on a per-atom basis than chlorine in chemical reactions that destroy ozone in the stratosphere. The gases with small ODP values generally have short atmospheric lifetimes or fewer chlorine and bromine atoms. The production and consumption of all principal halogen source gases by humans are regulated under the provisions of the Montreal Protocol (see Q15).

Fluorine and iodine. Fluorine and iodine are also halogen atoms. Many of the source gases in Figure Q7-1 also contain fluorine atoms in addition to chlorine or bromine. After the source gases undergo conversion in the stratosphere (see Q6), the fluorine content of these gases is left in chemical forms that do not cause ozone depletion. Iodine is a component of several gases that are naturally emitted from the oceans. Although iodine can

Table Q7-1. Atmospheric lifetimes, emissions, and Ozone Depletion Potentials of halogen source gases. [a]

Halogen Source Gas	Atmospheric Lifetime (years)	Global Emissions in 2003 [b]	Ozone Depletion Potential (ODP) [d]
Chlorine			
CFC-12	100	101-144	1
CFC-113	85	1-15	1
CFC-11	45	60-126	1
Carbon tetrachloride (CCl$_4$)	26	58-131	0.73
HCFCs	1-26	312-403	0.02-0.12
Methyl chloroform (CH$_3$CCl$_3$)	5	~20	0.12
Methyl chloride	1.0	1700-13600	0.02
Bromine			
Halon-1301	65	~3	16
Halon-1211	16	7-10	7.1
Methyl bromide (CH$_3$Br)	0.7	160-200	0.51
Very short-lived gases (e.g., CHBr$_3$)	< 0.5	c	c

[a] Includes both human activities and natural sources.
[b] Emission in gigagrams per year (1 gigagram = 10^9 grams = 1000 metric tons).
[c] Estimates are uncertain for most species.
[d] Values are calculated for emissions of equal mass for each gas.

participate in ozone destruction reactions, these iodine-containing source gases generally have very short life-times and, as a result, most are removed in the troposphere before they reach the stratosphere.

Other gases. Other gases that influence stratospheric ozone abundances also have increased in the stratosphere as a result of human activities. Important examples are methane (CH$_4$) and nitrous oxide (N$_2$O), which react in the stratosphere to form water vapor and reactive hydrogen, and nitrogen oxides, respectively. These reactive products also participate in the production and loss balance of stratospheric ozone (see Q2). The overall effect of increases in these other gases on ozone is much smaller than that caused by increases in chlorine- and bromine-containing gases from human activities (see Q18).

Heavier-Than-Air CFCs

CFCs and other halogen source gases reach the stratosphere despite the fact that they are "heavier than air." All the principal source gases are emitted and accumulate in the lower atmosphere (troposphere). The distributions of gases in the troposphere and stratosphere are not controlled by the molecular weight of the gases because air is in continual motion in these regions as a result of winds and convection. Air motions ensure that most source gases become horizontally and vertically well mixed throughout the troposphere in a matter of months. It is this well-mixed air that enters the lower stratosphere from upward air motions in tropical regions, bringing with it source gas molecules emitted from a wide variety of locations on Earth's surface.

Atmospheric measurements confirm that halogen source gases with long atmospheric lifetimes are well mixed in the troposphere and are present in the stratosphere (see Figure Q8-2). The amounts found in these regions are consistent with the emissions estimates reported by industries and governments. Measurements also show that gases that are "lighter than air," such as hydrogen (H$_2$) and methane (CH$_4$), are also well mixed in the troposphere, as expected. Only at altitudes well above the troposphere and stratosphere (above 85 kilometers (53 miles)), where much less air is present, does the influence of winds and convection diminish to the point where heavy gases begin to separate from lighter gases as a result of gravity.

Q8: What are the reactive halogen gases that destroy stratospheric ozone?

Emissions from human activities and natural processes include large sources of chlorine- and bromine-containing gases that eventually reach the stratosphere. When exposed to ultraviolet radiation from the Sun, these halogen source gases are converted to more reactive gases also containing chlorine and bromine. Important examples of the reactive gases that destroy stratospheric ozone are chlorine monoxide (ClO) and bromine monoxide (BrO). These reactive gases participate in "catalytic" reaction cycles that efficiently destroy ozone. Volcanoes can emit some chlorine-containing gases, but these gases are ones that readily dissolve in rainwater and ice and are usually "washed out" of the atmosphere before they can reach the stratosphere.

Reactive gases containing the halogens chlorine and bromine lead to the chemical destruction of stratospheric ozone. Halogen-containing gases present in the stratosphere can be divided into two groups: *halogen source gases* and *reactive halogen gases*. The source gases are emitted at Earth's surface by natural processes and by human activities (see Q7). Once in the stratosphere, the halogen source gases chemically convert at different rates to form the reactive halogen gases. The conversion occurs in the stratosphere instead of the troposphere because solar UV radiation is more intense in the stratosphere.

Reactive halogen gases. The chemical conversion of halogen source gases, which involves ultraviolet sunlight and other chemical reactions, produces a number of reactive halogen gases. These reactive gases contain all of the chlorine and bromine atoms originally present in the source gases.

The most important reactive chlorine- and bromine-containing gases that form in the stratosphere are shown in Figure Q8-1. Away from polar regions, the most abundant are hydrogen chloride (HCl) and chlorine nitrate ($ClONO_2$). These two gases are considered *reservoir* gases because they do not react directly with ozone but can be converted to the most reactive forms that do chemically destroy ozone. The *most reactive* forms are chlorine monoxide (ClO) and bromine monoxide (BrO), and chlorine and bromine atoms (Cl and Br). A large fraction of available stratospheric bromine is generally in the form of BrO, whereas usually only a small fraction of stratospheric chlorine is in the form of ClO. In polar regions, the reservoirs $ClONO_2$ and HCl undergo a further conversion on polar stratospheric clouds to form ClO (see Q10). In that case, ClO becomes a large fraction of available reactive chlorine.

Reactive chlorine observations. Reactive chlorine gases have been observed extensively in the stratosphere with both local and remote measurement techniques. The measurements from space at middle latitudes displayed in

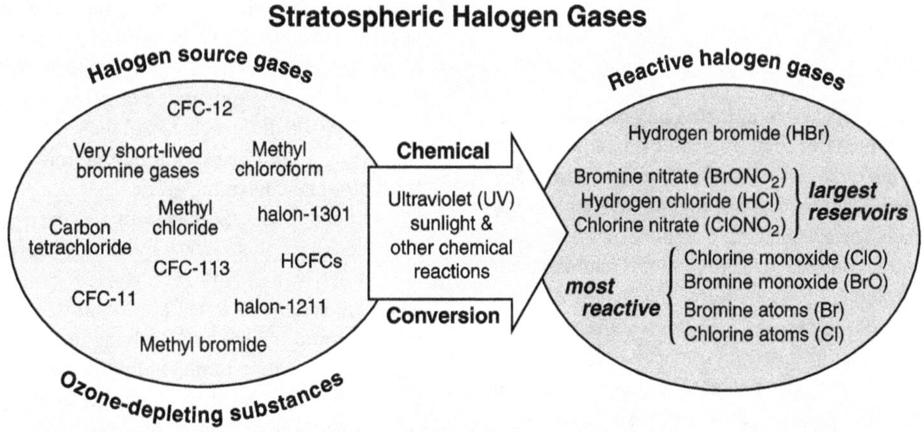

Stratospheric Halogen Gases

Figure Q8-1. Conversion of halogen source gases. Halogen source gases (also known as ozone-depleting substances) are chemically converted to reactive halogen gases primarily in the stratosphere. The conversion requires ultraviolet sunlight and a few other chemical reactions. The short-lived gases undergo some conversion in the troposphere. The reactive halogen gases contain all the chlorine and bromine originally present in the source gases. The reactive gases separate into reservoir gases, which do not destroy ozone, and reactive gases, which participate in ozone destruction cycles (see Q9).

Figure Q8-2 are representative of how the amounts of chlorine-containing gases change between the surface and the upper stratosphere. Available chlorine (see red line in Figure Q8-2) is the sum of chlorine contained in halogen source gases and the reactive gases HCl, ClONO$_2$, ClO, and other minor gases. Available chlorine is constant within a few percent from the surface to 47 kilometers (31 miles) altitude. In the troposphere, available chlorine is contained almost entirely in the source gases described in Figure Q7-1. At higher altitudes, the source gases become a smaller fraction of available chlorine as they are converted to reactive chlorine gases. At the highest altitudes, available chlorine is all in the form of reactive chlorine gases.

Measurements of Reactive Chlorine from Space
November 1994 (35° – 49°N)

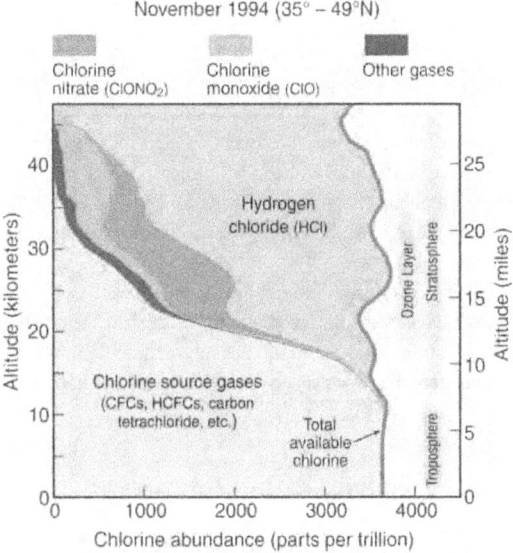

Figure Q8-2. Reactive chlorine gas observations. The abundances of chlorine source gases and reactive chlorine gases as measured from space are displayed with altitude for a midlatitude location. In the troposphere (below about 10 kilometers), all chlorine is contained in the source gases. In the stratosphere, reactive chlorine gases increase with altitude as chlorine source gases decrease. This is a consequence of chemical reactions involving ultraviolet sunlight (see Figure Q8-1). The principal reactive gases formed are HCl, ClONO$_2$, and ClO. Summing the source gases with the reactive gases gives *total available chlorine*, which is nearly constant with altitude up to 47 km. In the ozone layer, HCl and ClONO$_2$ are the most abundant reactive chlorine gases. (The unit "parts per trillion" is defined in the caption of Figure Q7-1.)

In the altitude range of the ozone layer at midlatitudes, as shown in Figure Q8-2, the reactive chlorine gases HCl and ClONO$_2$ account for most of available chlorine. ClO, the most reactive gas in ozone depletion, is a small fraction of available chlorine. This small value limits the amount of ozone destruction that occurs outside of polar regions.

Reactive chlorine in polar regions. Reactive chlorine gases in polar regions in summer look similar to the altitude profiles shown in Figure Q8-2. In winter, however, the presence of polar stratospheric clouds (PSCs) causes further chemical changes (see Q10). PSCs convert HCl and ClONO$_2$ to ClO when temperatures are near minimum values in the winter Arctic and Antarctic stratosphere. In that case, ClO becomes the principal reactive chlorine species in sunlit regions and ozone loss becomes very rapid. An example of the late-winter ClO and ozone distributions is shown in Figure Q8-3 for the Antarctic stratosphere. These space-based measurements show that ClO abundances are high in the lower stratosphere over a region that exceeds the size of the Antarctic continent (greater than 13 million square kilometers or 5 million square miles). The peak abundance of ClO exceeds 1500 parts per trillion, which is much larger than typical midlatitude values shown in Figure Q8-2 and represents a large fraction of reactive chlorine in that altitude region. Because high ClO amounts cause rapid ozone loss (see Q9), ozone depletion is found in regions of elevated ClO (see Figure Q8-3).

Reactive bromine observations. Fewer measurements are available for reactive bromine gases in the lower stratosphere than for reactive chlorine, in part because of the lower abundance of bromine. The most widely observed bromine gas is bromine monoxide (BrO). Recent observations have shown that measured BrO abundances in the stratosphere are larger than expected from the conversion of the halons and methyl bromide to BrO, suggesting a significant contribution from the very short-lived bromine-containing gases.

Other sources. Some reactive halogen gases are also produced at Earth's surface by natural processes and by human activities. However, because reactive halogen gases are soluble in water, almost all become trapped in the lower atmosphere by dissolving in rainwater and ice, and ultimately are returned to Earth's surface before they can reach the stratosphere. For example, reactive chlorine is present in the atmosphere as sea salt (sodium chloride) produced by evaporation of ocean spray. Because sea salt dissolves in water, this chlorine is removed and does not reach the stratosphere in appreciable quantities. Another ground-level source is emission of chlorine gases from swimming pools, household bleach, and other uses.

Satellite Observations in the Lower Stratosphere

30 August 1996

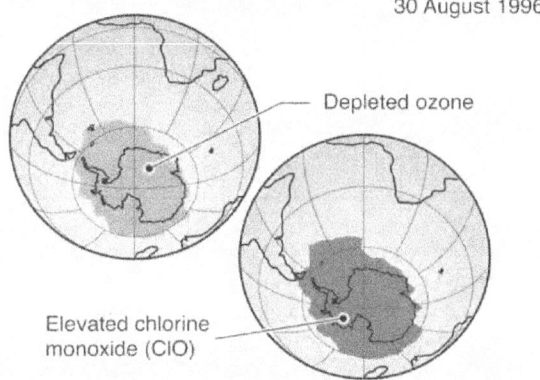

Depleted ozone

Elevated chlorine monoxide (CIO)

Figure Q8-3. Antarctic chlorine monoxide and ozone. Satellite instruments monitor ozone and reactive chlorine gases in the global stratosphere. Results are shown here for Antarctic winter for a narrow altitude region within the ozone layer. In winter, chlorine monoxide (CIO) reaches high values (1500 parts per trillion) in the ozone layer, much higher than observed anywhere else in the stratosphere because CIO is produced by reactions on polar stratospheric clouds (see Q10). These high CIO values in the lower stratosphere last for 1 to 2 months, cover an area that at times exceeds that of the Antarctic continent, and efficiently destroy ozone in sunlit regions in late winter/early spring. Ozone values measured simultaneously within the ozone layer show very depleted values.

When released to the atmosphere, this chlorine is rapidly converted to forms that are soluble in water and removed. The Space Shuttle and other rocket motors release reactive chlorine gases directly in the stratosphere: in this case, the quantities are very small in comparison with other tropospheric sources.

Volcanoes. Volcanic plumes generally contain large quantities of chlorine in the form of hydrogen chloride (HCl). Because the plumes also contain a considerable amount of water vapor, the HCl is efficiently scavenged by rainwater and ice and removed from the atmosphere. As a result, most of the HCl in the plume does not enter the stratosphere. After large recent eruptions, the increase in HCl in the stratosphere has been small compared with the total amount of chlorine in the stratosphere from other sources.

Replacing the Loss of Ozone in the Stratosphere

The idea is sometimes put forth that humans could replace the loss of global stratospheric ozone by making ozone and transporting it to the stratosphere. Ozone amounts in the stratosphere reflect a balance between continual production and destruction by mostly naturally occurring reactions (see Q2). The addition of chlorine and bromine to the stratosphere from human activities has increased ozone destruction and lowered stratospheric ozone amounts. Adding manufactured ozone to the stratosphere would upset the existing balance. As a consequence, most added ozone would be destroyed in chemical reactions within weeks to months as the balance was restored. So, it is not practical to consider replacing the loss of global stratospheric ozone because the replacement effort would need to continue indefinitely, or as long as increased chlorine and bromine amounts remained.

Other practical difficulties in replacing stratospheric ozone are the large amounts of ozone required and the delivery method. The total amount of atmospheric ozone is approximately 3,000 megatons (1 megaton = 1 billion kilograms) with most residing in the stratosphere. The replacement of the average global ozone loss of about 4% would require 120 megatons of stratospheric ozone to be distributed throughout the layer located many kilometers above Earth's surface. The energy required to produce this amount of ozone would be a significant fraction of the electrical power generated in the United States, which is now approximately 5 trillion kilowatt hours. Processing and storing requirements for ozone, which is explosive and toxic in large quantities, would increase the energy requirement. In addition, methods suitable to deliver and distribute large amounts of ozone to the stratosphere have not been demonstrated. Concerns for a global delivery system would include further significant energy use and unforeseen environmental consequences.

Q9: What are the chlorine and bromine reactions that destroy stratospheric ozone?

Reactive gases containing chlorine and bromine destroy stratospheric ozone in "catalytic" cycles made up of two or more separate reactions. As a result, a single chlorine or bromine atom can destroy many hundreds of ozone molecules before it reacts with another gas, breaking the cycle. In this way, a small amount of reactive chlorine or bromine has a large impact on the ozone layer. Certain ozone destruction reactions become most effective in polar regions because the reactive gas chlorine monoxide reaches very high levels there in the late winter/early spring season.

Stratospheric ozone is destroyed by reactions involving *reactive halogen gases*, which are produced in the chemical conversion of *halogen source gases* (see Figure Q8-1). The most reactive of these gases are chlorine monoxide (ClO), bromine monoxide (BrO), and chlorine and bromine atoms (Cl and Br). These gases participate in three principal reaction cycles that destroy ozone.

Cycle 1. Ozone destruction Cycle 1 is illustrated in Figure Q9-1. The cycle is made up of two basic reactions: ClO + O and Cl + O$_3$. The net result of Cycle 1 is to convert one ozone molecule and one oxygen atom into two oxygen molecules. In each cycle, chlorine acts as a *catalyst* because ClO and Cl react and are reformed. In this way, one Cl atom participates in many cycles, destroying many ozone molecules. For typical stratospheric conditions at middle or low latitudes, a single chlorine atom can destroy hundreds of ozone molecules before it happens to react with another gas, breaking the catalytic cycle.

Polar Cycles 2 and 3. The abundance of ClO is greatly increased in polar regions during winter as a result of reactions on the surfaces of polar stratospheric cloud (PSC) particles (see Q10). Cycles 2 and 3 (see Figure Q9-2) become the dominant reaction mechanisms for polar ozone loss because of the high abundances of ClO and the relatively low abundance of atomic oxygen (which limits the rate of ozone loss by Cycle 1). Cycle 2 begins with the self-reaction of ClO. Cycle 3, which begins with the reaction of ClO with BrO, has two reaction pathways to produce either Cl and Br or BrCl. The net result of both cycles is to destroy two ozone molecules and create three oxygen molecules. Cycles 2 and 3 account for most of the ozone loss observed in the Arctic and Antarctic stratospheres in the late winter/early spring

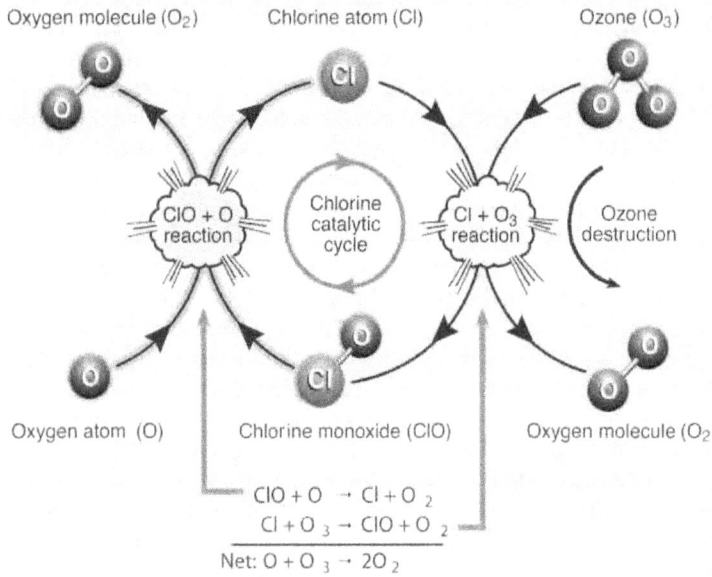

Ozone Destruction Cycle 1

Oxygen molecule (O$_2$)   Chlorine atom (Cl)   Ozone (O$_3$)

ClO + O reaction

Chlorine catalytic cycle

Cl + O$_3$ reaction

Ozone destruction

Oxygen atom (O)   Chlorine monoxide (ClO)   Oxygen molecule (O$_2$)

$$ClO + O \rightarrow Cl + O_2$$
$$Cl + O_3 \rightarrow ClO + O_2$$
$$\text{Net: } O + O_3 \rightarrow 2O_2$$

Figure Q9-1. Ozone destruction Cycle 1. The destruction of ozone in Cycle 1 involves two separate chemical reactions. The net or overall reaction is that of atomic oxygen with ozone, forming two oxygen molecules. The cycle can be considered to begin with either ClO or Cl. When starting with ClO, the first reaction is ClO with O to form Cl. Cl then reacts with (and thereby destroys) ozone and reforms ClO. The cycle then begins again with another reaction of ClO with O. Because Cl or ClO is reformed each time an ozone molecule is destroyed, chlorine is considered a catalyst for ozone destruction. Atomic oxygen (O) is formed when ultraviolet sunlight reacts with ozone and oxygen molecules. Cycle 1 is most important in the stratosphere at tropical and middle latitudes, where ultraviolet sunlight is most intense.

Ozone Destruction Cycles

Cycle 2	Cycle 3
$ClO + ClO \rightarrow (ClO)_2$	$ClO + BrO \rightarrow Cl + Br + O_2$
	or $\left(\begin{array}{c} ClO + BrO \rightarrow BrCl + O_2 \\ BrCl + sunlight \rightarrow Cl + Br \end{array}\right)$
$(ClO)_2 + sunlight \rightarrow ClOO + Cl$	
$ClOO \rightarrow Cl + O_2$	$Cl + O_3 \rightarrow ClO + O_2$
$2(Cl + O_3 \rightarrow ClO + O_2)$	$Br + O_3 \rightarrow BrO + O_2$
Net: $2O_3 \rightarrow 3O_2$	Net: $2O_3 \rightarrow 3O_2$

Figure Q9-2. Polar ozone destruction Cycles 2 and 3. Significant destruction of ozone occurs in polar regions because ClO abundances reach large values. In this case, the cycles initiated by the reaction of ClO with another ClO (Cycle 2) or the reaction of ClO with BrO (Cycle 3) efficiently destroy ozone. The net reaction in both cases is two ozone molecules forming three oxygen molecules. The reaction of ClO with BrO has two pathways to form the Cl and Br product gases. Ozone destruction Cycles 2 and 3 are catalytic, as illustrated for Cycle 1 in Figure Q9-1, because chlorine and bromine gases react and are reformed in each cycle. Sunlight is required to complete each cycle and to help form and maintain ClO abundances.

season (see Q11 and Q12). At high ClO abundances, the rate of ozone destruction can reach 2 to 3% per day in late winter/early spring.

Sunlight requirement. Sunlight is required to complete and maintain Cycles 1 through 3. Cycle 1 requires sunlight because atomic oxygen is formed only with ultraviolet sunlight. Cycle 1 is most important in the stratosphere at tropical and middle latitudes, where ultraviolet sunlight is most intense.

Cycles 2 and 3 require visible sunlight to complete the reaction cycles and to maintain ClO abundances. In the continuous darkness of winter in the polar stratospheres, reaction Cycles 2 and 3 cannot occur. It is only in late winter/early spring when sunlight returns to the polar regions that these cycles can occur. Therefore, the greatest destruction of ozone occurs in the partially to fully sunlit periods after midwinter in the polar stratospheres. The

visible sunlight needed in Cycles 2 and 3 is not sufficient to form ozone because this process requires ultraviolet sunlight. In the stratosphere in the late winter/early spring period, ultraviolet sunlight is weak because Sun angles are low. As a result, ozone is destroyed by Cycles 2 and 3 in the sunlit winter stratosphere but is not produced in significant amounts.

Other reactions. Global ozone abundances are controlled by many reactions that both produce and destroy ozone (see Q2). Chlorine and bromine catalytic reactions are but one group of ozone destruction reactions. Reactive hydrogen and reactive nitrogen gases, for example, are involved in other catalytic ozone-destruction cycles that also occur in the stratosphere. These reactions occur naturally in the stratosphere and their importance has not been as strongly influenced by human activities as have reactions involving halogens.

Q10: Why has an "ozone hole" appeared over Antarctica when ozone-depleting gases are present throughout the stratosphere?

Ozone-depleting gases are present throughout the stratospheric ozone layer because they are transported great distances by atmospheric air motions. The severe depletion of the Antarctic ozone layer known as the "ozone hole" occurs because of the special weather conditions that exist there and nowhere else on the globe. The very low temperatures of the Antarctic stratosphere create ice clouds called polar stratospheric clouds (PSCs). Special reactions that occur on PSCs and the relative isolation of polar stratospheric air allow chlorine and bromine reactions to produce the ozone hole in Antarctic springtime.

The severe depletion of stratospheric ozone in Antarctic winter is known as the "ozone hole" (see Q11). Severe depletion first appeared over Antarctica because atmospheric conditions there increase the effectiveness of ozone destruction by reactive halogen gases (see Q8). The formation of the Antarctic ozone hole requires abundant reactive halogen gases, temperatures low enough to form polar stratospheric clouds (PSCs), isolation of air from other stratospheric regions, and sunlight.

Distributing halogen gases. Halogen source gases emitted at Earth's surface are present in comparable abundances throughout the stratosphere in both hemispheres even though most of the emissions occur in the Northern Hemisphere. The abundances are comparable because most source gases have no important natural removal processes in the lower atmosphere and because winds and warm-air convection redistribute and mix air efficiently throughout the troposphere. Halogen gases (in the form of source gases and some reactive products) enter the stratosphere primarily from the tropical upper troposphere. Atmospheric air motions then transport them upward and toward the poles in both hemispheres.

Low temperatures. The severe ozone destruction represented by the ozone hole requires that low tempera-

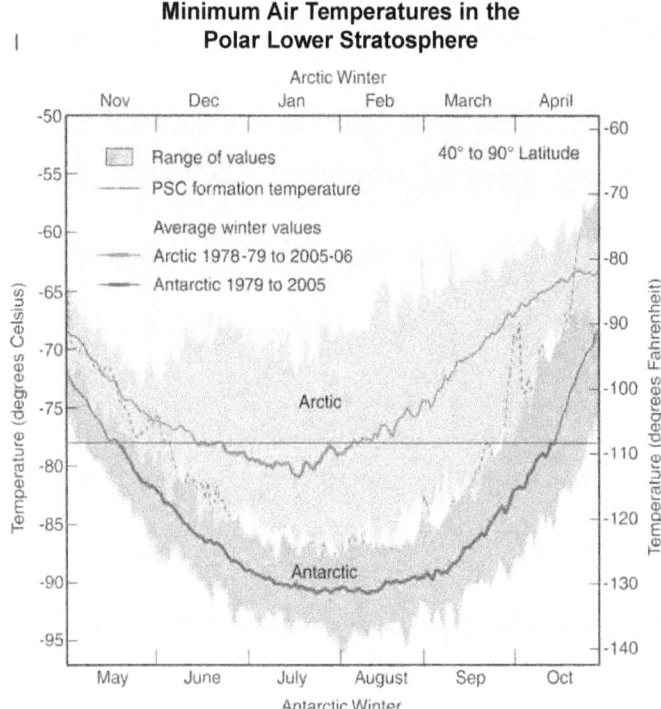

Minimum Air Temperatures in the Polar Lower Stratosphere

Figure Q10-1. Arctic and Antarctic temperatures. Stratospheric air temperatures in both polar regions reach minimum values in the lower stratosphere in the winter season. Average minimum values over Antarctica are as low as –90°C in July and August in a typical year. Over the Arctic, average minimum values are near –80°C in January and February. Polar stratospheric clouds (PSCs) are formed when winter minimum temperatures fall below the formation temperature (about –78°C). This occurs on average for 1 to 2 months over the Arctic and 5 to 6 months over Antarctica (see heavy red and blue lines). Reactions on PSCs cause the highly reactive chlorine gas ClO to be formed, which increases the destruction of ozone (see Q9). The range of winter minimum temperatures found in the Arctic is much greater than in the Antarctic. In some years, PSC formation temperatures are not reached in the Arctic, and significant ozone depletion does not occur. In the Antarctic, PSCs are present for many months, and severe ozone depletion now occurs in each winter season.

tures be present over a range of stratospheric altitudes, over large geographical regions, and for extended time periods. Low temperatures are important because they allow polar stratospheric clouds (PSCs) to form. Reactions on the surfaces of the cloud particles initiate a remarkable increase in the most reactive halogen gases (see below and Q8). Temperatures are lowest in the stratosphere over both polar regions in winter. In the Antarctic winter, minimum temperatures are generally lower and less variable than in the Arctic winter (see Figure Q10-1). Antarctic temperatures also remain below the PSC formation temperature for much longer periods during winter. This occurs, in part, because there are significant meteorological differences between the hemispheres, resulting from the differences in the distributions of land, ocean, and mountains at middle and high latitudes. The winter temperatures are low enough for PSCs to form for nearly the entire Antarctic winter but usually only for part of every Arctic winter.

Isolated conditions. Air in the polar stratospheric regions is relatively isolated from other stratospheric regions for long periods in the winter months. The isolation comes about because of strong winds that encircle the poles, preventing substantial motion of air in or out of the polar stratospheres. The isolation is much more effective in the Antarctic than the Arctic. Once chemical changes occur in the cold air as a result of the presence of PSCs, the changes remain for many weeks to months.

Polar stratospheric clouds (PSCs). Polar stratospheric clouds cause changes in the relative abundances of reactive chlorine gases. Reactions occur on the surfaces of PSC particles that convert the reservoir forms of reactive chlorine gases, $ClONO_2$ and HCl, to the most reactive form, ClO (see Figure Q8-1). ClO increases from a small fraction of available reactive chlorine gases to nearly all that is available (see Q8). With increased ClO, additional catalytic cycles involving ClO and BrO become active in the chemical destruction of ozone when sunlight is available (see Q9).

PSCs form when stratospheric temperatures fall below about −78°C (−108°F) in polar regions (see Figure Q10-1). As a result, PSCs are often found over large areas of the winter polar regions and over a significant altitude range. At low polar temperatures, nitric acid (HNO_3) and water condense on preexisting sulfur-containing particles to form solid and liquid PSC particles. At even lower temperatures, ice particles also form. PSC particles grow large enough and are numerous enough that cloud-like features can be observed from the ground under certain conditions, particularly when the Sun is near the horizon (see Figure Q10-2). PSCs are often found near mountain

ranges in polar regions because the motion of air over the mountains can cause local cooling of stratospheric air.

When temperatures increase by early spring, PSCs no longer form and the production of ClO ends. Without continued ClO production, ClO amounts decrease as other chemical reactions reform $ClONO_2$ and HCl. As a result, the intense period of ozone depletion ends.

PSC removal. Once formed, PSC particles move downward because of gravity. The largest particles move down several kilometers or more in the stratosphere during the low-temperature winter/spring period. Because most PSCs contain nitric acid, their downward motion removes nitric acid from regions of the ozone layer. That process is called *denitrification*. With less nitric acid, the

Arctic Polar Stratospheric Clouds

Figure Q10-2. Polar stratospheric clouds. This photograph of an Arctic polar stratospheric cloud (PSC) was taken from the ground at Kiruna, Sweden (67°N), on 27 January 2000. PSCs form during winters in the Arctic and Antarctic stratospheres. The particles grow from the condensation of water and nitric acid (HNO_3). The clouds often can be seen with the human eye when the Sun is near the horizon. Reactions on PSCs cause the highly reactive chlorine gas ClO to be formed, which is very effective in the chemical destruction of ozone (see Q9).

highly reactive chlorine gas ClO remains chemically active for a longer period, thereby increasing chemical ozone destruction. Denitrification occurs each winter in the Antarctic and in some, but not all, Arctic winters, because PSC formation temperatures are required over an extensive time period.

Discovering the role of PSCs. The formation of PSCs has been recognized for many years from ground-based observations. However, the geographical and altitude extent of PSCs in both polar regions was not known fully until PSCs were observed by a satellite instrument in the late 1970s. The role of PSCs in converting reactive chlorine gases to ClO was not understood until after the discovery of the Antarctic ozone hole in 1985. Our understanding of the PSC role developed from laboratory studies of their surface reactivity, computer modeling studies of polar stratospheric chemistry, and sampling of PSC particles and reactive chlorine gases, such as ClO, in the polar stratospheric regions.

The Discovery of the Antarctic Ozone Hole

The first decreases in Antarctic total ozone were observed in the early 1980s over research stations located on the Antarctic continent. The measurements were made with ground-based Dobson spectrophotometers (see box in Q5). The observations showed unusually low total overhead ozone during the late winter/early spring months of September, October, and November. Total ozone was lower in these months compared with previous observations made as early as 1957. The early published reports came from the British Antarctic Survey and the Japan Meteorological Agency. The results became more widely known in the international community after three scientists from the British Antarctic Survey published them in the journal *Nature* in 1985. Soon after, satellite measurements confirmed the spring ozone depletion and further showed that in each late winter/early spring season starting in the early 1980s, the depletion extended over a large region centered near the South Pole. The term "ozone hole" came about from satellite images of total ozone that showed very low values encircling the Antarctic continent each spring (see Q11). Currently, the formation and severity of the Antarctic "ozone hole" are documented each year by a combination of satellite, ground-based, and balloon observations of ozone.

III. STRATOSPHERIC OZONE DEPLETION

Q11: How severe is the depletion of the Antarctic ozone layer?

Severe depletion of the Antarctic ozone layer was first observed in the early 1980s. Antarctic ozone depletion is seasonal, occurring primarily in late winter and early spring (August-November). Peak depletion occurs in early October when ozone is often completely destroyed over a range of altitudes, reducing overhead total ozone by as much as two-thirds at some locations. This severe depletion creates the "ozone hole" in images of Antarctic total ozone made from space. In most years the maximum area of the ozone hole far exceeds the size of the Antarctic continent.

The severe depletion of Antarctic ozone, known as the "ozone hole," was first observed in the early 1980s. The depletion is attributable to chemical destruction by reactive halogen gases, which increased in the stratosphere in the latter half of the 20th century (see Q16). Conditions in the Antarctic winter stratosphere are highly suitable for ozone depletion because of (1) the long periods of extremely low temperatures, which promote polar stratospheric cloud (PSC) formation; (2) the abundance of reactive halogen gases, which chemically destroy ozone; and (3) the isolation of stratospheric air during the winter, which allows time for chemical destruction to occur (see Q10). The severity of Antarctic ozone depletion can be seen using satellite observations of total ozone, ozone altitude profiles, and long-term average values of polar total ozone.

Antarctic ozone hole. The most widely used images of Antarctic ozone depletion are those from space-based measurements of total ozone. Satellite images made during Antarctic winter and spring show a large region centered near the South Pole in which total ozone is highly depleted (see Figure Q11-1). This region has come to be called the "ozone hole" because of the near-circular contours of low ozone values in the images. The area of the ozone hole is defined here as the area contained within the 220-Dobson unit (DU) contour in total ozone maps (light blue color in Figure Q11-1). The maximum area has reached 25 million square kilometers (about 10 million square miles) in recent years, which is nearly twice the area of the Antarctic continent (see Figure Q11-2). Minimum values of total ozone inside the ozone hole averaged in late September have reached below 100 DU, which is well below normal springtime values of about 200 DU (see Figure Q11-2).

Altitude profiles of Antarctic ozone. Ozone within the "ozone hole" is also measured using balloonborne instruments (see Q5). Balloon measurements show changes within the ozone layer, the vertical region that contains the highest ozone abundances in the stratosphere. At geographic locations where the lowest total ozone

values occur in ozone hole images, balloon measurements show that the chemical destruction of ozone is complete over a vertical region of several kilometers. Balloon measurements shown in Figure Q11-3 give an example of such depletion over South Pole, Antarctica, on 2 October 2001. The altitude region of total depletion (14-20 kilometers) in the profile corresponds to the region of lowest

Antarctic Ozone Hole

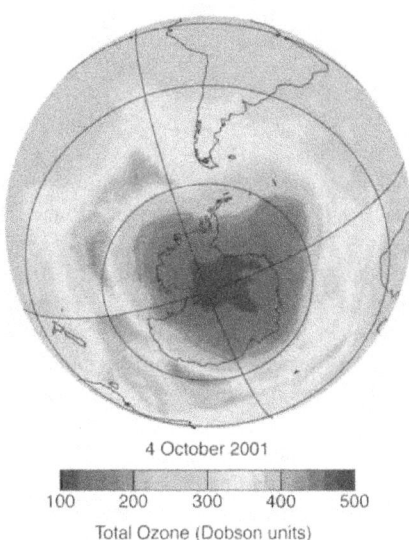

4 October 2001

100   200   300   400   500

Total Ozone (Dobson units)

Figure Q11-1. Antarctic "ozone hole." Total ozone values are shown for high southern latitudes as measured by a satellite instrument. The dark blue and purple regions over the Antarctic continent show the severe ozone depletion or "ozone hole" now found during every spring. Minimum values of total ozone inside the ozone hole are close to 100 Dobson units (DU) compared with normal springtime values of about 200 DU (see Q4). In late spring or early summer (November-December) the ozone hole disappears in satellite images as ozone-depleted air is displaced and mixed with ozone-rich air transported poleward from outside the ozone hole.

Antarctic Ozone Depletion

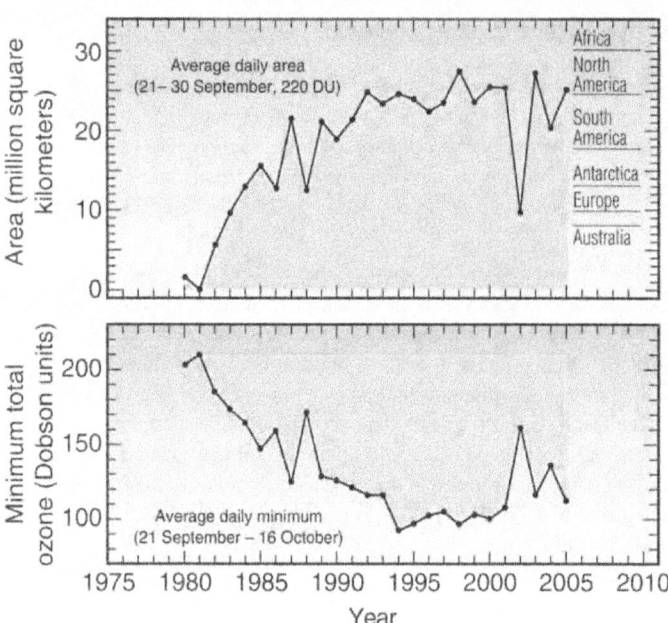

Figure Q11-2. Antarctic ozone hole features. Values are shown for key parameters of the Antarctic ozone hole: the area enclosed by the 220-DU total ozone contour and the minimum total ozone amount, as determined from space-based observations. The values are averaged for each year near the peak of ozone depletion, as defined by the dates shown in each panel. The ozone hole areas are contrasted to the areas of continents in the upper panel. The intensity of ozone depletion gradually increased beginning in 1980. In the 1990s, the depletion reached fairly steady values, except for the anomalously low depletion in 2002 (see Figure Q11-box). The intensity of Antarctic ozone depletion will decrease as part of the ozone recovery process (see Q19 and Q20).

winter temperatures and highest chlorine monoxide (ClO) abundances. The average South Pole ozone profiles for the decades 1962-1971 and 1992-2001 (see Figure Q11-3) show how reactive halogen gases have dramatically altered the ozone layer. For the 1960s, the ozone layer is clearly evident in the October average profile and has a peak near 16 kilometers. For the 1990s, minimum average values in the center of the layer have fallen by 90% from the earlier values.

Long-term total ozone changes. Low winter temperatures and isolated conditions occur each year in the Antarctic stratosphere, but significant spring ozone depletion has been observed every year only since the early 1980s. In prior years, the amounts of reactive halogen gases in the stratosphere were insufficient to cause significant depletion. Satellite observations can be used to examine how ozone depletion has changed with time in both polar regions for the last three decades. Changes in ozone hole areas and minimum Antarctic ozone amounts are shown in Figure Q11-2. Depletion has increased since 1980 to become fairly stable in the 1990s and early 2000s, with the exception of 2002 (see Q11-box). Total ozone averaged over the Antarctic region in late winter/early spring shows similar features (Figure Q12-1). Average values decreased steadily through the 1980s and 1990s,

reaching minimum values that were 37% less than in pre-ozone-hole years (1970-1982). The year-to-year changes in the average values reflect variations in the meteorological conditions, which affect the extent of low polar temperatures and the transport of air into and out of the Antarctic winter stratosphere (see Figure Q11-box). However, essentially all of the ozone depletion in the Antarctic in most years is attributable to chemical loss from reactive halogen gases.

Restoring ozone in spring. The depletion of Antarctic ozone occurs primarily in the late winter/early spring season. In spring, temperatures in the lower polar stratosphere eventually warm, thereby ending PSC formation as well as the most effective chemical cycles that destroy ozone (see Q10). The transport of air between the polar stratosphere and lower latitudes also increases during this time, ending winter isolation. This allows ozone-rich air to be transported to polar regions, displacing air in which ozone has been severely depleted. This displaced air is mixed at lower latitudes with more abundant ozone-rich air. As a result, the ozone hole disappears by December and Antarctic ozone amounts remain near normal until the next winter season.

Polar Ozone Depletion

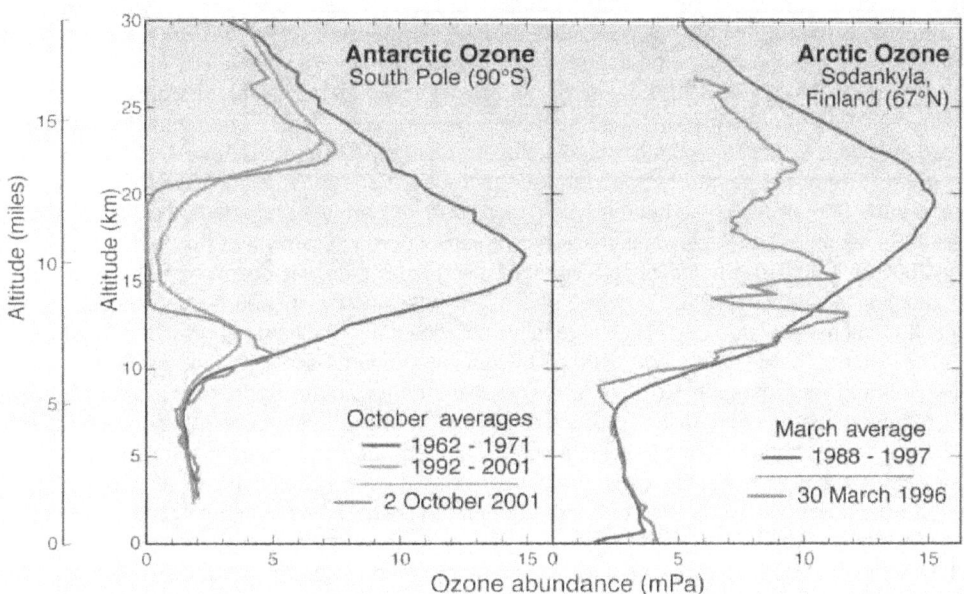

Figure Q11-3. Arctic and Antarctic ozone distribution. The stratospheric ozone layer resides between about 10 and 50 kilometers (6 to 31 miles) above Earth's surface over the globe. Long-term observations of the ozone layer with balloonborne instruments allow the winter Antarctic and Arctic regions to be compared. In the Antarctic at the South Pole, halogen gases have destroyed ozone in the ozone layer beginning in the 1980s. Before that period, the ozone layer was clearly present, as shown here using average ozone values from balloon observations made between 1962 and 1971. In more recent years, as shown here for 2 October 2001, ozone is destroyed completely between 14 and 20 kilometers (8 to 12 miles) in the Antarctic in spring. Average October values in the ozone layer now are reduced by 90% from pre-1980 values. The Arctic ozone layer is still present in spring as shown by the average March profile obtained over Finland between 1988 and 1997. However, March Arctic ozone values in some years are often below normal average values as shown here for 30 March 1996. In such years, winter minimum temperatures are generally below PSC formation temperatures for long periods. Ozone abundances are shown here with the unit "milli-Pascals" (mPa), which is a measure of absolute pressure (100 million mPa = atmospheric sea-level pressure).

The Anomalous 2002 Antarctic Ozone Hole

The 2002 Antarctic ozone hole showed features that surprised scientists. They considered it anomalous at the time because the hole had much less area as viewed from space and much less ozone depletion as measured by minimum column ozone amounts when compared with values in several preceding years (see Figure Q11-box). The 2002 ozone hole area and minimum ozone values stand out clearly in displays of the year-to-year changes in these quantities (see Figure Q11-2). The smaller area was unexpected because the conditions required to deplete ozone, namely low temperatures and available reactive halogen gases, are not expected to have large year-to-year variations. Ozone was being depleted in August and early September 2002, but the hole *broke apart* into two separate depleted regions during the last week of September. The depletion in these two regions was significantly less than was observed inside either the 2001 or 2003 ozone holes, but still substantially greater than was observed in the early 1980s.

The anomalous behavior in 2002 occurred because of specific atmospheric air motions that sometimes occur in polar regions, not large decreases in reactive chlorine and bromine amounts in the Antarctic stratosphere. The Antarctic stratosphere was warmed by very strong, large-scale weather systems in 2002 that originated in the lower atmosphere (troposphere) at midlatitudes in late September. In late September, Antarctic temperatures are generally very low (see Q10) and ozone destruction rates are near their peak values. These tropospheric systems traveled poleward and upward into the stratosphere, upsetting the circumpolar wind flow and warming the lower stratosphere where ozone depletion was ongoing. The higher-than-normal impact of these weather disturbances during the critical time period for ozone loss reduced the total loss of ozone in 2002.

The warming in 2002 was unprecedented in Antarctic meteorological observations. Warming events are difficult to predict because of their complex formation conditions.

Large Antarctic ozone depletion returned in 2003 through 2005, in a manner similar to that observed from the mid-1990s to 2001 (see Figures Q11-box and Q11-2). The high ozone depletion found since the mid-1990s, with the exception of 2002, is expected to be typical of coming years. A significant, sustained reduction of Antarctic ozone depletion, defined as ozone recovery, requires the removal of halogen source gases from the stratosphere (see Q19 and Q20).

Antarctic Ozone Hole

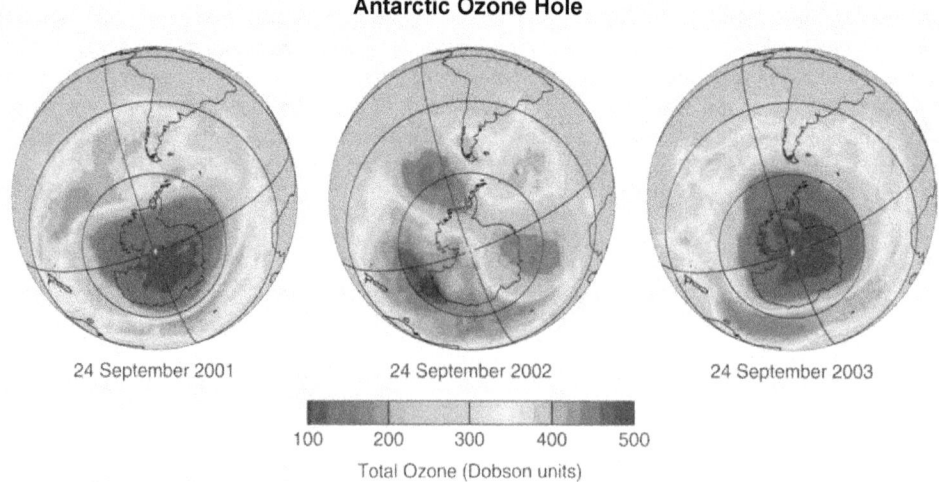

24 September 2001   24 September 2002   24 September 2003

100   200   300   400   500

Total Ozone (Dobson units)

Figure Q11-Box. Anomalous 2002 ozone hole. Views from space of the Antarctic ozone hole as observed on 24 September in each of three consecutive years. The hole split and elongated in 2002, reducing the total depletion of ozone observed that year in comparison with 2001 and 2003. The anomalous depletion in 2002 is attributable to an early warming of the polar stratosphere caused by air disturbances originating in midlatitudes, rather than to large changes in the amounts of reactive chlorine and bromine in the Antarctic stratosphere.

Q12: Is there depletion of the Arctic ozone layer?

Yes, significant depletion of the Arctic ozone layer now occurs in some years in the late winter/early spring period (January-April). However, the maximum depletion is less severe than that observed in the Antarctic and is more variable from year to year. A large and recurrent "ozone hole," as found in the Antarctic stratosphere, does not occur in the Arctic.

Significant ozone depletion in the Arctic stratosphere occurs in cold winters because of reactive halogen gases. The depletion, however, is much less than the depletion that now occurs in every Antarctic winter and spring. Although Arctic depletion does not generally create persistent "ozone hole"-like features in Arctic total ozone maps, depletion is observed in altitude profiles of ozone and in long-term average values of polar ozone.

Altitude profiles of Arctic ozone. Arctic ozone is measured using a variety of instruments (see Q5), as for the Antarctic (see Q11). These measurements show changes within the ozone layer, the vertical region that contains the highest ozone abundances in the stratosphere. Figure Q11-2 shows an example of balloonborne measurements of a depleted ozone profile in the Arctic region on 30 March 1996, and contrasts the depletion with that found in the Antarctic. The 30 March spring profile shows much less depletion than the 2 October spring profile in the Antarctic. In general, some reduction in the Arctic ozone layer occurs each late winter/early spring season. However, complete depletion each year over a broad vertical layer, as is now common in the Antarctic stratosphere, is not found in the Arctic.

Long-term total ozone changes. Satellite and ground-based observations can be used to examine the average total ozone abundances in the Arctic region for the last three decades and to contrast them with Antarctic abundances (see Figure Q12-1). Decreases from the pre-ozone-hole average values (1970-1982) were observed in the Arctic beginning in the 1980s, when similar changes were occurring in the Antarctic. The decreases have reached a maximum of about 30% but have remained smaller than those found in the Antarctic since the mid-1980s. The year-to-year changes in the Arctic and Antarctic average ozone values reflect annual variations in meteorological conditions that affect the extent of low polar temperatures and the transport of air into and out of the polar stratosphere. The effect of these variations is generally greater for the Arctic than the Antarctic.

Figure Q12-1. Average polar ozone. Total ozone in polar regions is measured by well-calibrated satellite instruments. Shown here is a comparison of average springtime total ozone values found between 1970 and 1982 (solid and dashed red lines) with those in later years. Each point represents a monthly average in October in the Antarctic or in March in the Arctic. After 1982, significant ozone depletion is found in most years in the Arctic and all years in the Antarctic. The largest average depletions have occurred in the Antarctic since 1990. The ozone changes are the combination of chemical destruction and natural variations. Variations in meteorological conditions influence the year-to-year changes in depletion,

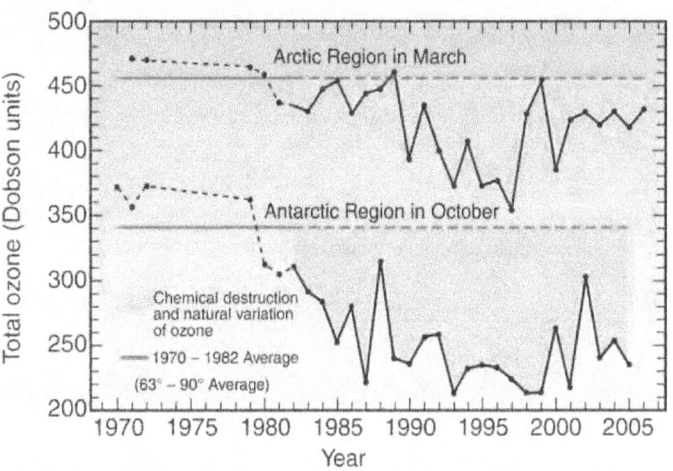

particularly in the Arctic. Essentially all of the decrease in the Antarctic and usually most of the decrease in the Arctic each year are attributable to chemical destruction by reactive halogen gases. Average total ozone values over the Arctic are naturally larger at the beginning of each winter season because more ozone is transported poleward each season in the Northern Hemisphere than in the Southern Hemisphere.

Arctic vs. Antarctic. The Arctic winter stratosphere is generally warmer than its Antarctic counterpart (see Figure Q10-1). Higher temperatures reduce polar stratospheric cloud (PSC) formation, which slows the conversion of reactive chlorine gases to form ClO and, as a consequence, reduces the amount of ozone depletion (see Q10). Furthermore, the temperature and wind conditions are much more variable in the Arctic from winter to winter and within a winter season than in the Antarctic. Large year-to-year differences occur in Arctic minimum temperatures and the duration of PSC-forming temperatures into early spring. In a few Arctic winters, minimum temperatures are not low enough for PSCs to form. These factors combine to cause ozone depletion to be variable in the Arctic from year to year, with some years having little to no ozone depletion.

As in the Antarctic, depletion of ozone in the Arctic is confined to the late winter/early spring season. In spring, temperatures in the lower stratosphere eventually warm, thereby ending PSC formation as well as the most effective chemical cycles that destroy ozone. The subsequent transport of ozone-rich air into the Arctic stratosphere displaces ozone-depleted air. As a result, ozone layer abundances are restored to near-normal values until the following winter.

High Arctic total ozone. A significant difference exists between the Northern and Southern Hemispheres in how ozone-rich stratospheric air is transported into the polar regions from lower latitudes during fall and winter. In the northern stratosphere, the poleward and downward transport of ozone-rich air is stronger. As a result, total ozone values in the Arctic are considerably higher than in the Antarctic at the beginning of each winter season (see Figure Q12-1).

Q13: How large is the depletion of the global ozone layer?

The ozone layer has been depleted gradually since 1980 and now is about an average of 4% lower over the globe. The average depletion exceeds the natural variability of the ozone layer. The ozone loss is very small near the equator and increases with latitude toward the poles. The larger polar depletion is primarily a result of the late winter/early spring ozone destruction that occurs there each year.

Stratospheric ozone has decreased over the globe since the 1980s. The depletion, which in the period 1997-2005 averaged about 4% (see Figure Q13-1), is larger than natural variations in ozone. The observations shown in Figure Q13-1 have been smoothed to remove regular ozone changes that are due to seasonal and solar effects (see Q14). The increase in reactive halogen gases in the stratosphere is considered to be the primary cause of the average depletion. The lowest ozone values in recent years occurred following the 1991 eruption of Mt. Pinatubo, which increased the number of sulfur-containing particles in the stratosphere. The particles remain in the stratosphere for several years, increasing the effectiveness of reactive halogen gases in destroying ozone (see Q14).

Observed ozone depletion varies significantly with latitude on the globe (see Figure Q13-1). The largest losses occur at the highest southern latitudes as a result of the severe ozone loss over Antarctica each late winter/early spring period. The next largest losses are observed in the Northern Hemisphere, caused in part by late winter/early spring losses over the Arctic. Ozone-depleted air over both polar regions is dispersed away from the poles during and after each winter/spring period. Ozone depletion also occurs directly at latitudes between the equator and polar regions but is smaller because of the presence of lower amounts of reactive halogen gases (see Q8).

Tropical regions. There has been little or no depletion of total ozone in the tropics (between about 20° latitude north and south of the equator in Figure Q13-1). In

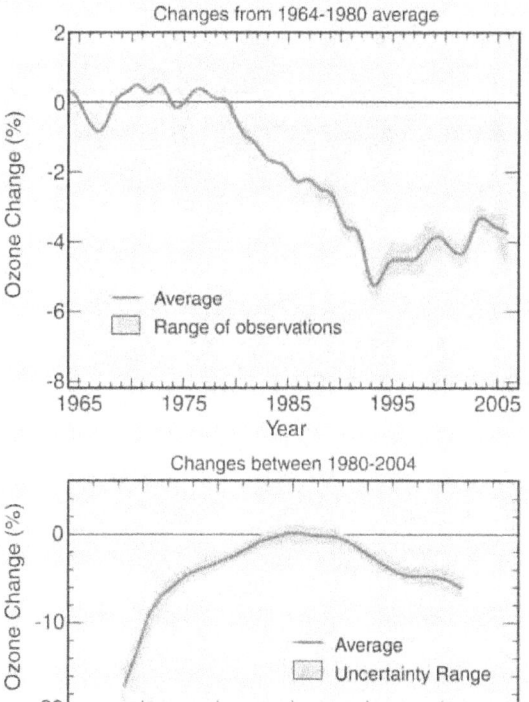

Global Total Ozone Change

Changes from 1964-1980 average

Changes between 1980-2004

Figure Q13-1. Global total ozone changes. Satellite observations show a decrease in global total ozone values over more than two decades. The top panel compares global ozone values (annual averages) with the average from the period 1964 to 1980. Seasonal and solar effects have been removed from the data. On average, global ozone decreased each year between 1980 and the early 1990s. The decrease worsened during the few years when volcanic aerosol from the Mt. Pinatubo eruption in 1991 remained in the stratosphere. Now global ozone is about 4% below the 1964-to-1980 average. The bottom panel compares ozone changes between 1980 and 2004 for different latitudes. The largest decreases have occurred at the highest latitudes in both hemispheres because of the large winter/spring depletion in polar regions. The losses in the Southern Hemisphere are greater than those in the Northern Hemisphere because of the Antarctic ozone hole. Long-term changes in the tropics are much smaller because reactive halogen gases are less abundant in the tropical lower stratosphere.

this region of the lower stratosphere, air has only recently (less than 18 months) been transported from the lower atmosphere. As a result, the conversion of halogen source gases to reactive halogen gases is very small. Because of the low abundance of reactive gases, total ozone depletion in this region is very small. In contrast, stratospheric air in polar regions has been in the stratosphere for an average of 4 to 7 years; therefore, the abundance of reactive halogen gases is much larger.

Seasonal changes. The magnitude of global ozone depletion also depends on the season of the year. In comparison with the 1964-1980 averages, total ozone averaged for 2002-2005 is about 3% lower in northern middle latitudes (35°N-60°N) and about 6% lower at southern middle latitudes (35°S-60°S). The seasonality of these changes is also somewhat different in the two hemispheres. In the summer/autumn periods, the decline in total ozone is about 2% in the Northern Hemisphere and 5% in the Southern Hemisphere. In winter/spring, the decline is about 5-6% in both hemispheres.

Q14: Do changes in the Sun and volcanic eruptions affect the ozone layer?

Yes, factors such as changes in solar radiation, as well as the formation of stratospheric particles after volcanic eruptions, do influence the ozone layer. However, neither factor can explain the average decreases observed in global total ozone over the last two decades. If large volcanic eruptions occur in the coming decades, ozone depletion will increase for several years after the eruption.

Changes in solar radiation and increases in stratospheric particles from volcanic eruptions both affect the abundance of stratospheric ozone, but they have not caused the long-term decreases observed in total ozone.

Solar changes. The formation of stratospheric ozone is initiated by ultraviolet (UV) radiation coming from the Sun (see Figure Q2-1). As a result, an increase in the Sun's radiation output increases the amount of ozone in Earth's atmosphere. The Sun's radiation output and sunspot number vary over the well-known 11-year solar cycle.

Observations over several solar cycles (since the 1960s) show that global total ozone levels vary by 1 to 2% between the maximum and minimum of a typical cycle. Changes in solar output at a wavelength of 10.7 cm, although much larger than changes in total solar output, are often used to show when periods of maximum and minimum total output occur (see Figure Q14-1). The Sun's output has gone through maximum values around 1969, 1980, 1991, and 2002. In 2006, the solar output was decreasing towards a minimum.

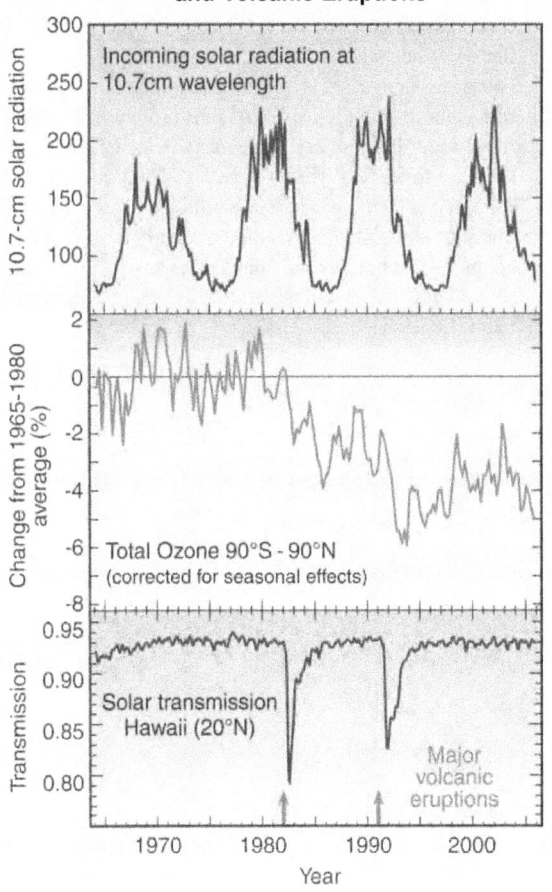

The Solar Cycle, Global Ozone, and Volcanic Eruptions

Figure Q14-1. Solar changes and volcanoes. Total ozone values have decreased beginning in the early 1980s (see middle panel). The ozone values shown are 3-month averages corrected for seasonal effects. Incoming solar radiation, which produces ozone in the stratosphere, changes on a well-recognized 11-year cycle. The amount of solar radiation at a wavelength of 10.7-cm is often used to document the 11-year cycle (see top panel). A comparison of the top and middle panels indicates that the cyclic changes in solar output cannot account for the long-term decreases in ozone. Volcanic eruptions occurred frequently in the 1965 to 2005 period. The largest recent eruptions are El Chichón (1982) and Mt. Pinatubo (1991) (see red arrows in bottom panel). Large volcanic eruptions are monitored by the decreases in solar transmission to Earth's surface that occur because new particles are formed in the stratosphere from volcanic sulfur emissions (see bottom panel). These particles increase ozone depletion only temporarily because they do not remain in the stratosphere for more than a few years. A comparison of the middle and bottom panels indicates that large volcanic eruptions also cannot account for the long-term decreases found in global total ozone.

Over the last two decades, average total ozone has decreased over the globe. Average values in recent years show about 4% depletion from pre-1980 values (see Figure Q14-1). The ozone values shown are 3-month averages corrected for seasonal effects but not for solar effects. Over the same period, changes in solar output show the expected 11-year cycle but do not show a decrease with time. For this reason, the long-term decreases in global ozone cannot result from changes in solar output alone. Most examinations of long-term ozone changes presented in this and previous international scientific assessments quantitatively account for the influence of the 11-year solar cycle.

Past volcanoes. Large volcanic eruptions inject sulfur gases directly into the stratosphere, causing new sulfate particles to be formed. The particles initially form in the stratosphere above and downwind of the volcano location and then often spread throughout the hemisphere or globally as air is transported by stratospheric winds. The presence of volcanic particles in the stratosphere is shown in observations of solar transmission through the atmosphere. When large amounts of particles are present in the stratosphere, transmission of solar radiation is significantly reduced. The large eruptions of El Chichón (1982) and Mt. Pinatubo (1991) are recent examples of events that temporarily reduced solar transmission (see Figure Q14-1).

Laboratory measurements and stratospheric observations have shown that chemical reactions on the surface of volcanically produced particles increase ozone destruction by increasing the amounts of the highly reactive chlorine gas, chlorine monoxide (ClO). The amount of ClO produced is proportional to the total abundance of reactive chlorine in the stratosphere (see Figure Q16-1). Ozone depletion increases as a consequence of increased ClO. The most recent large eruption was that of Mt. Pinatubo, which resulted in up to a 10-fold increase in the number of particles available for surface reactions. Both El Chichón and Mt. Pinatubo increased global ozone depletion for a few years (see Figure Q14-1). After a few years, however, the effect of volcanic particles on ozone is diminished by their gradual removal from the stratosphere by natural air circulation. Because of particle removal, the two large volcanic eruptions of the last two decades cannot account for the long-term decreases observed in ozone over the same period.

Future volcanoes. Observations and atmospheric models indicate that the record-low ozone levels observed in 1992-1993 resulted from the large number of particles produced by the Mt. Pinatubo eruption, combined with the relatively large amounts of reactive halogen gases present in the stratosphere in the 1990s. If the Mt. Pinatubo eruption had occurred before 1980, changes to global ozone would have been much smaller than observed in 1992-1993 because the abundances of reactive halogen gases in the stratosphere were smaller. In the early decades of the 21st century, the abundance of halogen source gases will still be substantial in the global atmosphere (see Figure Q16-1). If large volcanic eruptions occur in these early decades, ozone depletion will increase for several years. If an eruption larger than Mt. Pinatubo occurs, ozone losses could be larger than previously observed and persist longer. Only later in the 21st century when halogen gas abundances have declined close to pre-1980 values will the effect of volcanic eruptions on ozone be lessened.

IV. CONTROLLING OZONE-DEPLETING GASES

Q15: Are there regulations on the production of ozone-depleting gases?

Yes, the production of ozone-depleting gases is regulated under a 1987 international agreement known as the "Montreal Protocol on Substances that Deplete the Ozone Layer" and its subsequent Amendments and Adjustments. The Protocol, now ratified by over 190 nations, establishes legally binding controls on the national production and consumption of ozone-depleting gases. Production and consumption of all principal halogen-containing gases by developed and developing nations will be significantly phased out before the middle of the 21ˢᵗ century.

Montreal Protocol. In 1985, a treaty called the *Vienna Convention for the Protection of the Ozone Layer* was signed by 20 nations in Vienna. The signing nations agreed to take appropriate measures to protect the ozone layer from human activities. The Vienna Convention supported research, exchange of information, and future protocols. In response to growing concern, the *Montreal Protocol on Substances that Deplete the Ozone Layer* was signed in 1987 and, following country ratification, entered into force in 1989. The Protocol established legally binding controls for developed and developing nations on the production and consumption of halogen source gases known to cause ozone depletion. National consumption of a halogen gas is defined as the amount that production and imports of a gas exceed its export to other nations.

Amendments and Adjustments. As the scientific basis of ozone depletion became more certain after 1987 and substitutes and alternatives became available to replace the principal halogen source gases, the Montreal Protocol was strengthened with Amendments and Adjustments. These revisions put additional substances under regulation, accelerated existing control measures, and prescribed phaseout dates for the production and con-

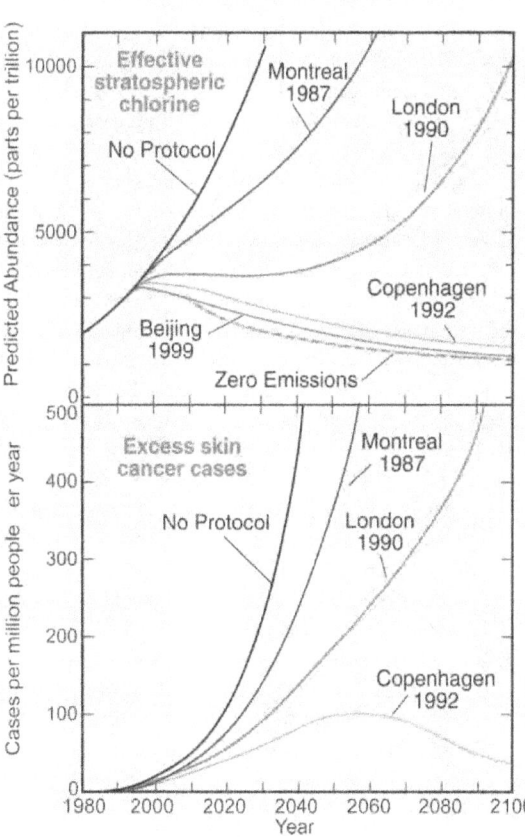

Effect of the Montreal Protocol

Figure Q15-1. Effect of the Montreal Protocol. The purpose of the Montreal Protocol is to achieve reductions in stratospheric abundances of chlorine and bromine. The reductions follow from restrictions on the production and consumption of manufactured halogen source gases. Projections of the future abundance of *effective stratospheric chlorine* (see Q16) are shown in the top panel assuming (1) no Protocol regulations, (2) only the regulations in the original 1987 Montreal Protocol, and (3) additional regulations from the subsequent Amendments and Adjustments. The city names and years indicate where and when changes to the original 1987 Protocol provisions were agreed upon. Effective stratospheric chlorine as used here accounts for the combined effect of chlorine and bromine gases. Without the Protocol, stratospheric halogen gases are projected to increase significantly in the 21ˢᵗ century. The "zero emissions" line shows a hypothetical case of stratospheric abundances if all emissions were reduced to zero beginning in 2007. The lower panel shows how excess skin cancer cases (see Q17) might increase with no regulation and how they might be reduced under the Protocol provisions. (The unit "parts per trillion" is defined in the caption of Figure Q7-1.)

sumption of certain gases. The initial Protocol called for only a slowing of chlorofluorocarbon (CFC) and halon production. The 1990 London Amendments to the Protocol called for a phaseout of the production and consumption of the most damaging ozone-depleting substances in developed nations by 2000 and in developing nations by 2010. The 1992 Copenhagen Amendments accelerated the date of the phaseout to 1996 in developed nations. Further controls on ozone-depleting substances were agreed upon in later meetings in Vienna (1995), Montreal (1997), and Beijing (1999).

Montreal Protocol projections. Future stratospheric abundances of effective stratospheric chlorine (see Q16) can be calculated based on the provisions of the Montreal Protocol. The concept of *effective stratospheric chlorine* accounts for the combined effect on ozone of chlorine- and bromine-containing gases. The results are shown in Figure Q15-1 for the following cases:

- No Protocol and continued production increases of 3% per year (business-as-usual scenario).
- Continued production and consumption as allowed by the Protocol's original provisions agreed upon in Montreal in 1987.
- Restricted production and consumption as outlined in the subsequent Amendments and Adjustments as decided in London in 1990, Copenhagen in 1992, and Beijing in 1999.
- Zero emissions of ozone-depleting gases starting in 2007.

In each case, production of a gas is assumed to result in its eventual emission to the atmosphere. Without the Montreal Protocol and with continued production and use of CFCs and other ozone-depleting gases, effective stratospheric chlorine is projected to have increased tenfold by the mid-2050s compared with the 1980 value. Such high values likely would have increased global ozone depletion far beyond that currently observed. As a result, harmful UV-B radiation would have also increased substantially at Earth's surface, causing a rise in excess skin cancer cases (see Q17 and lower panel of Figure Q15-1).

The 1987 provisions of the Montreal Protocol alone would have only slowed the approach to high effective chlorine values by one or more decades in the 21st century. Not until the 1992 Copenhagen Amendments and Adjustments did the Protocol projections show a *decrease* in future effective stratospheric chlorine values. Now, with full compliance to the Montreal Protocol and its Amendments and Adjustments, use of the major human-produced ozone-depleting gases will ultimately be phased out and effective stratospheric chlorine will slowly decay, reaching pre-1980 values in the mid-21st century (see Q16).

Zero emissions. Effective chlorine values in the coming decades will be influenced by emissions of halogen source gases produced in those decades, as well as the emission of currently existing gases that are now being used or stored in various ways. Examples of long-term storage are CFCs in refrigeration equipment and foams, and halons in fire-fighting equipment. Some continued production and consumption of ozone-depleting gases is allowed, particularly in developing nations, under the agreements. As a measure of the contribution of these continued emissions to the effective chlorine value, the "zero emissions" case is included in Figure Q15-1. In this hypothetical case, all emissions of ozone-depleting gases are set to zero beginning in 2007. The reductions in effective stratospheric chlorine below the values expected with the 1999 Beijing agreement would be relatively small.

HCFC substitute gases. The Montreal Protocol provides for the transitional use of hydrochlorofluorocarbons (HCFCs) as substitute compounds for principal halogen source gases such as CFC-12. HCFCs differ chemically from most other halogen source gases in that they contain hydrogen (H) atoms in addition to chlorine and fluorine atoms. HCFCs are used for refrigeration, for blowing foams, and as solvents, which were primary uses of CFCs. HCFCs are 88 to 98% less effective than CFC-12 in depleting stratospheric ozone because they are chemically removed primarily in the troposphere (see Q18). This removal partially protects stratospheric ozone from the halogens contained in HCFCs. In contrast, CFCs and many other halogen source gases are chemically inert in the troposphere and, hence, reach the stratosphere without being significantly removed. Because HCFCs still contribute to the chlorine abundance in the stratosphere, the Montreal Protocol requires a gradual phaseout of HCFC consumption in developed and developing nations that will be complete in 2040.

HFC substitute gases. Hydrofluorocarbons (HFCs) are also used as substitute compounds for CFCs and other halogen source gases. HFCs contain only hydrogen, fluorine, and carbon atoms. Because HFCs contain no chlorine or bromine, they do not contribute to ozone depletion (see Q18). As a consequence, the Montreal Protocol does not regulate the HFCs. However, HFCs (as well as all halogen source gases) are radiatively active gases that contribute to human-induced climate change as they accumulate in the atmosphere (see Q18). HFCs are included in the group of gases listed in the Kyoto Protocol of the United Nations Framework Convention on Climate Change (UNFCCC).

Q16: Has the Montreal Protocol been successful in reducing ozone-depleting gases in the atmosphere?

Yes, as a result of the Montreal Protocol, the total abundance of ozone-depleting gases in the atmosphere has begun to decrease in recent years. If the nations of the world continue to follow the provisions of the Montreal Protocol, the decrease will continue throughout the 21st century. Some individual gases, such as halons and hydrochlorofluorocarbons (HCFCs), are still increasing in the atmosphere but will begin to decrease in the next decades if compliance with the Protocol continues. Around midcentury, the effective abundance of ozone-depleting gases should fall to values that were present before the Antarctic "ozone hole" began to form in the early 1980s.

Effective stratospheric chlorine. The Montreal Protocol has been successful in slowing and reversing the increase of ozone-depleting gases (halogen source gases) in the atmosphere. An important measure of its success is the change in the value of *effective stratospheric chlorine.* Effective stratospheric chlorine values are a measure of the potential for ozone depletion in the stratosphere, obtained by summing over adjusted amounts of all chlorine and bromine gases. The adjustments account for the different rates of decomposition of the gases and the greater per-atom effectiveness of bromine in depleting ozone (see Q7). Although chlorine is much more abundant in the stratosphere than bromine (160 times) (see Figure Q7-1), bromine atoms are about 60 times more effective than chlorine atoms in chemically destroying ozone molecules. Increases in effective stratospheric chlorine in the past decades have caused ozone depletion. Accordingly, ozone is expected to recover in the future as effective stratospheric chlorine values decrease.

Effective stratospheric chlorine changes. In the latter half of the 20th century up until the 1990s, effective stratospheric chlorine values steadily increased (see Figure Q16-1). Values are derived from individual halogen source gas abundances obtained from measurements, historical estimates of abundance, and projections of future abundance. As a result of the Montreal Protocol regulations, the long-term increase in effective stratospheric chlorine slowed, reached a peak, and began to decrease in the 1990s. This initial decrease means that the potential for stratospheric ozone depletion has begun to lessen as a result of the Montreal Protocol. The decrease in effective chlorine is projected to continue throughout the 21st century if all nations continue to comply with the provisions of the Protocol. The decrease will continue because, as emissions become small, natural destruction processes gradually remove halogen-containing gases from the global atmosphere. Reduction of effective stratospheric chlorine amounts to 1980 values or lower will require many decades because the lifetimes of halogen source gas molecules in the atmosphere range up to 100 years (see Figure Q16-1 and Table Q7-1).

Individual halogen source gas reductions. The reduction in the atmospheric abundance of a gas in response to regulation depends on a number of factors that include (1) how rapidly gas reserves are used and released to the atmosphere, (2) the lifetime for the removal of the gas from the atmosphere, and (3) the total amount of the gas that has already accumulated in the atmosphere.

The regulation of human-produced halogen source gases under the Montreal Protocol is considered separately for each class of one or more gases and is based on several factors. The factors include (1) the effectiveness of each class in depleting ozone in comparison with other halogen source gases, (2) the availability of suitable substitute gases for domestic and industrial use, and (3) the impact of regulation on developing nations.

Methyl chloroform and CFCs. The largest reduction in the abundance of a halogen source gas has occurred for methyl chloroform (CH_3CCl_3) (see Figure Q16-1). The implementation of the Montreal Protocol caused global production of methyl chloroform to be reduced to near zero. Atmospheric abundances subsequently dropped rapidly because methyl chloroform has a short atmospheric lifetime (about 5 years). Methyl chloroform is used mainly as a solvent and has no significant long-term storage following production. The reduction in effective chlorine in the 1990s came primarily from the reduction in methyl chloroform abundance in the atmosphere. Significant emissions reductions have also occurred for the chlorofluorocarbons CFC-11, CFC-12, and CFC-113 starting in the 1990s. As a result, the atmospheric amounts of these gases have all peaked, and CFC-11 and CFC-113 abundances have decreased slightly (see Figure Q16-1). As emissions of CFCs are reduced, their atmospheric abundances will decrease more slowly than methyl chloroform because of longer CFC atmospheric lifetimes (see Table Q7-1) and because CFCs escape very slowly to the atmosphere from their use in refrigeration

Past and Expected Future Abundances of Atmospheric Halogen Source Gases

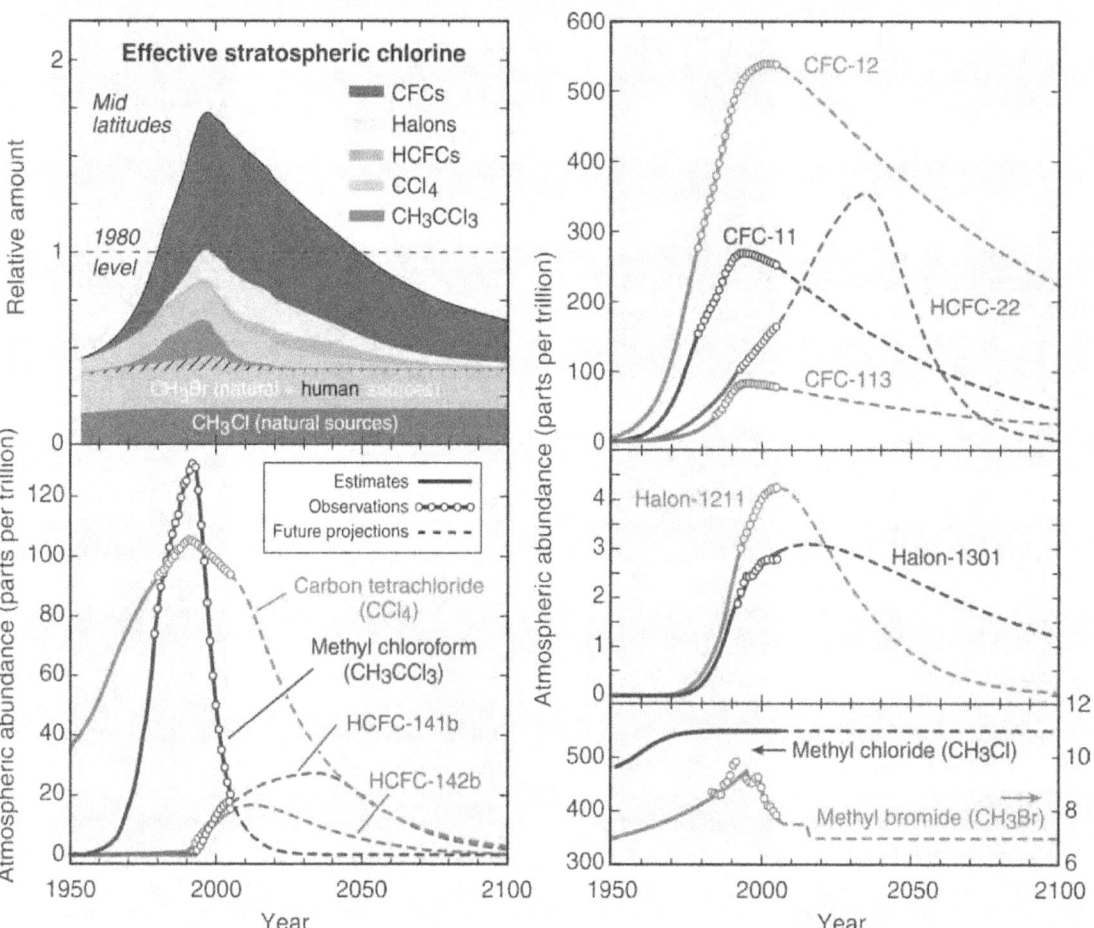

Figure Q16-1. Halogen source gas changes. The rise in effective stratospheric chlorine values in the 20th century has slowed and reversed in the last decade (top left panel). Effective stratospheric chlorine values are a measure of the potential for ozone depletion in the stratosphere, obtained by summing over adjusted amounts of all chlorine and bromine gases. Effective stratospheric chlorine levels as shown here for midlatitudes will return to 1980 values around 2050. The return to 1980 values will occur around 2065 in polar regions. In 1980, ozone was not significantly depleted by the chlorine and bromine then present in the stratosphere. A decrease in effective stratospheric chlorine abundance follows reductions in emissions of individual halogen source gases. Overall emissions and atmospheric concentrations have decreased and will continue to decrease given international compliance with the Montreal Protocol provisions (see Q15). The changes in the atmospheric abundance of individual gases at Earth's surface shown in the panels were obtained using a combination of direct atmospheric measurements, estimates of historical abundance, and future projections of abundance. The past increases of CFCs, along with those of CCl_4 and CH_3CCl_3, have slowed significantly and most have reversed in the last decade. HCFCs, which are used as CFC substitutes, will continue to increase in the coming decades. Some halon abundances will also continue to grow in the future while current halon reserves are depleted. Smaller relative decreases are expected for CH_3Br in response to production and use restrictions because it has substantial natural sources. CH_3Cl has large natural sources and is not regulated under the Montreal Protocol. (See Figure Q7-1 for chemical names and formulas. The unit "parts per trillion" is defined in the caption of Figure Q7-1.)

and foam products.

HCFC substitute gases. The Montreal Protocol allows for the use of hydrochlorofluorocarbons (HCFCs) as short-term substitutes for CFCs. As a result, the abundances of HCFC-22, HCFC-141b, and HCFC-142b continue to grow in the atmosphere (see Figure Q16-1). HCFCs pose a lesser threat to the ozone layer than CFCs because they are partially destroyed in the troposphere by chemical processes, thus reducing the overall effectiveness of their emissions in destroying stratospheric ozone. Under the Montreal Protocol, HCFC consumption will reach zero in developed nations by 2030 and in developing nations by 2040 (see Q15). Thus, the future projections in Figure Q16-1 show HCFC abundances reaching a peak in the first decades of the 21st century and steadily decreasing thereafter.

Halons. The atmospheric abundances of halon-1211 and halon-1301 account for a significant fraction of bromine from all source gases (see Figure Q7-1) and continue to grow despite the elimination of production in developed nations in 1994 (see Figure Q16-1). The growth in abundance continues because substantial reserves are held in fire-extinguishing equipment and are gradually being released, and production and consumption are still allowed in developing nations. Atmospheric halon abundances can be expected to remain high well into the 21st century because of their long lifetimes and continued release.

Methyl chloride and methyl bromide. Both methyl chloride (CH_3Cl) and methyl bromide (CH_3Br) are distinct among principal halogen source gases because a substantial fraction of their emissions is associated with natural processes (see Q7). The average atmospheric abundance of methyl chloride, which is not regulated under the Montreal Protocol, will remain fairly constant throughout this century if natural sources remain unchanged. At century's end, methyl chloride is expected to account for a large fraction of remaining effective stratospheric chlorine because the abundances of other gases, such as the CFCs, are expected to be greatly reduced (see Figure Q16-1). The abundance of methyl bromide, which is regulated under the Protocol, has already decreased in recent years and is projected to decrease further as a result of production phaseouts in developed and developing countries. For the later decades of the century, methyl bromide abundances are shown as nearly constant in Figure Q16-1. However, these abundances are uncertain because the amounts of exempted uses of methyl bromide under the Montreal Protocol are not known for future years.

V. IMPLICATIONS OF OZONE DEPLETION

Q17: Does depletion of the ozone layer increase ground-level ultraviolet radiation?

Yes, ultraviolet radiation at Earth's surface increases as the amount of overhead total ozone decreases, because ozone absorbs ultraviolet radiation from the Sun. Measurements by ground-based instruments and estimates made using satellite data have confirmed that surface ultraviolet radiation has increased in regions where ozone depletion is observed.

The depletion of stratospheric ozone leads to an increase in surface ultraviolet radiation. The increase occurs primarily in the ultraviolet-B (UV-B) component of the Sun's radiation. UV-B is defined as radiation in the wavelength range of 280 to 315 nanometers. Changes in UV-B at the surface have been observed directly and can be estimated from ozone changes.

Surface UV-B radiation. The amount of ultraviolet radiation reaching Earth's surface depends in large part on the amount of ozone in the atmosphere. Ozone molecules in the stratosphere absorb UV-B radiation, thereby significantly reducing the amount of this radiation that reaches Earth's surface (see Q3). If total ozone amounts are reduced in the stratosphere, then the amount of UV radiation reaching Earth's surface generally increases. This relationship between total ozone and surface UV radiation has been studied at a variety of locations with direct measurements of both ozone and UV. The actual amount of UV reaching a location depends on a large number of additional factors, including the position of the Sun in the sky, cloudiness, and air pollution. In general, surface UV at a particular location on Earth changes throughout the day and with season as the Sun's position in the sky changes.

Long-term surface UV changes. Satellite observations of long-term global ozone changes can be used to estimate changes in global surface UV that have occurred over the past two decades. These changes are of interest because UV radiation can cause harm to humans, other life forms, and materials (see Q3). The amount of UV that produces an "erythemal" or sunburning response in humans is often separately evaluated. Long-term changes in sunburning UV at a particular location have been estimated from the changes in total ozone at that location. The results show that average erythemal UV has increased due to ozone reduction by up to a few percent per decade between 1979 and 1998 over a wide range of latitudes (see Figure Q17-1). The largest increases are found at high polar latitudes in both hemispheres. As expected, the increases occur where decreases in total ozone are observed to be the largest

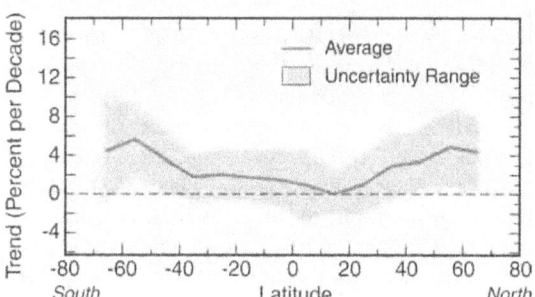

Changes in Surface Ultraviolet Radiation

Figure Q17-1. Changes in surface UV radiation. Ultraviolet (UV) radiation at Earth's surface has increased over much of the globe since 1979. Also known as "erythemal radiation," sunburning UV is harmful to humans and other life forms. The increases shown here for 1979-1998 are estimated from observed decreases in ozone and the relationship between ozone and surface UV established at some surface locations. The estimates are based on the assumption that all other factors that influence the amount of UV radiation reaching the Earth's surface, such as aerosol abundances and cloudiness, are unchanged. The estimated changes in ultraviolet radiation in the tropics are the smallest because observed ozone changes are the smallest there.

(see Figure Q13-1). The smallest changes in erythemal UV are in the tropics, where long-term total ozone changes are smallest.

UV Index changes. The "UV Index" is a measure of daily surface UV levels that is relevant to the effects of UV on human skin. The UV Index is used internationally to increase public awareness about the detrimental effects of UV on human health and to guide the need for protective measures. The UV Index is essentially the amount of erythemal irradiance as measured on a horizontal surface. The daily maximum UV Index varies with location and season, as shown for three locations in

Figure Q17-2. The highest daily values generally occur at the lowest latitudes (tropics) and in summer when the midday Sun is closest to overhead. Values in San Diego, California, for example, normally are larger year round than those found in Barrow, Alaska, which is at higher latitude. At a given latitude, UV Index values increase in mountainous regions. The UV Index becomes zero in periods of continuous darkness found during winter at high-latitude locations.

An illustrative example of how polar ozone depletion increases the maximum daily UV Index is shown in Figure Q17-2. Normal UV Index values for Palmer, Antarctica, in spring were estimated from satellite measurements made during the period 1978-1983, before the appearance of the "ozone hole" over Antarctica (see red dotted line). In the last decade (1991-2001), severe and persistent ozone depletion in spring has increased the UV Index well above normal values for several months (see thick red line). Now, spring UV Index values in Palmer, Antarctica (64°S), sometimes equal or exceed even the peak summer values measured in San Diego, California (32°N).

Other causes of long-term UV changes. The sur-

face UV values may also change as a result of other human activities or climate change. Long-term changes in cloudiness, aerosols, pollution, and snow or ice cover will cause long-term changes in surface UV. At some ground sites, measurements indicate that long-term changes in UV have resulted from changes in one or more of these factors. The impact of some of the changes can be complex. For example, an increase in cloud cover usually results in a reduction of UV radiation below the clouds, but can increase radiation above the clouds (in mountainous regions).

UV changes and skin cancer. Skin cancer cases in humans are expected to increase with the amount of UV reaching Earth's surface. Atmospheric scientists working together with health professionals can estimate how skin cancer cases will change in the future. The estimates are based on knowing how UV increases as total ozone is depleted and how total ozone depletion changes with effective stratospheric chlorine (see Q16). Estimates of future excess skin cancer cases are shown in Figure Q15-1 using future estimates of effective stratospheric chlorine based on the 1992 and earlier Montreal Protocol provisions and assuming that other factors (besides

Figure Q17-2. Changes in UV Index. The maximum daily UV Index is a measure of peak sunburning UV that occurs during the day at a particular location. UV-B, which is absorbed by ozone, is an important component of sunburning UV. The UV Index varies with latitude and season, and with the Sun's elevation in the local sky. The highest values of the maximum daily UV Index occur in the tropics, where the midday Sun is highest throughout the year and where total ozone values are lowest. The lowest average UV Index values occur at high latitudes. As an example, the figure compares the seasonal UV Index at three locations. The UV Index is higher throughout the year in San Diego, a low-latitude location, than in Barrow, a high-latitude location. Index values are zero at high latitudes in winter when darkness is continuous. The effect

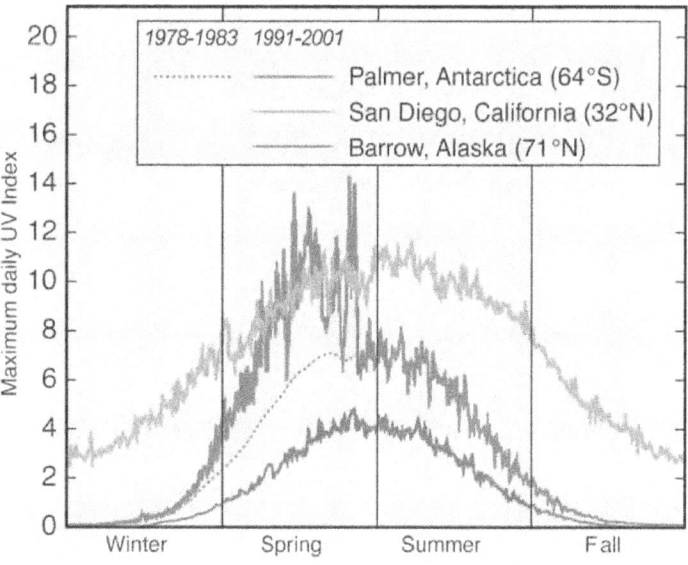

Seasonal Changes in the UV Index

of Antarctic ozone depletion is demonstrated by comparing the Palmer and San Diego data in the figure. Normal values estimated for Palmer are shown for the 1978-1983 period before the "ozone hole" occurred each season (see red dotted line). In the decade 1991-2001, Antarctic ozone depletion has increased the maximum UV Index value at Palmer throughout spring (see yellow shaded region). Values at Palmer now sometimes equal or exceed those measured in spring and even in the summer in San Diego, which is located at much lower latitude.

ozone) affecting surface UV are unchanged in the future. The cases are those that would occur in a population with the UV sensitivity and age distribution such as that of the United States. The cases counted are those in *excess* of the number that occurred in 1980 before ozone depletion was observed (about 2000 per million population), with the assumption that the population's sun exposure remains unchanged. The case estimates include the fact that skin cancer in humans occurs long after the exposure to sunburning UV. The results illustrate that,

with current Protocol provisions, excess skin cancer cases are predicted to increase in the early to middle decades of the 21st century. By century's end, with the expected decreases in halogen source gas emissions, the number of excess cases is predicted to return close to 1980 values. Without the provisions of the Protocol, excess skin cancer cases would have been expected to increase substantially throughout the century.

Q18: Is depletion of the ozone layer the principal cause of climate change?

No, ozone depletion itself is not the principal cause of climate change. However, because ozone absorbs solar radiation and is a greenhouse gas, ozone changes and climate change are linked in important ways. Stratospheric ozone depletion and increases in global tropospheric ozone that have occurred in recent decades both contribute to climate change. These contributions to climate change are significant but small compared with the total contribution from all other greenhouse gases. Ozone and climate change are indirectly linked because both ozone-depleting gases and substitute gases contribute to climate change.

Radiative forcing of climate change. Human activities and natural processes have led to the accumulation in the atmosphere of several long-lived and radiatively active gases known as "greenhouse gases." Ozone is a greenhouse gas, along with carbon dioxide (CO_2), methane (CH_4), nitrous oxide (N_2O), and halogen source gases. The accumulation of these gases in Earth's atmosphere changes the balance between incoming solar radiation and outgoing infrared radiation. Greenhouse gases generally change the balance by absorbing outgoing radiation, leading to a warming at Earth's surface. This change in Earth's radiative balance is called a *radiative forcing of climate change.*

A summary of radiative forcings resulting from the increases in long-lived greenhouse gases in the industrial era is shown in Figure Q18-1. All forcings shown relate to human activities. Positive forcings generally lead to *warming* and negative forcings lead to *cooling* of Earth's surface. The accumulation of carbon dioxide represents the largest forcing term. Carbon dioxide concentrations are increasing in the atmosphere primarily as the result of burning coal, oil, and natural gas for energy and transportation; and from cement manufacturing. The atmospheric abundance of carbon dioxide is currently about 35% above what it was 250 years ago, in preindustrial times. In other international assessments, much of the observed surface warming over the last 50 years has been linked to increases in carbon dioxide and other greenhouse gas concentrations caused by human activities.

Stratospheric and tropospheric ozone. Stratospheric and tropospheric ozone both absorb infrared radiation emitted by Earth's surface, effectively trapping heat in the atmosphere. Stratospheric ozone also significantly absorbs solar radiation. As a result, increases or decreases

Figure Q18-1. Radiative forcing of climate change from atmospheric gas changes. Human activities since the start of the Industrial Era (around 1750) have caused increases in the abundances of several long-lived gases, changing the radiative balance of Earth's atmosphere. These gases, known as "greenhouse gases," result in radiative forcings, which can lead to climate change. Other international assessments have shown that the largest radiative forcings come from carbon dioxide, followed by methane, tropospheric ozone, the halogen-containing gases (see Figure Q7-1), and nitrous oxide. Ozone increases in the troposphere result from pollution associated with human activities. All these forcings are positive, which leads to a warming of Earth's sur-

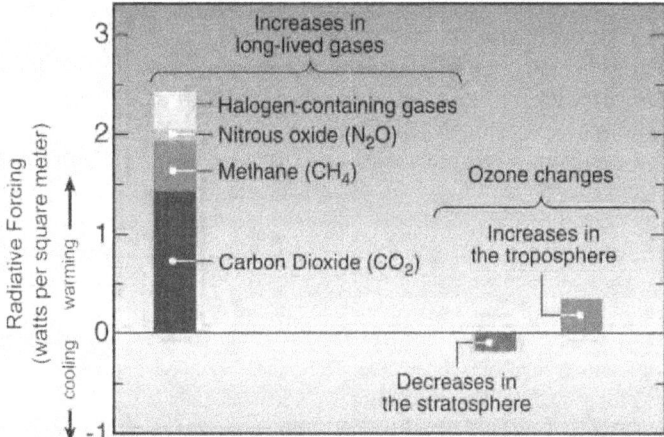

Radiative Forcing of Climate Change from Atmospheric Gas Changes (1750-2000)

face. In contrast, stratospheric ozone depletion represents a small negative forcing, which leads to cooling of Earth's surface. In the coming decades, halogen gas abundances and stratospheric ozone depletion are expected to be reduced along with their associated radiative forcings. The link between these two forcing terms is an important aspect of the radiative forcing of climate change.

in stratospheric or tropospheric ozone cause radiative forcings and represent direct links of ozone to climate change. In recent decades, stratospheric ozone has decreased due to rising chlorine and bromine amounts in the atmosphere, while troposphere ozone in the industrial era has increased due to pollution from human activities (see Q3). Stratospheric ozone depletion causes a negative radiative forcing, while increases in tropospheric ozone cause a positive radiative forcing (see Figure Q18-1). The radiative forcing due to tropospheric ozone increases is currently larger than that associated with stratospheric ozone depletion. The negative forcing from ozone depletion represents an offset to the positive forcing from the halogen source gases, which cause ozone depletion.

Halogen source gases and HFCs. An important link between ozone depletion and climate change is the radiative forcing from halogen source gases and hydrofluorocarbons (HFCs). Halogen source gases are the cause of ozone depletion (see Q7) and HFCs are substitute gases (see Q15). Both groups of gases cause radiative forcing in the atmosphere, but with a wide range of effectiveness. The principal gases in each group are intercompared in Figure Q18-2 (top panel) using their "*ozone depletion potentials*" (ODPs) and "*global warming potentials*" (GWPs), which indicate the effectiveness of each gas in causing ozone depletion and climate change, respectively. The ODPs of CFC-11 and CFC-12, and the GWP of CO_2 all are assigned a value of 1.0. For ozone depletion, the halons are the most effective gases (for equal mass amounts) and HFCs cause no ozone depletion (see Q7). For climate change, all gases make a contribution, with CFC-12 and HFC-23 having the largest effect (for equal mass amounts). Montreal Protocol actions (see Q15) that have led to reductions in CFC concentrations and increases in HCFC and HFC concentrations have also reduced the total radiative forcing from these gases. It is important to note that, despite a GWP that is small in comparison to many other greenhouse gases, CO_2 is the most important greenhouse gas related to human activities because its atmospheric abundance is so much greater than the abundance of other gases.

The relative importance of total emissions of halogen source gases and HFCs to ozone depletion and climate change is illustrated for a single year (2004) of emissions in the bottom panel of Figure Q18-2. The values displayed are proportional to the product of 2004 annual global emissions and the ODP or GWP. The results in the lower panel are shown relative to CFC-11, because it is often used as a reference gas. The comparison shows that the importance of CFC emissions in 2004 to future ozone depletion exceeds that of the halons, despite the higher halon ODP values, because CFC emissions are larger. Similarly, the contributions of CFC and HCFC-22 emissions in 2004 to climate change are currently larger than the halon or HFC contributions. These 2004 results represent only incremental contributions of these gases to either ozone depletion or climate change. The overall contribution of a gas depends on its total accumulation in the atmosphere, which in turn depends on its long-term emission history and atmospheric lifetime (see Q7 and Q16). In the case of ozone depletion, the relative contributions of the halogen source gases can be compared through their respective contributions to effective stratospheric chlorine (see Q16).

As a group, the principal halogen source gases represent a positive direct radiative forcing in the Industrial Era that is comparable to the forcing from methane, the second most important greenhouse gas. In the coming decades, the abundances of these ozone-depleting gases and their associated positive radiative forcings are expected to decrease (see Q16). Future growth in HFC emissions, while uncertain, will contribute a positive forcing that will counter the decrease from ozone-depleting gases. Finally, reductions in ozone-depleting gases will be followed by reductions in stratospheric ozone depletion and its associated *negative* radiative forcing.

Impact of climate change on ozone. Certain changes in Earth's climate could affect the future of the ozone layer. Stratospheric ozone is influenced by changes in temperatures and winds in the stratosphere. For example, lower temperatures and stronger polar winds could both affect the extent and severity of winter polar ozone depletion. While the Earth's surface is expected to warm in response to the net positive radiative forcing from greenhouse gas increases, the stratosphere is expected to cool. A cooler stratosphere would extend the time period over which polar stratospheric clouds (PSCs) are present in polar regions and, as a result, might increase winter ozone depletion. In the upper stratosphere at altitudes above PSC formation regions, a cooler stratosphere is expected to increase ozone amounts and, hence, hasten recovery, because lower temperatures favor ozone production over loss (see Q2). Similarly, changes in atmospheric composition that lead to a warmer climate may also alter ozone amounts (see Q20).

Evaluation of Selected Ozone-Depleting Substances and Substitute Gases

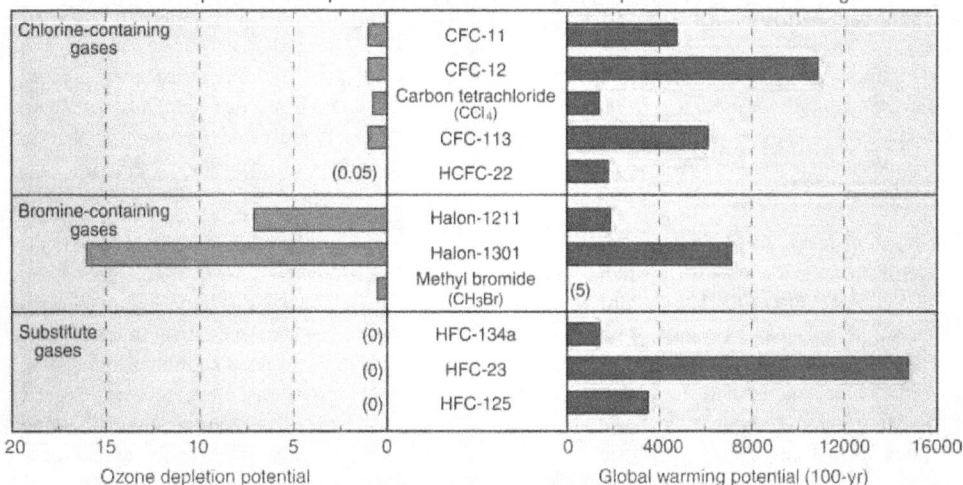

Relative importance of equal mass amounts for ozone depletion and climate change

Relative importance of 2004 emissions for ozone depletion and climate change

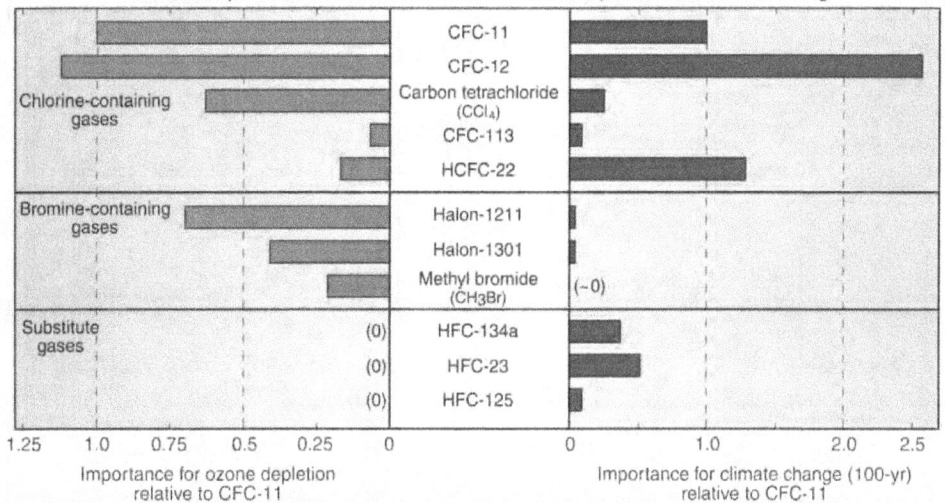

Figure Q18-2. Evaluation of ozone-depleting gases and their substitutes. Ozone-depleting gases (halogen source gases) and their substitutes can be compared via their ozone depletion potentials (ODPs) and global warming potentials (GWPs). The GWPs are evaluated for a 100-yr time interval after emission. The CFCs, halons, and HCFCs are ozone-depleting gases (see Q7) and HFCs, used as substitute or replacement gases, do not destroy ozone. The ODPs of CFC-11 and CFC-12, and the GWP of CO_2 have values of 1.0 by definition. Larger ODPs or GWPs indicate greater potential for ozone depletion or climate change, respectively. The top panel compares ODPs and GWPs for emissions of equal mass amounts of each gas. The ODPs of the halons far exceed those of the CFCs. HFCs have zero ODPs. All gases have non-zero GWPs that span a wide range of values. The bottom panel compares the contributions of the 2004 emissions of each gas, using CFC-11 as the reference gas. Each bar represents the product of a global emission value and the respective ODP or GWP factor. The comparison shows that 2004 emissions of ozone-depleting gases currently contribute more than substitute gas emissions to both ozone depletion and climate change. Future projections guided by Montreal Protocol provisions suggest that the contributions of ozone-depleting gases to climate change will decrease, while those of the substitute gases will increase.

VI. STRATOSPHERIC OZONE IN THE FUTURE

Q19: How will recovery of the ozone layer be identified?

Scientists expect to identify the recovery of the ozone layer with detailed ozone measurements in the atmosphere and with global models of ozone amounts. Increases in global ozone and reductions in the extent and severity of the Antarctic "ozone hole" will be important factors in gauging ozone recovery. Natural variations in ozone amounts will limit how soon recovery can be detected with future ozone measurements.

Recovery process. Identifying the recovery of the ozone layer from depletion associated with halogen gases will rely on comparisons of the latest ozone values with values measured in the past. Because of its importance, ozone will likely be measured continuously in the future using a variety of techniques and measurement platforms (see Q5). Atmospheric computer models will be used to predict future abundances of ozone and attribute observed changes to ozone-depleting gases and other factors.

The recovery process is schematically shown for *global* ozone in Figure Q19-1. Ozone has declined from pre-1980 amounts due to past increases in halogen gases in the stratosphere (see Q16). In the future, as the overall decline in these gases continues in response to Montreal Protocol provisions, global ozone is expected to recover, approaching or exceeding pre-1980 values (see Q20). Ozone recovery attributable to decreases in ozone-depleting gases can be described, in general, as a process

Recovery Stages of Global Ozone

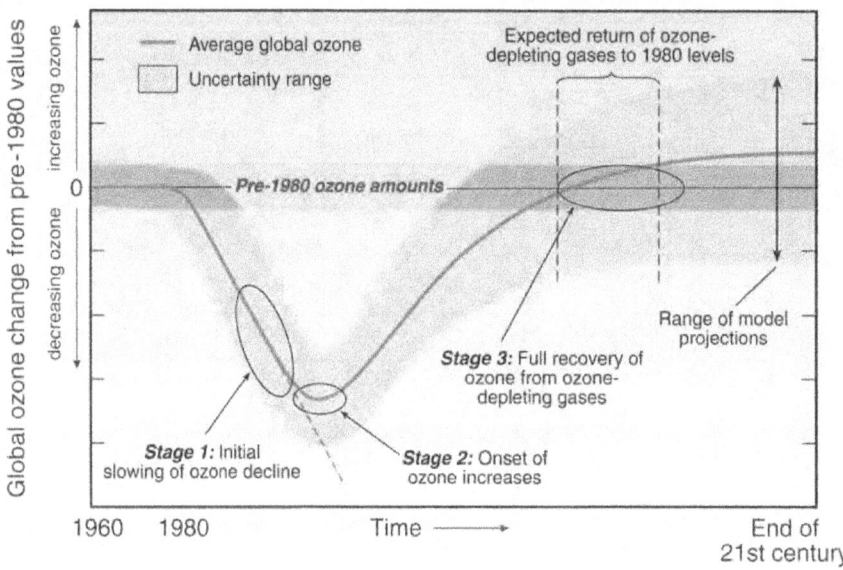

Figure Q19-1. Recovery stages of global ozone. Significant ozone depletion from the release of ozone-depleting gases in human activities first became recognized in the 1980s. The Montreal Protocol provisions are expected to further reduce and eliminate these gases in the atmosphere in the coming decades, thereby leading to the return of ozone amounts to near pre-1980 values. The timeline of the recovery process is schematically illustrated with three stages identified. The large uncertainty range illustrates natural ozone variability in the past and potential uncertainties in global model projections of future ozone amounts. When ozone reaches the full recovery stage, global ozone values may be above or below pre-1980 values, depending on other changes in the atmosphere (see Q20).

involving three stages:

(1) The **initial slowing of ozone decline,** identified as the occurrence of a statistically significant reduction in the rate of decline in ozone.

(2) The **onset of ozone increases (turnaround),** identified as the occurrence of statistically significant increases in ozone above previous minimum values.

(3) The **full recovery of ozone from ozone-depleting gases,** identified as when ozone is no longer significantly affected by ozone-depleting gases from human activities.

Each recovery stage is noted in Figure Q19-1. The red line and shaded region in the figure indicate the expected average value and the uncertainty range, respectively, in global ozone amounts. The large uncertainty range illustrates natural ozone variability in the past and potential uncertainties in global model projections of future ozone amounts.

In the full recovery of global ozone, the milestone of the return of ozone to pre-1980 levels is considered important because prior to 1980 ozone was not significantly affected by human activities. As a consequence, this milestone is useful, for example, to gauge when the adverse impacts of enhanced surface ultraviolet (UV) radiation on human health and ecosystems caused by ozone-depleting

substances are likely to become negligible. The uncertainty range in model results indicates that ozone amounts may be below or above pre-1980 values when ozone has fully recovered from the effects of ozone-depleting gases from human activities (see Q20). The wide range of uncertainty for global ozone in the final stage of recovery represents, in part, the difficulty in accurately forecasting the effects of future changes in climate and atmospheric composition on the abundance of ozone (see Q20).

Natural factors. Stratospheric ozone is influenced by two important natural factors, namely, changes in the output of the Sun and volcanic eruptions (see Q14). Evaluations of ozone recovery include the effects of these natural factors. The solar effect on ozone is expected to be predictable based on the well-established 11-year cycle of solar output. The uncertainty range in Figure Q19-1 includes solar changes. Volcanic eruptions are particularly important because they enhance ozone depletion caused by reactive halogen gases, but cannot be predicted. The occurrence of a large volcanic eruption in the next decades when effective stratospheric chlorine levels are still high (see Figure Q16-1) may obscure progress in overall ozone recovery by temporarily increasing ozone depletion. The natural variation of ozone amounts also limits how easily small improvements in ozone abundances can be detected.

Q20: When is the ozone layer expected to recover?

Substantial recovery of the ozone layer is expected near the middle of the 21ˢᵗ century, assuming global compliance with the Montreal Protocol. Recovery will occur as chlorine- and bromine-containing gases that cause ozone depletion decrease in the coming decades under the provisions of the Protocol. However, the influence of changes in climate and other atmospheric parameters could accelerate or delay ozone recovery, and volcanic eruptions in the next decades could temporarily reduce ozone amounts for several years.

Halogen source gas reductions. Ozone depletion caused by human-produced chlorine and bromine gases is expected to gradually disappear by about the middle of the 21ˢᵗ century as the abundances of these gases decline in the stratosphere. The decline in *effective stratospheric chlorine* will follow the reductions in emissions that are expected to continue under the provisions of the Montreal Protocol and its Adjustments and Amendments (see Figure Q16-1). The emission reductions are based on the assumption of full compliance by the developed and developing nations of the world. The slowing of increases in atmospheric abundances and the initial decline of several halogen gases have already been observed (see Figure Q16-1). One gas, methyl chloroform, has already decreased by about 90% from its peak value. Natural chemical and transport processes limit the rate at which halogen gases are removed from the stratosphere. The

atmospheric lifetimes of the halogen source gases range up to 100 years (see Table Q7-1). Chlorofluorocarbon-12 (CFC-12), with its 100-year lifetime, will require about 200 to 300 years before it is removed (less than 5% remaining) from the atmosphere (see Figure Q16-1). At midlatitudes, effective stratospheric chlorine is not expected to reach pre-1980 values until about 2050.

Ozone projections. Computer models of the atmosphere are used to assess past changes in the global ozone distribution and to project future changes. Two important measures of ozone considered by scientists are global total ozone averaged between 60°N and 60°S latitudes, and minimum ozone values in the Antarctic "ozone hole." Both measures show ongoing ozone depletion that began in the 1980s (see Figure Q20-1). The model projections indicate that for 60°N-60°S total ozone, the first two stages of recovery (slowing of the decline and turnaround

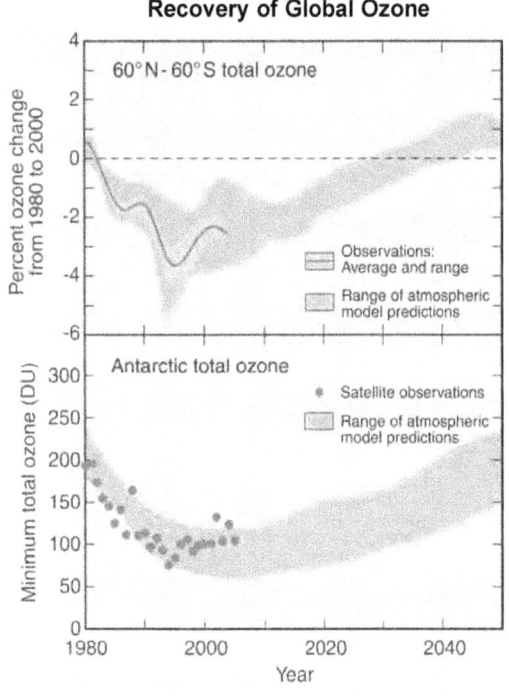

Recovery of Global Ozone

Figure Q20-1. Global ozone recovery predictions. Observed values of midlatitude total ozone (top panel) and September-October minimum total ozone values, over Antarctica (bottom panel) have decreased beginning in the early 1980s. As halogen source gas emissions decrease in the 21ˢᵗ century, ozone values are expected to recover by increasing toward pre-1980 values. Atmospheric computer models that account for changes in halogen gases and other atmospheric parameters are used to predict how ozone amounts will increase. These model results show that full recovery is expected in midlatitudes by 2050, or perhaps earlier. Recovery in the Antarctic will occur somewhat later. The range of model projections comes from the use of several different models of the future atmosphere.

(see Q19)) will be reached before 2020. Full recovery, with ozone reaching or exceeding pre-1980 values, is expected to occur by the middle of the 21st century. The range of projections comes from several computer models of the atmosphere. Some of these models indicate that recovery of 60°N-60°S total ozone may come well before midcentury.

Models predict that Antarctic ozone depletion will also reach the first two stages of recovery by 2020, but somewhat more slowly than 60°N-60°S total ozone. Full recovery could occur by mid-century but some models show later recovery, between 2060 and 2070. Declines in effective stratospheric chlorine amounts will occur later over the Antarctic than at lower latitudes because air in the Antarctic stratosphere is older than air found at lower latitudes. As a result, reductions in halogen loading to pre-1980 values will occur 10-15 years later in the Antarctic stratosphere than in the mid-latitude stratosphere.

A different atmosphere in 2050. By the middle of the 21st century, halogen amounts in the stratosphere are expected to be similar to those present in 1980 before the onset of significant ozone depletion (see Figure Q16-1). However, climate and other atmospheric factors will not be the same in 2050 as in 1980, and this could cause ozone abundances in 2050 to be somewhat different from those observed in 1980. Stratospheric ozone abundances are affected by a number of natural and human-caused factors in addition to the atmospheric abundance of halogen gases. Important examples are stratospheric temperatures and air motions, volcanic eruptions, solar activity, and changes in atmospheric composition. Separating the effects of these factors is challenging because of the complexity of atmospheric processes affecting ozone.

The ozone recovery projections in Figures Q19-1 and Q20-1 attempt to take these various factors into account.

For example, since 1980 human activities have increased the atmospheric abundance of important greenhouse gases, including carbon dioxide, methane, and nitrous oxide. Other international assessments have shown that the accumulation of these gases is linked to the warmer surface temperatures and lower stratospheric temperatures observed within recent decades. Warmer surface temperatures could change the emission rates of naturally occurring halogen source gases. Lower temperatures in the upper stratosphere (at about 40 kilometers (25 miles) altitude) accelerate ozone recovery because ozone destruction reactions proceed at a slower rate. In contrast, reduced temperatures in the polar lower stratosphere during winter might increase the occurrence of polar stratospheric clouds (PSCs) and, therefore, enhance chemical ozone destruction (see Q10). Further increases of stratospheric water vapor, such as those that have occurred over the last two decades, could also increase PSC occurrences and associated ozone destruction. Therefore, a cooler, wetter polar stratosphere could delay polar ozone recovery beyond what would be predicted for the 1980 atmosphere. Increased abundances of methane and nitrous oxide due to human activities also cause some change in the overall balance of the chemical production and destruction of global stratospheric ozone. Finally, one outcome that cannot be included precisely in models is the occurrence of one or more large volcanic eruptions in the coming decades. Large eruptions would increase stratospheric sulfate particles for several years, temporarily reducing global ozone amounts (see Q14).

As a consequence of these potential changes, the return of effective stratospheric chlorine and ozone to pre-1980 levels may not occur at the same time. In some regions of the stratosphere, ozone may remain below pre-1980 values after effective chlorine has declined to pre-1980 levels.

GLOSSARY AND ACRONYMS

GLOSSARY

aerosol
A suspension of very fine particles (solid or liquid) in air.

anthropogenic
Human-caused.

bank
The amount of a chemical that has been produced but has not yet been emitted or chemically altered. They exist either in reserve storage or in current applications (*i.e.*, refrigerators, air conditioners, fire extinguishers, etc.).

catalytic reaction
Acceleration (increase in rate) of a chemical reaction by means of a substance, called a catalyst. Chlorine acts as a catalyst in the destruction of ozone in the stratosphere.

chlorofluorocarbons (CFCs)
Halocarbons containing only chlorine, fluorine, and carbon atoms. CFCs are ozone-depleting substances (ODSs).

climate forcing
Changes that affect the energy balance of the planet and that consequently "force" the climate to change (see also radiative forcing). Examples of climate forcing include changes in atmospheric carbon dioxide, or suspended particulates (see aerosols), or energy from the sun.

CO_2-equivalent
The amount of carbon dioxide (CO_2) that would cause the same amount of radiative forcing as a given amount of another greenhouse gas.

column ozone
The total amount of ozone in a vertical column above the Earth's surface. Column ozone is measured in Dobson units (DU).

consumption
Used here as defined by the Montreal Protocol: The magnitude of an ozone-depleting substance produced or imported, minus the amount of that substance that is exported.

Equivalent Effective Chlorine (EECl)
An index used to approximately quantify overall changes in reactive halogen trends based on the measured mix of ODSs in the troposphere. It accounts for the number of halogen atoms in the ODS molecule, their relative efficiency at destroying ozone, and the rate at which the ODSs decompose in the stratosphere to release the halogen atoms.

Equivalent Effective Stratospheric Chlorine (EESC)
An index related to EECl that considers the time lags associated with transporting air from the troposphere to the stratosphere. EESC roughly estimates the effect of ODSs on stratospheric ozone. The EESC index is used to estimate the time evolution of ozone-depleting halogen (chlorine, bromine) in the atmosphere. It is often used to estimate when the cumulative effect of all ODSs on ozone will return to a level attained at some earlier time, or to evaluate the relative effects of potential policy scenarios.

Global Warming Potential
An index comparing the cumulative radiative forcing (*i.e.*, climate impact) of a unit mass of a greenhouse gas relative to the same quantity of a unit mass of a reference gas (usually CO_2) over some time period (usually 100 years).

100-year GWP
The Global Warming Potential of a chemical integrated over a 100-year time horizon relative to CO_2. When applied as a weighting factor to emissions or production of other chemicals, the resulting quantity provides a CO_2-equivalent emission or production.

greenhouse gases
Gases including water vapor, carbon dioxide, methane, nitrous oxide, and halocarbons that trap infrared heat, warming the air near the surface and in the lower levels of the atmosphere.

halocarbons
Chemical compounds containing carbon atoms, and one or more atoms of the halogens chlorine (Cl), bromine (Br), or iodine (I). Halocarbons include chlorofluorocarbons (CFCs), hydrochlorofluorocarbons (HCFCs), hydrofluorocarbons (HFCs), perfluorocarbons (PFCs), and halons.

halogen

A member of the family of elements that includes fluorine, chlorine, bromine, and iodine. In the context of ozone depletion, chlorine (Cl), bromine (Br), and iodine (I) are the halogens of greatest relevance. "Halogenated" compounds are chemicals that contain one or more halogen atoms in their molecular structure. Fully halogenated halocarbons contain only carbon and halogen atoms, whereas partially halogenated halocarbons also contain hydrogen (H) atoms.

halon

Fully halogenated halocarbons that contain bromine and fluorine. Halons are used in fire-extinguishing applications.

mixing ratio (by mass)

The ratio of the mass of a specific substance in a volume of air to the total mass of that volume of air.

ozone-depleting substance (ODS)

A substance known to deplete the stratospheric ozone layer. This includes CFCs, halons, HCFCs, and other halocarbons. Most ODSs are regulated by the Montreal Protocol, though some with very short lifetimes (*e.g.*, $CHBr_3$) or small anthropogenic sources (*e.g.*, CH_3Cl) are not.

ozone depletion

Chemical destruction of the stratospheric ozone layer by substances produced by human activities.

Ozone Depletion Potential

An index comparing the amount of global ozone destroyed by a particular ozone-depleting substance per unit mass compared to the amount destroyed by a reference gas (CFC-11) per unit mass.

production

The magnitude of ODS or substitute chemical produced by industry.

radiative forcing

Broadly defined as the difference between the incoming radiation energy and the outgoing radiation energy in the climate system. If more energy is incoming than outgoing, it tends to warm the climate (and is a planetary energy imbalance). A source of radiative forcing might be more solar energy, or more greenhouse gases for example. (This term is used in a more specific manner in IPCC).

stratosphere

The highly stratified region of the atmosphere above the troposphere extending from about 10 km (ranging from 9 km in high latitudes to 16 km in the tropics on average) to about 50 km. In the stratosphere, temperatures increase with height.

substitutes for ozone-depleting substances

Chemicals that are used in place of the ozone-depleting substances that are regulated by the Montreal Protocol. These include hydrochlorofluorocarbons (HCFCs), which are ozone-depleting substances but not as potent as those they replace, and perfluorocarbons (PFCs), which do not deplete ozone but are greenhouse gases.

troposphere

The lowest part of the atmosphere from the surface to about 10 km in altitude in mid-latitudes (ranging from 9 km in high latitudes to 16 km in the tropics on average) where clouds and "weather" phenomena occur. In the troposphere, temperatures generally decrease with height.

ACRONYMS AND ABBREVIATIONS

A1	baseline halocarbon scenario
ACE	Atmospheric Chemistry Experiment
AFEAS	Alternative Fluorocarbons Environmental Acceptability Study
AGAGE	Advanced Global Atmospheric Gases Experiment
AMTRAC	Atmospheric Model with Transport and Chemistry
ATMOS	Atmospheric Trace Molecule Spectroscopy
Br	atomic bromine
BrO	bromine monoxide
BrO$_x$	reactive bromine
BrONO$_2$	bromine nitrate
Br$_y$	inorganic bromine
C	carbon
°C	degree celsius
CALIPSO	Cloud-Aerosol Lidar and Infrared Pathfinder Satellite Observation
CCl$_4$	carbon tetrachloride
C$_2$Cl$_4$	tetrachloroethene
CCM	Chemistry Climate Model
CCSP	Climate Change Science Program
CFC	chlorofluorocarbon
CFC-11	trichlorofluoromethane
CFC-12	dichlorodifluoromethane
CH$_4$	methane
CHBr$_3$	tribromomethane
CHCl$_3$	trichloromethane (chloroform)
CH$_2$Br$_2$	dibromomethane
CH$_2$Cl$_2$	dichloromethane or methylene chloride
CH$_3$Br	methyl bromide
CH$_3$Cl	methyl chloride
CH$_3$CCl$_3$	methyl chloroform
Cl	chlorine
ClO	chlorine monoxide
ClO$_x$	reactive chlorine
ClONO$_2$	chlorine nitrate
ClOOCl	chlorine monoxide dimer
CMDL	Climate Monitoring and Diagnostics Laboratory (NOAA)
CO$_2$	carbon dioxide
CTM	chemical transport model
CUE	Critical Use Exemption
DU	Dobson units
EECl	Equivalent Effective Chlorine
EESC	Equivalent Effective Stratospheric Chlorine
EP	Earth-Probe TOMS
EPA	U.S. Environmental Protection Agency
ESRL	Earth System Research Laboratory (NOAA)
Gg	gigagram, or one billion grams
GHG	greenhouse gas
GMD	Global Monitoring Division (NOAA/ESRL)
GOMOS	Global Ozone Monitoring by Occultation of Stars
Gt	gigaton, or one billion metric tons
GtCO$_2$	gigatons of carbon dioxide
GWP	Global Warming Potential
H	hydrogen
HALOE	Halogen Occultation Experiment
HCFC	hydrochlorofluorocarbon
HCl	hydrogen chloride (or hydrochloric acid)
HFC	hydrofluorocarbon
H$_2$	hydrogen gas
H$_2$O	water
HO$_x$	reactive hydrogen
hPa	hectopascal
ILAS	Improved Limb Atmospheric Spectrometer
IPCC	Intergovernmental Panel on Climate Change
IR	infrared
K	kelvin (unit of temperature)
kJ/m^2	kilojoules per meter squared
km	kilometer, or one thousand meters
Kt	kiloton, or one thousand metric tons
LOSU	level of scientific understanding
MAM	March-April-May
MBTOC	Methyl Bromide Technical Options Committee
MLS	Microwave Limb Sounder
MSU	Microwave Sounding Unit
Mt	megaton, or one million metric tons
N	nitrogen
NASA	National Aeronautics and Space Administration
NIH	National Institutes of Health
NH	Northern Hemisphere
NRL	Naval Research Laboratory
NO	nitric oxide
NO$_x$	reactive nitrogen
NO$_2$	nitrogen dioxide
N$_2$O	nitrous oxide
NOAA	National Oceanic and Atmospheric Administration
NOCAR	NOAA and NCAR model
O	oxygen atom
O$_2$	molecular oxygen
O$_3$	ozone
O$_x$	odd oxygen
ODP	Ozone Depletion Potential
ODS	ozone-depleting substance

OH	hydroxyl radical
OMI	Ozone Monitoring Instrument
ppb	parts per billion
ppmv	parts per million by volume
ppt	parts per trillion
PSC	polar stratospheric cloud
POAM	Polar Ozone and Aerosol Measurement
QBO	quasi-biennial oscillation
QPS	quarantine and pre-shipment
RAF	radiation amplification factor
RF	radiative forcing
SAGE	Stratospheric Aerosol and Gas Experiment
SAM	Stratospheric Aerosol Monitor
SBUV	Solar Backscatter Ultraviolet
SH	Southern Hemisphere
SLIMCAT	Single-Layer Isentropic Model of Chemistry and Transport
SO$_2$	sulfur dioxide
SPARC	Stratospheric Processes and their Role in Climate (WCRP)
SRES	Special Report on Emissions Scenarios (IPCC)
SROC	Special Report on Ozone and Climate (IPCC)
2-D	two-dimensional
3-D	three-dimensional
TEAP	Technology and Economic Assessment Panel (UNEP)
TOMS	Total Ozone Mapping Spectrometer
UNEP	United Nations Environment Programme
UNFCCC	United Nations Framework Convention on Climate Change
UV	ultraviolet radiation
UV-vis	ultraviolet/visible camera
UVA	ultraviolet–A radiation
UVB	ultraviolet–B radiation
UVC	ultraviolet–C radiation
V$_{PSC}$	volume of polar stratospheric clouds
VSL	very short-lived
VSLS	very short-lived substances
WCRP	World Climate Research Programme
WG1-AR4	Working Group I–Fourth Assessment Report (IPCC)
W per m^2	watts per meter squared
WMGHG	well-mixed greenhouse gases
WMO	World Meteorological Organization

REFERENCES

EXECUTIVE SUMMARY REFERENCES

Fahey, D.W., 2007: Twenty questions and answers about the ozone layer: 2006 update. In: *Scientific Assessment of Ozone Depletion: 2006.* Global Ozone Research and Monitoring Project report no. 50. World Meteorological Organization, Geneva, Switzerland, [50 pp.] Included as Appendix A of this report; also at <http://www.esrl.noaa.gov/csd/assessments/2006/report.html>

IPCC (Intergovernmental Panel on Climate Change), 2007: *Climate Change 2007: The Physical Science Basis.* Contribution of Working Group I to the Fourth Assessment Report of the Intergovernmental Panel on Climate Change. [Solomon, S., D. Qin, M. Manning, Z. Chen, M. Marquis, K.B. Averyt, M. Tignor, and H.L. Miller (eds.)]. Cambridge University Press, UK, and New York, 996 pp. <http://www.ipcc.ch>

WMO (World Meteorological Organization), 2003: *Scientific Assessment of Ozone Depletion: 2002.* Global Ozone Research and Monitoring Project report no. 47. World Meteorological Organization, Geneva, Switzerland, 498 pp.

WMO (World Meteorological Organization), 2007: *Scientific Assessment of Ozone Depletion: 2006.* Global Ozone Research and Monitoring Project report no. 50. World Meteorological Organization, Geneva, Switzerland, 572 pp. <http://www.esrl.noaa.gov/csd/assessments/2006/>

CHAPTER 1 REFERENCES

Fahey, D.W., 2007: Twenty questions and answers about the ozone layer: 2006 update. In: *Scientific Assessment of Ozone Depletion: 2006.* Global Ozone Research and Monitoring Project report no. 50. World Meteorological Organization, Geneva, Switzerland, [50 pp.] Included as Appendix A of this report; also at <http://www.esrl.noaa.gov/csd/assessments/2006/report.html>

IPCC (Intergovernmental Panel on Climate Change), 2007: *Climate Change 2007: The Physical Science Basis.* Contribution of Working Group I to the Fourth Assessment Report of the Intergovernmental Panel on Climate Change [Solomon, S., D. Qin, M. Manning, Z. Chen, M. Marquis, K.B. Averyt, M. Tignor, and H.L. Miller (eds.)]. Cambridge University Press, Cambridge, UK, and New York, 996 pp. <http://www.ipcc.ch>

IPCC/TEAP (Intergovernmental Panel on Climate Change/Technology and Economic Assessment Panel), 2005: *IPCC/TEAP Special Report on Safeguarding the Ozone Layer and the Global Climate System: Issues Related to Hydrofluorocarbons and Perfluorocarbons.* [Metz, B., L. Kuijpers, S. Solomon, S.O. Andersen, O. Davidson, J. Pons, D. de Jager, T. Kestin, M. Manning, and L.A. Meyer (eds.)]. Cambridge University Press, Cambridge, UK, and New York, 478 pp.

WMO (World Meteorological Organization), 2007: *Scientific Assessment of Ozone Depletion: 2006.* Global Ozone Research and Monitoring Project report no. 50. World Meteorological Organization, Geneva, Switzerland, 572 pp. <http://www.esrl.noaa.gov/csd/assessments/2006/report.html>

CHAPTER 2 REFERENCES

AFEAS (Alternative Fluorocarbons Environmental Acceptability Study), 2007: Data tables at <http://www.afeas.org/>

Campbell, N. and R. Shende (coordinating lead authors), M. Bennett, O. Blinova, R. Derwent, A. McCulloch, M. Yamabe, J. Shevlin, and T. Vink, 2005: HFCs and PFCs: current and future supply, demand and emissions, plus emissions of CFCs, HCFCs and halons. In: *IPCC/TEAP Special Report on Safeguarding the Ozone Layer and the Global Climate System: Issues Related to Hydrofluorocarbons and Perfluorocarbons* [Metz, B., L. Kuijpers, S. Solomon, S.O. Andersen, O. Davidson, J. Pons, D. de Jager, T. Kestin, M. Manning, and L.A. Meyer (eds.)]. Cambridge University Press, Cambridge, UK, and New York, pp. 403-436.

Clerbaux, C. and D.M. Cunnold (lead authors), J. Anderson, A. Engel, P.J. Fraser, E. Mahieu, A. Manning, S.A. Montzka, R. Nassar, R. Prinn, S. Reimann, C.P. Rinsland, P. Simmonds, D. Verdonik, R. Weiss, D. Wuebbles, and Y. Yokouchi, 2007: Long-lived compounds. In: *Scientific Assessment of Ozone Depletion: 2006.* Global Ozone Research and Monitoring Project report no. 50. World Meteorological Organization, Geneva, Switzerland, pp. 1.1-1.63.

Daniel, J.S. and G.J.M. Velders (lead authors), A.R. Douglass, P.M.D. Forster, D.A. Hauglustaine, I.S.A. Isaksen, L.J.M. Kuijpers, A. McCulloch, and T.J. Wallington, 2007: Halocarbon scenarios, ozone depletion potentials, and global warming potentials. In: *Scientific Assessment of Ozone Depletion: 2006.* Global Ozone Research and Monitoring Project report no. 50. World Meteorological Organization, Geneva, Switzerland, pp. 8.1-8.39.

Daniel, J.S., S. Solomon, and D.L. Albritton, 1995: On the evaluation of halocarbon radiative forcing and global warming potentials. *Journal of Geophysical Research*, **100(D1)**, 1271-1285.

Dorf, M., J.H. Butler, A. Butz, C. Camy-Peyret, M.P. Chipperfield, L. Kritten, S. Montzka, B. Simmes, F. Weidner, and K. Pfeilsticker, 2006: Long-term observations of stratospheric bromine reveal slow down in growth. *Geophysical Research Letters*, **33**, L24803, doi:10.1029/2006GL027714.

Forster, P., V. Ramaswamy, P. Artaxo, T. Berntsen, R. Betts, D.W. Fahey, J. Haywood, J. Lean, D.C. Lowe, G. Myhre, J. Nganga, R. Prinn, G. Raga, M. Schulz, and R. Van Dorland, 2007: Changes in atmospheric constituents and in radiative forcing. In: *Climate Change 2007: The Physical Basis*. Contribution of Working Group I to the Fourth Assessment Report of the Intergovernmental Panel on Climate Change [Solomon, S., D. Qin, M. Manning, Z. Chen, M. Marquis, K.B. Averyt, M. Tignor, and H.L. Miller (eds.)]. Cambridge University Press, Cambridge, UK, and New York, pp. 129-234.

IPCC (Intergovernmental Panel on Climate Change), 2001: *Climate Change 2001: The Scientific Basis*. Contribution of Working Group I to the Third Assessment Report of the Intergovernmental Panel on Climate Change [Houghton, J.T., Y. Ding, D.J. Griggs, M. Noguer, P.J. van der Linden, X. Dai, K. Maskell, and C.A. Johnson (eds.)]. Cambridge University Press, Cambridge, UK, and New York, 881 pp.

IPCC (Intergovernmental Panel on Climate Change), 2007: *Climate Change 2007: The Physical Science Basis*. Contribution of Working Group I to the Fourth Assessment Report of the Intergovernmental Panel on Climate Change. [Solomon, S., D. Qin, M. Manning, Z. Chen, M. Marquis, K.B. Averyt, M. Tignor, and H.L. Miller (eds.)]. Cambridge University Press, UK, and New York, 996 pp. <http://www.ipcc.ch>

IPCC/TEAP (Intergovernmental Panel on Climate Change/Technology and Economic Assessment Panel), 2005: *IPCC/TEAP Special Report on Safeguarding the Ozone Layer and the Global Climate System: Issues Related to Hydrofluorocarbons and Perfluorocarbons*. [Metz, B., L. Kuijpers, S. Solomon, S.O. Andersen, O. Davidson, J. Pons, D. de Jager, T. Kestin, M. Manning, and L.A. Meyer (eds.)]. Cambridge University Press, Cambridge, UK, and New York, 478 pp.

Joshi, M., K. Shine, M. Ponater, N. Stuber, R. Sausen, and L. Li, 2003: A comparison of climate response to different radiative forcings in three general circulation models: towards an improved metric of climate change. *Climate Dynamics*, **20(7-8)**, 843-854.

Law, K.S. and W.T. Sturges (lead authors), D.R. Blake, N.J. Blake, J.B. Burkholder, J.H. Butler, R.A. Cox, P.H. Haynes, K. Kreher, C. Mari, K. Pfeilsticker, J.M.C. Plane, R.J

Salawitch, C. Schiller, B.-M. Sinnhuber, R. von Glasow, N.J. Warwick, D.J. Wuebbles, and S.A. Yvon-Lewis, 2007: Halogenated and very short-lived substances. In: *Scientific Assessment of Ozone Depletion: 2006*. Global Ozone Research and Monitoring Project report no. 50. World Meteorological Organization, Geneva, Switzerland, pp. 2.1-2.57.

McCulloch, A., P.M. Midgley, and D.A. Fisher, 1994: Distribution of emissions of chlorofluorocarbons (CFCs) 11, 12, 113, 114, and 115 among reporting and non-reporting countries in 1986. *Atmospheric Environment*, **28(16)**, 2567-2582.

Midgley, P.M. and A. McCulloch, 1999: Properties and applications of industrial halocarbons. In: *The Handbook of Environmental Chemistry Vol. 4. Part E, Reactive Halogen Compounds in the Atmosphere* [Fabian, P. and O.N. Singh (eds.)]. Springer-Verlag, Berlin and New York, chapter 5.

Newman, P.A., E.R. Nash, S.R. Kawa, S.A. Montzka, and S.M. Schauffler, 2006: When will the Antarctic ozone hole recover? *Geophysical Research Letters*, **33**, L12814, doi:10.1029/2005GL025232.

Newman, P.A., J.S. Daniel, D.W. Waugh, and E.R. Nash, 2007: A new formulation of equivalent effective stratospheric chlorine (EESC). *Atmospheric Chemistry and Physics*, **7(17)**, 4537-4552.

Ramaswamy, V., O. Boucher, J. Haigh, D. Hauglustaine, J. Haywood, G. Myhre, T. Nakajima, G.Y. Shi, and S. Solomon, 2001: Radiative forcing of climate change. In: *Climate Change 2001: The Scientific Basis*. Contribution of Working Group I to the Third Assessment Report of the Intergovernmental Panel on Climate Change [Houghton, J.T., Y. Ding, D.J. Griggs, M. Noguer, P.J. van der Linded, X. Dai, K. Maskell, and C.A. Johnson (eds.)]. Cambridge University Press, Cambridge, UK, and New York, pp. 349-416.

Rand, S. and M. Yamabe (coordinating lead authors), N. Campbell, J. Hu, P. Lapin, A. McCulloch, A. Merchant, K. Mizuno, J. Owens, and P. Rollet, 2005: Non-medical aerosols, solvents, and HFC-23. In: *IPCC/TEAP Special Report on Safeguarding the Ozone Layer and the Global Climate System: Issues Related to Hydrofluorocarbons and Perfluorocarbons* [Metz, B., L. Kuijpers, S. Solomon, S.O. Andersen, O. Davidson, J. Pons, D. de Jager, T. Kestin, M. Manning, and L.A. Meyer (eds.)]. Cambridge University Press, Cambridge, UK, and New York, pp. 379-402.

Schauffler, S.M., E.L. Atlas, S.G. Donnelly, A. Andrews, S.A. Montzka, J.W. Elkins, D.F. Hurst, P.A. Romashkin, G.S. Dutton, and V. Stroud, 2003: Chlorine budget and partitioning during SOLVE. *Journal of Geophysical Research*, **108 (D5)**, 4173, doi:10.1029/2001JD002040.

Shine, K.P., L.K. Gohar, M.D. Hurley, G. Marston, D. Martin, P.G. Simmonds, T.J. Wallington, and M. Watkins, 2005:

Perfluorodecalin: global warming potential and first detection in the atmosphere. *Atmospheric Environment*, **39(9)**, 1759-1763.

Solomon, S. and D.L. Albritton, 1992: Time-dependent ozone depletion potentials for short- and long-term forecasts. *Nature*, **357(6373)**, 33-37.

Solomon, S., M. Mills, L.E. Heidt, W.H. Pollock, and A.F. Tuck, 1992: On the evaluation of ozone depletion potentials. *Journal of Geophysical Research*, **97(D1)**, 825-842.

UNEP (United Nations Environment Programme), 2007: Ozone Secretariat ozone depleting substances production and consumption data tables on the web at <http://ozone.unep.org/Data_Reporting/>

UNEP/CTOC (United Nations Environment Programme/Chemical Technical Options Committee), 2007: 2006 *Report of the Chemical Technical Options Committee: 2006 Assessment*. UN Ozone Secretariat, Nairobi, Kenya, 55 pp. <http://ozone.unep.org/teap/Reports/CTOC/ctoc_assessment_report06.pdf>

UNEP/MBTOC (United Nations Environment Programme/Methyl Bromide Technical Options Committee), 2007: *2006 Report of the Methyl Bromide Technical Options Committee: 2006 Assessment*. UN Ozone Secretariat, Nairobi, Kenya, 453 pp. <http://ozone.unep.org/teap/Reports/MBTOC/MBTOC-2006-Assessment%20Report.pdf>

UNEP/TEAP (United Nations Environment Programme/Technology and Economic Assessment Panel), 2006: *Task Force on Emissions Discrepancies Report*. UN Ozone Secretariat, Nairobi, Kenya, 80 pp. <http://ozone.unep.org/teap/Reports/TEAP_Reports/TEAP-Discrepancy-report.pdf>

U.S. EPA (Environmental Protection Agency), 2007: *Inventory of U.S. Greenhouse Gas Emissions and Sinks, 1990-2005*. EPA #430-R-07-002. Environmental Protection Agency, Washington DC, 393 pp. <http://epa.gov/climatechange/emissions/usginv_archive.html>

Velders, G.J.M., S.O. Andersen, J.S. Daniel, D.W. Fahey, and M. McFarland, 2007: The importance of the Montreal Protocol in protecting climate. *Proceedings of the National Academy of Sciences*, **104(12)**, 4814-4819.

Wuebbles, D.J., 1983: Chlorocarbon emission scenarios: potential impact on stratospheric ozone. *Journal of Geophysical Research*, **88(C2)**, 1433-1443.

WMO (World Meteorological Organization), 2003: *Scientific Assessment of Ozone Depletion: 2002*. Global Ozone Research and Monitoring Project report no. 47. World Meteorological Organization, Geneva, Switzerland, 498 pp.

WMO (World Meteorological Organization), 2007: *Scientific Assessment of Ozone Depletion: 2006*. Global Ozone Research and Monitoring Project report no. 50. World Meteorological Organization, Geneva, Switzerland, 572 pp. <http://www.esrl.noaa.gov/csd/assessments/2006/>

CHAPTER 3 REFERENCES

Allen, D.R., R.M. Bevilacqua, G.E. Nedoluha, C.E. Randall, and G.L. Manney, 2003: Unusual stratospheric transport and mixing during the 2002 Antarctic winter. *Geophysical Research Letters*, **30(12)**, 1599, doi: 10.1029/2003GL017117.

Andrews, D.G., J.R. Holton, and C.B. Leovy, 1987: *Middle Atmosphere Dynamics*. Academic Press, Orlando, FL, 489 pp.

Bais, A. and D. Lubin (lead authors), A. Arola, G. Bernhard, M. Blumthaler, N. Chubarova, C. Erlick, H.P. Gies, N. Krotkov, K. Lantz, B. Mayer, R.L. McKenzie, R.D. Piacentini, G. Seckmeyer, J.R. Slusser, and C.S. Zerefos, 2007: Surface ultraviolet radiation: past, present, and future. In: *Scientific Assessment of Ozone Depletion: 2006*. Global Ozone Research and Monitoring Project report no. 50. World Meteorological Organization, Geneva, Switzerland, pp. 7.1-7.54.

Brönnimann, S. and L.L. Hood, 2003: Frequency of low-ozone events over northwestern Europe in 1952-1963 and 1990-2000. *Geophysical Research Letters*, **30(21)**, 2118, doi:10.1029/2003GL018431.

Canty, T., E.D. Rivière, R.J. Salawitch, G. Berthet, J.-B. Renard, K. Pfeilsticker, M. Dorf, A. Butz, H. Bösch, R.M. Stimpfle, D.M. Wilmouth, E.C. Richard, D.W. Fahey, P.J. Popp, M.R. Schoeberl, L.R. Lait, and T.P. Bui, 2005: Nighttime OClO in the winter Arctic vortex. *Journal of Geophysical Research*, **110**, D01301, doi:10.1029/2004JD005035.

Cede, A., M. Kowalewski, S. Kazadzis, A. Bais, N. Kouremeti, M. Blumthaler, and J. Herman, 2006: Solar zenith angle effect for direct-sun measurements of Brewer spectrophotometers due to polarization. *Geophysical Research Letters*, **33**, L02806, doi:10.1029/2005GL024860.

Crutzen, P.J. and F. Arnold, 1986: Nitric-acid cloud formation in the cold Antarctic stratosphere: A major cause for the springtime 'ozone hole.' *Nature*, **324(6098)**, 651-655.

Dhomse, S., M. Weber, I. Wohltmann, M. Rex, and J.P. Burrows, 2006: On the possible causes of recent increases in northern hemisphere total ozone from a statistical analysis of satellite data from 1979 to 2003. *Atmospheric Chemistry and Physics*, **6(5)**, 1165-1180.

Díaz, S., D. Nelson, G. Deferrari, and C. Camilión, 2003: Estimated and measured DNA, plant-chromosphere and erythemal-weighted irradiance at Barrow and South Pole (1979-2000). *Agricultural and Forest Meteorology*, **120(1-4)**, 69-82.

Diffey, B.L., 1991: Solar ultraviolet radiation effects on biological systems. *Review in Physics in Medicine and Biology*, **36(3)**, 299-328.

Eyring, V., N. Butchart, D.W. Waugh, H. Akiyoshi, J. Austin, S. Bekki, G.E. Bodeker, B.A. Boville, C. Brühl, M.P. Chipperfield, E. Cordero, M. Dameris, M. Deushi, V.E. Fioletov, S.M. Frith, R.R. Garcia, A. Gettelman, M.A. Giorgetta, V. Grewe, L. Jourdain, D.E. Kinnison, E. Mancini, E. Manzini, M. Marchand, D.R. Marsh, T. Nagashima, P.A. Newman, J.E. Nielsen, S. Pawson, G. Pitari, D.A. Plummer, E. Rozanov, M. Schraner, T.G. Shepherd, K. Shibata, R.S. Stolarski, H. Struthers, W. Tian, and M. Yoshiki, 2006: Assessment of temperature, trace species, and ozone in chemistry-climate model simulations of the recent past. *Journal of Geophysical Research*, **111**, D22308, doi:10.1029/2006JD007327.

Fahey, D.W., 2007: Twenty questions and answers about the ozone layer: 2006 update. In: *Scientific Assessment of Ozone Depletion: 2006*. Global Ozone Research and Monitoring Project report no. 50. World Meteorological Organization, Geneva, Switzerland, [50 pp.] Included as Appendix A of this report; also at <http://www.esrl.noaa.gov/csd/assessments/2006/report.html>

Feng, W., M.P. Chipperfield, H.K. Roscoe, J.J. Remedios, A.M. Waterfall, G.P. Stiller, N. Glatthor, M. Höpfner, and D.-Y. Wang, 2005: Three-dimensional model study of the Antarctic ozone hole in 2002 and comparison with 2000. *Journal of the Atmospheric Sciences*, **62(3)**, 822-837.

Feng, W., M.P. Chipperfield, S. Davies, P. von der Gathen, E. Kyrö, C.M. Volk, A. Ulanovsky, and G. Belyaev, 2007a: Large chemical ozone loss in 2004/2005 Arctic winter/spring. *Geophysical Research Letters*, **34**, L09803, doi:10.1029/2006GL029098.

Feng, W., M.P. Chipperfield, M. Dorf, K. Pfeilsticker, and P. Ricaud, 2007b: Mid-latitude ozone changes: studies with a 3-D CTM forced by ERA-40 analyses. *Atmospheric Chemistry and Physics*, **7(9)**, 2357-2369.

Fioletov, V.E. and W.F.J. Evans, 1997: The influence of ozone and other factors on surface radiation. In: *Ozone Science: A Canadian Perspective on the Changing Ozone Layer* [Wardle, D.I., J.B. Kerr, C.T. McElroy, and D.R. Francis (eds.)]. University of Toronto Press, Toronto, pp. 73-79.

Fioletov, V.E., L.J.B. McArthur, J.B. Kerr, and D.I. Wardle, 2001: Long-term variations of UV-B irradiance over Canada estimated from Brewer observations and derived from

ozone and pyranometer measurements. *Journal of Geophysical Research*, **106(D19)**, 23009-23027.

Fioletov, V.E., G.E. Bodeker, A.J. Miller, R.D. McPeters, and R. Stolarski, 2002: Global and zonal total ozone variations estimated from ground-based and satellite measurements: 1964-2000. *Journal of Geophysical Research*, **107(D22)**, 4647, doi:10.1029/2001JD001350.

Fioletov, V.E., M.G. Kimlin, N. Krotkov, L.J.B. McArthur, J.B. Kerr, D.I. Wardle, J.R. Herman, R. Meltzer, T.W. Mathews, and J. Kaurola, 2004: UV index climatology over North America from ground-based and satellite estimates. *Journal of Geophysical Research*, **109**, D22308, doi:10.1029/2004JD004820.

Frieler, K., M. Rex, R.J. Salawitch, T. Canty, M. Streibel, R.M. Stimpfle, K. Pfeilsticker, M. Dorf, D.K. Weisenstein, S. Godin-Beekmann, and P. von der Gathen, 2006: Toward a better quantitative understanding of polar stratospheric ozone loss. *Geophysical Research Letters*, **33**, L10812, doi:10.1029/2005GL025466.

Fromm, M., J. Alfred, and M. Pitts, 2003: A unified, long-term, high-latitude stratospheric aerosol and cloud database using SAM II, SAGE II, and POAM II/III data: algorithm description, database definition, and climatology. *Journal of Geophysical Research*, **108(D12)**, 4366, doi:10.1029/2002JD002772.

Fusco, A.C. and M.L. Salby, 1999: Interannual variations of total ozone and their relationship to variations of planetary wave activity. *Journal of Climate*, **12(6)**, 1619-1629.

Ghetti, F., G. Checcucci, and J. Bornman (eds.), 2006: *Environmental UV Radiation: Impact on Ecosystems and Human Health and Predictive Models*. NATO science series IV, earth and environmental sciences volume 57. Springer, Dordrecht, 288 pp.

Goutail, F., J.-P. Pommereau, F. Lefèvre, M. van Roozendael, S.B. Andersen, B.-A. Kåstad Høiskar, V. Dorokhov, E. Kyrö, M.P. Chipperfield, and W. Feng, 2005: Early unusual ozone loss during the Arctic winter 2002/2003 compared to other winters. *Atmospheric Chemistry and Physics*, **5(3)**, 665-677.

Grant, W.B., 2002: An estimate of premature cancer mortality in the U.S. due to inadequate doses of solar ultraviolet-B radiation. *Cancer*, **94(6)**, 1867-1875.

Grooß, J.-U., P. Konopka, and R. Müller, 2005: Ozone chemistry during the 2002 Antarctic vortex split. *Journal of the Atmospheric Sciences*, *62(3)*, 860-870.

Hadjinicolaou, P., J.A. Pyle, M.P. Chipperfield, and J.A. Kettleborough, 1997: Effect of interannual meteorological variability on mid-latitude O_3. *Geophysical Research Letters,* **24(23)**, 2993-2996.

Hadjinicolaou, P., A. Jrrar, J.A. Pyle, and L. Bishop, 2002: The dynamically driven long-term trend in stratospheric ozone over northern middle latitudes. *Quarterly Journal of the Royal Meteorological Society,* **128(583)**, 1393-1412.

Hadjinicolaou, P., J.A. Pyle, and N.R.P. Harris, 2005: The recent turnaround in stratospheric ozone over northern middle latitudes: a dynamical modeling perspective. *Geophysical Research Letters,* **32**, L12821, doi:10.1029/2005GL022476.

Herman, J.R. and E.A. Celarier, 1997: Earth surface reflectivity climatology at 340-380 nm from TOMS data. *Journal of Geophysical Research,* **102(D23)**, 28003-28011.

Herman, J.R., P.A. Newman, R.D. McPeters, A.J. Krueger, P.K. Bhartia, C.J. Seftor, O. Torres, G. Jaross, R.P. Cebula, D. Larko, and C. Wellemeyer, 1995: Meteor 3/Total Ozone Mapping Spectrometer observations of the 1993 ozone hole. *Journal of Geophysical Research,* **100(D2)**, 2973-2983.

Herman, J.R., S. McKenzie, S. Diaz, J. Kerr, S. Madronich, and G. Seckmeyer, 1999a: UV radiation at the Earth's surface. In: *Scientific Assessment of Ozone Depletion: 1998.* Global Ozone Research and Monitoring Project report no. 44. World Meteorological Organization, Geneva, Switzerland, chapter 9.

Herman, J.R., N.A. Krotkov, E.A. Celarier, D. Larko, and G. Labow, 1999b: The distribution of UV radiation at the Earth's surface from TOMS-measured UV-backscattered radiances. *Journal of Geophysical Research,* **104(D10)**, 12059-12076.

Herman, J.R., E. Celarier, and D. Larko, 2001a: UV 380 nm reflectivity of the Earth's surface, clouds, and aerosols. *Journal of Geophysical Research,* **106(D6)**, 5335-5351.

Herman, J.R., D. Larko, E. Celarier, and J. Ziemke, 2001b: Changes in the Earth's UV reflectivity from the surface, clouds and aerosols. *Journal of Geophysical Research,* **106(D6)**, 5353-5368.

Herman, J.R., G. Labow, N.C. Hsu, and D. Larko, 2008: Changes in cloud cover derived from reflectivity time series using SeaWiFS, N7-TOMS, EP-TOMS, SBUV-2, and OMI radiance data. *Journal of Geophysical Research,* in press, doi:10.1029/2007JD009508.

Hio, Y. and S. Yoden, 2005: Interannual variations of the seasonal March in the Southern Hemisphere stratosphere for 1979-2002 and characterization of the unprecedented year 2002. *Journal of the Atmospheric Sciences,* **62(3)**, 567-580.

Hofmann, D.J., S.J. Oltmans, J.M. Harris, B.J. Johnson, and J.A. Lathrop, 1997: Ten years of ozonesonde measurements at the south pole: implications for recovery of springtime Antarctic ozone. *Journal of Geophysical Research,* **102(D7)**, 8931-8943.

Holben, B.N., D. Tanre, A. Smirnov, T.F. Eck, I. Slutsker, N. Abuhassan, W.W. Newcomb, J. Schafer, B. Chatenet, F. Lavenue, Y.J. Kaufman, J. Vande Castle, A. Setzer, B. Markham, D. Clark, R. Frouin, R. Halthore, A. Karnieli, N.T. O'Neill, C. Pietras, R.T. Pinker, K. Voss, and G. Zibordi, 2001: An emerging ground-based aerosol climatology: aerosol optical depth from AERONET. *Journal of Geophysical Research,* **106(D11)**, 12067-12097.

Holick, M.F., 2004: Sunlight and vitamin D for bone health and prevention of autoimmune diseases, cancers, and cardiovascular disease. *American Journal of Clinical Nutrition,* **80(6)**, 1678S-1688S.

Hood, L.L. and B.E. Soukharev, 2005: Interannual variations of total ozone at northern midlatitudes correlated with stratospheric EP flux and potential vorticity. *Journal of the Atmospheric Sciences,* **62(10)**, 3724-3740.

Hood, L.L., J.P. McCormack, and K. Labitzke, 1997: An investigation of dynamical contributions to midlatitude ozone trends in winter. *Journal of Geophysical Research,* **102(D11)**, 13079-13093.

Hood, L.L., S. Rossi, and M. Beulen, 1999: Trends in lower stratospheric zonal winds, Rossby wave breaking behavior, and column ozone at northern midlatitudes. *Journal of Geophysical Research,* **104(D20)**, 24321-24339.

Hood, L.L., B.E. Soukharev, M. Fromm, and J. McCormack, 2001: Origin of extreme ozone minima at middle to high northern latitudes. *Journal of Geophysical Research,* **106(D18)**, 20925-20940.

Hoppel, K.H., R. Bevilacqua, G. Nedoluha, C. Deniel, F. Lefevre, J. Lumpe, M. Fromm, C. Randall, J. Rosenfield, and M. Rex, 2002: POAM III observations of Arctic ozone loss for the 1999/2000 winter. *Journal of Geophysical Research,* **107(D20)**, 8262, doi:10.1029/2001JD000476.

Hoppel, K., R. Bevilacqua, D. Allen, and G. Nedoluha, 2003: POAM III observations of the anomalous 2002 Antarctic ozone hole. *Geophysical Research Letters,* **30(7)**, 1394, doi:10.1029/2003GL016899.

Hoppel, K., G. Nedoluha, M. Fromm, A. Allen, R. Bevilacqua, J. Alfred, B. Johnson, and G. Konig-Langlo, 2005: Reduced ozone loss at the upper edge of the Antarctic ozone hole during 2001-2004. *Geophysical Research Letters,* **32**, L20816, doi:10.1029/2005GL023968.

IPCC/TEAP (Intergovernmental Panel on Climate Change/ Technology and Economic Assessment Panel), 2005: *IPCC/ TEAP Special Report on Safeguarding the Ozone Layer and the Global Climate System: Issues Related to Hydrofluoro-carbons and Perfluorocarbons* [Metz, B., L. Kuijpers, S. Solomon, S.O. Andersen, O. Davidson, J. Pons, D. de Jager, T. Kestin, M. Manning, and L.A. Meyer (eds.)]. Cambridge University Press, Cambridge, UK, and New York, 478 pp.

Kalliskota, S.J., J. Kaurola, P. Taalas, J.R. Herman, E. Celarier, and N. Krotkov, 2000: Comparison of daily UV doses estimated from Nimbus-7/TOMS measurements and ground-based spectroradiometric data. *Journal of Geophysical Research*, **105(D4)**, 5059-5067.

Koch, G., H. Wernli, C. Schwierz, J. Staehelin, and T. Peter, 2005: A composite study on the structure and formation of ozone miniholes and minihighs over central Europe. *Geophysical Research Letters*, **32**, L12810, doi:10.1029/2004GL022062.

Konopka, P., J.-U. Grooß, K.W. Hoppel, H.-M. Steinhorst, and R. Müller, 2005: Mixing and chemical ozone loss during and after the Antarctic polar vortex major warming in September 2002. *Journal of the Atmospheric Sciences*, **62(3)**, 848-859.

Krotkov, N.A., P.K. Bhartia, J.R. Herman, V. Fioletov, and J. Kerr, 1998: Satellite estimation of spectral surface UV irradiance in the presence of tropospheric aerosols 1: cloud-free case. *Journal of Geophysical Research*, **103(D8)**, 8779-8793.

Krotkov, N.A., J.R. Herman, P.K. Bhartia, Z. Ahmad, and V. Fioletov, 2001: Satellite estimation of spectral surface UV irradiance 2: effects of homogeneous clouds and snow. *Journal of Geophysical Research*, **106(D11)**, 11743-11759.

London, J., 1963: The distribution of total ozone in the Northern Hemisphere. *Beiträge zur Physik der Atmosphäre*, **36(3/4)**, 254-263.

Lucas, R., T. McMichael, W. Smith, and B. Armstrong, 2006: *Solar Ultraviolet Radiation: Global Burden of Disease from Solar Ultraviolet Radiation*. Environmental burden of disease series, no. 13. World Health Organization, Geneva, Switzerland, 87 pp. <http://www.who.int/quantifying_ehimpacts/publications/ebd13/en/index.html>

Madronich, S., 1993: The atmosphere and UV-B radiation at ground level. In: *Environmental UV Photobiology* [Björn, L.O. and A.R. Young (eds.)]. Plenum Press, New York, pp. 1-39.

Manney, G.L., J.L. Sabutis, D.R. Allen, W.A. Lahoz, A.A. Scaife, C.E. Randall, S. Pawson, B. Naujokat, and R. Swin-

bank, 2005: Simulations of dynamics and transport during the September 2002 Antarctic major warming. *Journal of the Atmospheric Sciences*, **62(3)**, 690-707.

Manney, G.L., M.L. Santee, L. Froidevaux, K. Hoppel, N.J. Livesey, and J.W. Waters, 2006: EOS MLS observations of ozone loss in the 2004-2005 Arctic winter. *Geophysical Research Letters*, **33**, L04802, doi: 10.1029/2005GL024494.

McKinlay, A.F. and B.L. Diffey, 1987: A reference action spectrum for ultraviolet induced erythema in human skin. In: *Human Exposure to Ultraviolet Radiation: Risks and Regulations* [Passchier, W.R. and B.F.M. Bosnjakovic (eds.)]. Elsevier, Amsterdam, 580 pp.

McPeters, R.D., G.J. Labow, and J.A. Logan, 2007: Ozone climatological profiles for satellite retrieval algorithms. *Journal of Geophysical Research*, **112**, D05308, doi:10.1029/2005JD006823.

Newman, P.A. and E.R. Nash, 2005: The unusual Southern Hemisphere stratosphere winter of 2002. *Journal of the Atmospheric Sciences*, **62(3)**, 614-628.

Newman, P.A. and W.J. Randel, 1988: Coherent ozone-dynamical changes during the southern hemisphere spring, 1979-1986. *Journal of Geophysical Research*, **93(D10)**, 12585-12606.

Newman, P.A., S.R. Kawa, and E.R. Nash, 2004: On the size of the Antarctic ozone hole. *Geophysical Research Letters*, **31**, L21104, doi:10.1029/2004GL020596.

Newman, P.A., E.R. Nash, S.R. Kawa, S.A. Montzka, and S.M. Schauffler, 2006: When will the Antarctic hole recover? *Geophysical Research Letters*, **33**, L12814, doi:10.1029/2005GL025232.

Newman, P.A., J.S. Daniel, D.W. Waugh, and E.R. Nash, 2007: A new formulation of equivalent effective stratospheric chlorine (EESC). *Atmospheric Chemistry and Physics*, **7(17)**, 4537-4552.

Orsolini, Y.J. and V. Limpasuvan, 2001: The North Atlantic Oscillation and the occurrences of ozone miniholes. *Geophysical Research Letters*, **28(21)**, 4099-4102.

Pope, F.D., J.C. Hansen, K.D. Bayes, R.R. Friedl, and S.P. Sander, 2007: Ultraviolet absorption spectrum of chlorine peroxide, ClOOCl. *Journal of Physical Chemistry A*, **111(20)**, 4322-4332.

Randel, W.J., F. Wu, and R. Stolarski, 2002: Changes in column ozone correlated with EP flux. *Journal of the Meteorological Society of Japan*, **80(4B)**, 849-862.

Reed, R.J., W.J. Campbell, L.A. Rasmussen, and D.G. Rogers, 1961: Evidence of a downward-propagating annual wind reversal in equatorial stratosphere. *Journal of Geophysical Research*, **66(3)**, 813-818.

Reid, S.J., A.F. Tuck, and G. Kiladis, 2000: On the changing abundance of ozone minima at northern midlatitudes. *Journal of Geophysical Research*, **105(D10)**, 12169-12180.

Rex, M., R.J. Salawitch, N.R.P. Harris, P. von der Gathen, G.O. Braathen, A. Schulz, H. Deckelmann, M. Chipperfield, B.M. Sinnhuber, E. Reimer, R. Alfier, R. Bevilacqua, K. Hoppel, M. Fromm, J. Lumpe, H. Küllmann, A. Kleinböhl, H. Bremer, M. von König, K. Künzi, D. Toohey, H. Vömel, E. Richard, K. Aikin, H. Jost, J.B. Greenblatt, M. Loewenstein, J.R. Podolske, C.R. Webster, G.J. Flesch, D.C. Scott, R.L. Herman, J.W. Elkins, E.A. Ray, F.L. Moore, D.F. Hurst, P. Romashkin, G.C. Toon, B. Sen, J.J. Margitan, P. Wennberg, R. Neuber, M. Allart, B.R. Bojkov, H. Claude, J. Davies, W. Davies, H. De Backer, H. Dier, V. Dorokhov, H. Fast, Y. Kondo, E. Kyrö, Z. Litynska, I.S. Mikkelsen, M.J. Molyneux, E. Moran, T. Nagai, H. Nakane, C. Parrondo, F. Ravegnani, P. Skrivankova, P. Viatte, and V. Yushkov, 2002: Chemical depletion of Arctic ozone in winter 1999/2000. *Journal of Geophysical Research*, **107(D20)**, 8276, doi:10.1029/2001JD000533.

Rex, M., R.J. Salawitch, P. von der Gathen, N.R.P. Harris, M.P. Chipperfield, and B. Naujokat, 2004: Arctic ozone loss and climate change. *Geophysical Research Letters*, **31**, L04116, doi: 10.1029/2003GL018844.

Rex, M., R.J. Salawitch, H. Deckelmann, P. von der Gathen, N.R.P. Harris, M.P. Chipperfield, B. Naujokat, E. Reimer, M. Allaart, S.B. Andersen, R. Bevilacqua, G.O. Braathen, H. Claude, J. Davies, H. De Backer, H. Dier, V. Dorokov, H. Fast, M. Gerding, S. Godin-Beekmann, K. Hoppel, B. Johnson, E. Kyrö, Z. Litynska, D. Moore, H. Nakane, M.C. Parrondo, A.D. Risley Jr., P. Skrivankova, R. Stübi, P. Viatte, V. Yushkov, and C. Zerefos, 2006: Arctic winter 2005: Implications for stratospheric ozone loss and climate change. *Geophysical Research Letters*, **33**, L23808, doi:10.1029/2006GL026731.

Ricaud, P., F. Lefèvre, G. Berthet, D. Murtagh, E.J. Llewellyn, G. Mégie, E. Kyrölä, G.W. Leppelmeier, H. Auvinen, C. Boonne, S. Brohede, D.A. Degenstein, J. de La Noë, E. Dupuy, L. El Amraoui, P. Eriksson, W.F.J. Evans, U. Frisk, R.L. Gattinger, F. Girod, C.S. Haley, S. Hassinen, A. Hauchecorne, C. Jimenez, E. Kyrö, N. Lautié, E. Le Flochmoën, N.D. Lloyd, J.C. McConnell, I.C. McDade, L. Nordh, M. Olberg, A. Pazmiño, S.V. Petelina, A. Sandqvist, A. Seppälä, C.E. Sioris, B.H. Solheim, J. Stegman, K. Strong, P. Taalas, J. Urban, J. von Savigny, F. von Scheele, and G. Witt, 2005: Polar vortex evolution during the 2002 Antarctic major warming as observed by the Odin satellite. *Journal of Geophysical Research*, **110**, D05302, doi:10.1029/2004JD005018.

Roscoe, H.K., A.E. Jones, and A.M. Lee, 1997: Midwinter start to Antarctic ozone depletion: Evidence from observations and models. *Science*, **278(5335)**, 93-96.

Roscoe, H.K., J.D. Shanklin, and S.R. Colwell, 2005: Has the Antarctic vortex split before 2002? *Journal of the Atmospheric Sciences*, **62(3)**, 581-588.

Salawitch, R.J., D.K. Weisenstein, L.J. Kovalenko, C.E. Sioris, P.O. Wennberg, K. Chance, M.K.W. Ko, and C.A. McLinden, 2005: Sensitivity of ozone to bromine in the lower stratosphere. *Geophysical Research Letters*, **32**, L05811, doi:10.1029/2004GL021504.

Salby, M.L. and P.F. Callaghan, 2002: Interannual changes of the stratospheric circulation: Relationship to ozone and tropospheric structure. *Journal of Climate*, **15(24)**, 3673-3685.

Salby, M.L. and P.F. Callaghan, 2004a: Systematic changes of northern hemisphere ozone and their relationship to random interannual changes. *Journal of Climate*, **17(23)**, 4512-4521.

Salby, M.L. and P.F. Callaghan, 2004b: Interannual changes of the stratospheric circulation: influence on the tropics and southern hemisphere. *Journal of Climate*, **17(5)**, 952-964.

Shepherd, T.G., 2007: Transport in the middle atmosphere. *Journal of the Meteorological Society of Japan*, **85B**, 165-191.

Singleton, C.S., C.E. Randall, V.L. Harvey, M.P. Chipperfield, W. Feng, G.L. Manney, L. Froidevaux, C.D. Boone, P.F. Bernath, K.A. Walker, C.T. McElroy, and K.W. Hoppel, 2007: Quantifying Arctic ozone loss during the 2004-2005 winter using satellite observations and a chemical transport model. *Journal of Geophysical Research*, **112**, D07304, doi:10.1029/2006JD007463.

Sinnhuber, B.-M., M. Weber, A. Amankwah, and J.P. Burrows, 2003: Total ozone during the unusual Antarctic winter of 2002. *Geophysical Research Letters*, **30(11)**, 1580, doi:10.1029/2002GL016798.

Smith, R.C., B.B. Prezelin, K.S. Baker, R.R. Bidigare, N.P. Boucher, T. Coley, D. Karentz, S. MacIntyre, H.A. Matlick, D. Menzies, M. Ondrusek, Z. Wan, and K.J. Waters, 1992: Ozone depletion: ultraviolet radiation and phytoplankton biology in Antarctic waters. *Science*, **255(5047)**, 952-959.

Solomon, S., R.R. Garcia, F.S. Rowland, and D.J. Wuebbles, 1986: On the depletion of Antarctic ozone. *Nature*, **321(6072)**, 755-758.

Solomon, S., R.W. Portmann, T. Sasaki, D.J. Hofmann, and D.W.J. Thompson, 2005: Four decades of ozonesonde measurements over Antarctica. *Journal of Geophysical Research*, **110(D7)**, doi: 10.1029/2005JD005917.

Staples, M., R. Marks, and G. Giles, 1998: Trends in the incidence of non-melanocytic skin cancer (NMSC) treated in Australia 1985-1995: are primary prevention programs starting to have an effect? *International Journal of Cancer*, **78(2)**, 144-148.

Steele, H.M., P. Hamill, M.P. McCormick, and T.J. Swissler, 1983: The formation of polar stratospheric clouds. *Journal of the Atmospheric Sciences*, **40(8)**, 2055-2068.

Steinbrecht, W., H. Claude, and U. Köhler, 1998: Correlations between tropopause height and total ozone: implications for long-term changes. *Journal of Geophysical Research*, **103(D15)**, 19183-19192.

Steinbrecht, W., H. Claude, F. Schönenborn, I.S. McDermid, T. Leblanc, S. Godin, T. Song, D.P.J. Swart, Y.J. Meijer, G.E. Bodeker, B.J. Connor, N. Kämpfer, K. Hocke, Y. Calisesi, N. Schneider, J. de la Noë, A.D. Parrish, I.S. Boyd, C. Brühl, B. Steil, M.A. Giorgetta, E. Manzini, L.W. Thomason, J.M. Zawodny, M.P. McCormick, J.M. Russell III, P.K. Bhartia, R.S. Stolarski, and S.M. Hollandsworth-Frith, 2006: Long-term evolution of upper stratospheric ozone at selected stations of the Network for the Detection of Stratospheric Change (NDSC). *Journal of Geophysical Research*, **111**, D10308, doi:10.1029/2005JD006454.

Stimpfle, R.M., D.M. Wilmouth, R.J. Salawitch, and J.G. Anderson, 2004: First measurements of ClOOCl in the stratosphere: the coupling of ClOOCl and ClO in the Arctic polar vortex. *Journal of Geoosical Research*, **109**, D03301, doi:10.1029/2003JD003811.

Stolarski, R.S. and S. Frith, 2006: Search for evidence of trend slow-down in the long-term TOMS/SBUV total ozone data record: the importance of instrument drift uncertainty. *Atmospheric Chemistry and Physics*, **6(12)**, 4057-4065.

Stolarski, R.S., R.D. McPeters, and P.A. Newman, 2005: The ozone hole of 2002 as measured by TOMS. *Journal of the Atmospheric Sciences*, **62(3)**, 716-720.

Tanskanen, A., A. Lindfors, A. Maatta, N. Krotkov, J. Herman, J. Kaurola, T. Koskela, K. Lakkala, V. Fioletov, G. Bernhard, R. McKenzie, Y. Kondo, M. O'Neill, H. Slaper, P. den Outer, A.F. Bais, and J. Tamminen, 2007: Validation of daily erythemal doses from Ozone Monitoring Instrument with ground-based UV measurement data. *Journal of Geophysical Research*, **112**, D24S44, doi:10.1029/2007JD008830.

Taylor, H.R., 1990: Cataracts and ultraviolet light. In: *Global Atmospheric Change and Public Health* [White, J.C. (ed.)]. Elsevier Science Publishing, New York, pp. 61-65.

Tilmes, S., R. Muller, A. Engel, M. Rex, and J.M. Russell, 2006: Chemical ozone loss in the Arctic and Antarctic stratosphere between 1992 and 2005. *Geophysical Research Letters*, **33**, L20812, doi:10.1029/2006GL026925.

Toon, O.B., P. Hamill, R.P. Turco, and J. Pinto, 1986: Condensation of HNO_3 and HCl in winter polar stratospheres. *Geophysical Research Letters*, **13(12)**, 1284-1287.

Trepte, C.R., R.E. Veiga, and M.P. McCormick, 1993: The poleward dispersal of Mount Pinatubo volcanic aerosol. *Journal of Geophysical Research*, **98(D10)**, 18563-18573.

U.S. EPA (Environmental Protection Agency), 1999: *The Benefits and Costs of the Clean Air Act 1990 to 2010, Appendix G: Stratospheric Ozone Assessment*. EPA-410-R-99-001. EPA Office of Air and Radiation, Washington, DC, pp. G.1-G.43.<http://www.epa.gov/oar/sect812/1990-2010/fullrept.pdf>

Vermeer, M., G.J. Schmieder, T. Yoshikawa, J-W. van den Berg, M.S. Metzman, J.R. Taylor, and J.W. Streilein, 1991: Effects of ultraviolet B light on cutaneous immune responses of humans with deeply pigmented skin. *Journal of Investigative Dermatology*, **97(4)**, 729-734.

von Hobe, M., R.J. Salawitch, T. Canty, H. Keller-Rudek, G.K. Moortgat, J.-U. Grooß, R. Müller, and F. Stroh, 2007: Understanding the kinetics of the ClO dimer cycle. *Atmospheric Chemistry and Physics*, **7(12)**, 3055-3069.

Weber, M., S. Dhomse, F. Wittrock, A. Richter, B.-M. Sinnhuber, and J.P. Burrows, 2003: Dynamical control of NH and SH winter/spring total ozone from GOME observations in 1995-2002. *Geophysical Research Letters*, **30(11)**, 1583, doi:10.1029/2002GL016799.

WMO (World Meteorological Organization), 1989: *Scientific Assessment of Stratospheric Ozone: 1989*. Global Ozone Research and Monitoring Project report no. 20. World Meteorological Organization, Geneva, Switzerland, 486 pp.

WMO (World Meteorological Organization), 1999: *Scientific Assessment of Ozone Depletion: 1998*. Global Ozone Research and Monitoring Project report no. 44. World Meteorological Organization, Geneva, Switzerland, 486 pp.

WMO (World Meteorological Organization), 2003: *Scientific Assessment of Ozone Depletion: 2002*. Global Ozone Research and Monitoring Project report no. 47. World Meteorological Organization, Geneva, Switzerland, 498 pp.

WMO (World Meteorological Organization), 2007: *Scientific Assessment of Ozone Depletion: 2006*. Global Ozone Research and Monitoring Project report no. 50. World Meteorological Organization, Geneva, Switzerland, 572 pp. <http://www.esrl.noaa.gov/csd/assessments/2006/>

CHAPTER 4 REFERENCES

Brasseur, G. and S. Solomon, 1986: *Aeronomy of the Middle Atmosphere*. D. Reidel, Dordrecht, Holland, and Boston, MA, 441 pp.

Butchart, N. and A.A. Scaife, 2001: Removal of chlorofluorocarbons by increased mass exchange between the stratosphere and the troposphere in a changing climate. *Nature*, **410(6830)**, 799-802.

Butchart, N., A.A. Scaife, M. Bourqui, J. de Grandpré, S.H.E. Hare, J. Kettleborough, U. Langematz, E. Manzini, F. Sassi, K. Shibata, D. Shindell, and M. Sigmond, 2006: Simulations of anthropogenic change in the strength of the Brewer-Dobson circulation. *Climate Dynamics*, **27(7-8)**, 727-741.

CCSP (Climate Change Science Program), 2006: *Temperature Trends in the Lower Atmosphere: Steps for Understanding and Reconciling Differences*. [Karl, T., S.J. Hassol, C.D. Miller, and W.L. Murray (eds.)]. Synthesis and assessment product 1.1. U.S. Climate Change Science Program, Washington, DC, 164 pp.

Coffey, M.T., 1996: Observations of the impact of volcanic activity on stratospheric chemistry. *Journal of Geophysical Research*, **101(D3)**, 6767-6780.

Dameris, M., V. Grewe, M. Ponater, R. Deckert, V. Eyring, F. Mager, S. Matthes, C. Schnadt, A. Stenke, B. Steil, C. Brühl, and M.A. Giorgetta, 2005: Long-term changes and variability in a transient simulation with a chemistry-climate model employing realistic forcing. *Atmospheric Chemistry and Physics*, **5(6)**, 2121-2145.

Daniel, J.S., S. Solomon, and D.L. Albritton, 1995: On the evaluation of halocarbon radiative forcing and global warming potentials. *Journal of Geophysical Research*, **100(D1)**, 1271-1285.

Douglass, A.R., R.S. Stolarski, S.E. Strahan, and B.C. Polansky, 2006: Sensitivity of Arctic ozone loss to polar stratospheric cloud volume and chlorine and bromine loading in a chemistry and transport model. *Geophysical Research Letters*, **33**, L17809, doi:10.1029/2006GL026492.

Dvortsov, V.L. and S. Solomon, 2001: Response of the stratospheric temperatures and ozone to past and future increases in stratospheric humidity. *Journal of Geophysical Research*, **106(D7)**, 7505-7514.

Eyring, V., N.R.P. Harris, M. Rex, T.G. Shepherd, D.W. Fahey, G.T. Amanatidis, J. Austin, M.P. Chipperfield, M. Dameris, P.M. De F. Forster, A. Gettleman, H.F. Graf, T. Nagashima, P.A. Newman, S. Pawson, M.J. Prather, J.A. Pyle, R.J. Salawitch, B.D. Santer, and D.W. Waugh, 2005: A strategy for process-oriented validation of coupled-chemistry-climate models. *Bulletin of the American Meteorological Society*, **86(8)**, 1117-1133.

Eyring V., N. Butchart, D.W. Waugh, H. Akiyoshi, J. Austin, S. Bekki, G.E. Bodeker, B.A. Boville, C. Brühl, M.P. Chipperfield, E. Cordero, M. Dameris, M. Deushi, V.E. Fioletov, S.M. Frith, R.R. Garcia, A. Gettelman, M.A. Giorgetta, V. Grewe, L. Jourdain, D.E. Kinnison, E. Mancini, E. Manzini, M. Marchand, D.R. Marsh, T. Nagashima, P.A. Newman, J.E. Nielsen, S. Pawson, G. Pitari, D.A. Plummer, E. Rozanov, M. Schraner, T.G. Shepherd, K. Shibata, R.S. Stolarski, H. Struthers, W. Tian, and M. Yoshiki, 2006: Assessment of temperature, trace species, and ozone in chemistry-climate model simulations of the recent past. *Journal of Geophysical Research*, **111**, D22308, doi:10.1029/2006JD007327.

Forster, P.M. and K.P. Shine, 1997: Radiative forcing and temperature trends from stratospheric ozone changes. *Journal of Geophysical Research*, **102(D9)**, 10841-10855.

Forster, P.M., G. Bodeker, R. Schofield, S. Solomon, and D. Thompson, 2007: Effects of ozone cooling in the tropical lower stratosphere and upper troposphere. *Geophysical Research Letters*, **34**, L23813, doi:10.1029/2007GL031994.

Gauss, M., G. Myhre, I.S.A. Isaksen, V. Grewe, G. Pitari, O. Wild, W.J. Collins, F.J. Dentener, K. Ellingsen, L.K. Gohar, D.A. Hauglustaine, D. Iachetti, F. Lamarque, E. Mancini, L.J. Mickley, M.J. Prather, J.A. Pyle, M.G. Sanderson, K.P. Shine, D.S. Stevenson, K. Sudo, S. Szopa, and G. Zeng, 2006: Radiative forcing since preindustrial times due to ozone changes in the troposphere and the lower stratosphere. *Atmospheric Chemistry and Physics*, **6(3)**, 575-599.

Gillett, N.P. and D.W.J. Thompson, 2003: Simulation of recent southern hemisphere climate change. *Science*, **302(5643)**, 273-275.

Hansen, J., M. Sato, R. Ruedy, L. Nazarenko, A. Lacis, G.A. Schmidt, G. Russell, I. Aleinov, M. Bauer, S. Bauer, N. Bell, B. Cairns, V. Canuto, M. Chandler, Y. Cheng, A. Del Genio, G. Faluvegi, E. Fleming, A. Friend, T. Hall, C. Jackman, M. Kelley, N. Kiang, D. Koch, J. Lean, J. Lerner, K. Lo, S. Menon, R. Miller, P. Minnis, T. Novakov, V. Oinas, Ja. Perlwitz, Ju. Perlwitz, D. Rind, A. Romanou, D. Shindell, P. Stone, S. Sun, N. Tausnev, D. Thresher, B. Wielicki, T. Wong, M. Yao, and S. Zhang, 2005: Efficacy of climate forcings. *Journal of Geophysical Research*, **110**, D18104, doi:10.1029/2005JD005776.

Holton, J.R., P.H. Haynes, M.E. McIntyre, A.R. Douglass, R.B. Rood, and L. Pfister, 1995: Stratosphere-troposphere exchange. *Reviews of Geophysics*, **33(4)**, 403-440.

IPCC (Intergovernmental Panel on Climate Change), 2001: *Climate Change 2001: The Scientific Basis*. Contribution of Working Group I to the Third Assessment Report of the Intergovernmental Panel on Climate Change [Houghton, J.T.,

Y. Ding, D.J. Griggs, M. Noguer, P.J. van der Linden, X. Dai, K. Maskell, and C.A. Johnson (eds.)]. Cambridge University Press, Cambridge, UK, and New York, 881 pp.

IPCC (Intergovernmental Panel on Climate Change), 2007: Summary for policymakers. In: *Climate Change 2007: The Physical Science Basis*. Contribution of Working Group I to the Fourth Assessment Report of the Intergovernmental Panel on Climate Change [Solomon, S., D. Qin, M. Manning, Z. Chen, M. Marquis, K.B. Averyt, M.Tignor, and H.L. Miller (eds.)]. Cambridge University Press, Cambridge, UK, and New York, pp. 1-18.

IPCC/TEAP (Intergovernmental Panel on Climate Change/ Technology and Economic Assessment Panel), 2005: *IPCC/ TEAP Special Report on Safeguarding the Ozone Layer and the Global Climate System: Issues Related to Hydrofluorocarbons and Perfluorocarbons* [Metz, B., L. Kuijpers, S. Solomon, S.O. Andersen, O. Davidson, J. Pons, D. de Jager, T. Kestin, M. Manning, and L.A. Meyer (eds.)]. Cambridge University Press, Cambridge, UK, and New York, 478 pp.

Isaksen, I.S.A. (ed.), 2003: *Ozone-Climate Interactions*. Air pollution research report no. 81. European Commission, Directorate-General for Research, Luxembourg, 143 pp. <http://xweb.geos.ed.ac.uk/%7Edstevens/publications/ isaksen_ozone_climate_ec03.pdf>

Jonsson, A.I., J. de Grandpré, V.I. Fomichev, J.C. McConnell, and S.R. Beagley, 2004: Doubled CO_2-induced cooling in the middle atmosphere: photochemical analysis of the ozone radiative feedback. *Journal of Geophysical Research*, **109**, D24103, doi:10.1029/2004JD005093.

Kiehl, J.T. and K.E. Trenberth, 1997: Earth's annual global mean energy budget. *Bulletin of the American Meteorological Society*, **78(2)**, 197-208.

Langematz, U., M. Kunze, K. Krüger, K. Labitzke, and G.L. Roff, 2003: Thermal and dynamical changes of the stratosphere since 1979 and their link to ozone and CO_2 changes. *Journal of Geophysical Research*, **108(D1)**, 4027, doi:10.1029/2002JD002069.

McCormick, M.P., L.W. Thomason, and C.R. Trepte, 1995: Atmospheric effects of the Mount Pinatubo eruption. *Nature*, **373(6513)**, 399-404.

Mears, C., M. Schabel, and F. Wentz, 2003: A reanalysis of the MSU channel 2 tropospheric temperature record. *Journal of Climate*, **16(22)**, 3650-3664.

Myhre, G., J.S. Nilsen, L. Gulstad, K.P. Shine, B. Rognerud, and I.S.A. Isaksen, 2007: Radiative forcing due to stratospheric water vapour from CH_4 oxidation. *Geophysical Research Letters*, **34**, L01807, doi:10.1029/2006GL027472.

Newman, P.A., S.R. Kawa, and E.R. Nash, 2004: On the size of the Antarctic ozone hole. *Geophysical Research Letters*, **31**, L21104, doi:10.1029/2004GL020596.

Oltmans, S.J., H. Vömel, D.J. Hofmann, K.H. Rosenlof, and D. Kley, 2000: The increase in stratospheric water vapor from balloonborne frostpoint hygrometer measurements at Washington D.C., and Boulder, Colorado. *Geophysical Research Letters*, **27(21)**, 3453-3456.

Pawson, S., K. Labitzke, and S. Leder, 1998: Stepwise changes in stratospheric temperature. *Geophysical Research Letters*, **25(12)**, 2157-2160.

Portmann, R.W. and S. Solomon, 2007: Indirect radiative forcing of the ozone layer during the 21st century. *Geophysical Research Letters*, **34**, L02813, doi:10.1029/2006GL028252.

Pyle, J., T. Shepherd, G. Bodeker, P. Canziani, M. Dameris, P. Forster, A. Gruzdev, R. Müller, N.J. Muthama, G. Pitari, and W. Randel, 2005: Ozone and climate: a review of interconnections. In: *IPCC/TEAP Special Report on Safeguarding the Ozone Layer and the Global Climate System: Issues Related to Hydrofluorocarbons and Perfluorocarbons* [Metz, B., L. Kuijpers, S. Solomon, S.O. Andersen, O. Davidson, J. Pons, D. de Jager, T. Kestin, M. Manning, and L.A. Meyer (eds.)]. Cambridge University Press, Cambridge, UK, and New York, pp. 83-132.

Ramaswamy, V. and M.D. Schwarzkopf, 2002: Effects of ozone and well-mixed gases on annual-mean stratospheric temperature trends. *Geophysical Research Letters*, **29(22)**, 2064, doi:10.1029/2002GL015141.

Ramaswamy, V., M.D. Schwarzkopf, W.J. Randel, B.D. Santer, B.J. Soden, and G.L. Stenchikov, 2006: Anthropogenic and natural influences in the evolution of lower stratospheric cooling. *Science*, **311(5764)**, 1138-1141.

Randel, W.J., F. Wu, S.J. Oltmans, K. Rosenlof, and G.E. Nedoluha, 2004: Interannual changes of stratospheric water vapor and correlations with tropical tropopause temperatures. *Journal of the Atmospheric Sciences*, **61(17)**, 2133-2148.

Randel, W.J., F. Wu, H. Vömel, G.E. Nedoluha, and P. Forster, 2006: Decreases in stratospheric water vapor after 2001: links to changes in the tropical tropopause and the Brewer-Dobson circulation. *Journal of Geophysical Research*, **112**, 10.1029/2006JD007339.

Rex, M., R.J. Salawitch, P. von der Gathen, N.R.P. Harris, M.P. Chipperfield, and B. Naujokat, 2004: Arctic ozone loss and climate change. *Geophysical Research Letters*, **31**, L04116, doi: 10.1029/2003GL018844.

Roscoe, H.K., 2001: The risk of large volcanic eruptions and the impact of this risk on future ozone depletion. *Natural Hazards*, **23(2-3)**, 231-246.

Rosenlof, K.H., S.J. Oltmans, D. Kley, J.M. Russell III, E.-W. Chiou, W.P. Chu, D.G. Johnson, K.K. Kelly, H.A. Michelsen, G.E. Nedoluha, E.E. Remsberg, G.C. Toon, and M.P. McCormick, 2001: Stratospheric water vapor increases over the past half-century. *Geophysical Research Letters*, **28(7)**, 1195-1198.

Santer, B.D., J.E. Penner, and P.W. Thorne, 2006: How well can the observed vertical temperature changes be reconciled with our understanding of the causes of these changes? In: *Temperature Trends in the Lower Atmosphere: Steps for Understanding and Reconciling Differences* [Karl, T., S.J. Hassol, C.D. Miller, and W.L. Murray (eds.)]. Synthesis and assessment product 1.1. U.S. Climate Change Science Program, Washington, DC, pp. 89-118.

Scherer, M., H. Vömel, S. Fueglistaler, S.J. Oltmans, and J. Staehlin, 2007: Trends and variability of midlatitude stratospheric water vapour deduced from the re-evaluated Boulder balloon series and HALOE. *Atmospheric Chemistry and Physics Discussions*, **7(5)**, 14511-14542.

Schwarzkopf, M.D. and V. Ramaswamy, 2002: Effects of changes in well-mixed gases and ozone on stratospheric seasonal temperatures. *Geophysical Research Letters*, **29(24)**, 2184, doi:10.1029/2002GL015759.

Seidel, D.J. and J.R. Lanzante, 2004: An assessment of three alternatives to linear trends for characterizing global atmospheric temperature changes. *Journal of Geophysical Research*, **109**, D14108, doi:10.1029/2003JD004414.

Seidel, D.J., R.J. Ross, J.K. Angell, and G.C. Reid, 2001: Climatological characteristics of the tropical tropopause as revealed by radiosondes. *Journal of Geophysical Research*, **106(D8)**, 7857-7878.

Shine, K.P., M.S. Bourqui, P.M.F. Forster, S.H.E. Hare, U. Langematz, P. Braesicke, V. Grewe, M. Ponater, C. Schnadt, C.A. Smith, J.D. Haigh, J. Austin, N. Butchart, D.T. Shindell, W.J. Randel, T. Nagashima, R.W. Portmann, S. Solomon, D.J. Seidel, J. Lanzante, S. Klein, V. Ramaswamy, and M.D. Schwarzkopf, 2003: A comparison of model-simulated trends in stratospheric temperatures. *Quarterly Journal of the Royal Meteorological Society*, **129(590)**, 1565-1588.

Solomon, S., R. Portmann, R. Garcia, W. Randel, F. Wu, R. Nagatani, J. Gleason, L. Thomason, L. Poole, and M. McCormick, 1998: Ozone depletion at mid-latitudes: coupling of volcanic aerosols and temperature variability to anthropogenic chlorine. *Geophysical Research Letters*, **25(11)**, 1871-1874.

SPARC (Stratospheric Processes and Their Role in Climate), 2000: *SPARC Assessment of Upper Tropospheric and Stratospheric Water Vapour.* [Kley, D., J.M. Russell III, and C. Phillips (eds.)]. SPARC report no. 2; WCRP no. 113; WMO/TD no. 1043. World Climate Research Programme,

[Geneva, Switzerland], 312 pp. <http://www.aero.jussieu.fr/~sparc/WAVASFINAL_000206/WWW_wavas/Cover.html>

Stenke, A. and V. Grewe, 2005: Simulation of stratospheric water vapor trends: impact on stratospheric ozone chemistry. *Atmospheric Chemistry and Physics*, **5(5)**, 1257-1272.

Tabazadeh, A. and R.P. Turco, 1993: Stratospheric chlorine injection by volcanic eruptions: HCl scavenging and implications for ozone. *Science*, **260(5111)**, 1082-1086.

Thompson, D.W.J. and S. Solomon, 2002: Interpretation of recent Southern Hemisphere climate change. *Science*, **296(5569)**, 895-899.

Tie, X. and G. Brasseur, 1995: The response of stratospheric ozone to volcanic eruptions: sensitivity to atmospheric chlorine loading. *Geophysical Research Letters*, **22(22)**, 3035-3038.

Tilmes, S., R. Muller, A. Engel, M. Rex, and J. Russell III, 2006: Chemical ozone loss in the Arctic and Antarctic stratosphere between 1992 and 2005. *Geophysical Research Letters*, **33**, L20812, doi:10.1029/2006GL026925.

UNFCCC (United Nations Framework Convention on Climate Change), 1997: *Kyoto Protocol to the United Nations Framework Convention on Climate Change.* United Nations, Geneva, Switzerland (and others). <http://unfccc.int/resource/docs/convkp/kpeng.html>

Velders, G.J.M., S.O. Andersen, J.S. Daniel, D.W. Fahey, and M. McFarland, 2007: The importance of the Montreal Protocol in protecting climate. *Proceedings of the National Academy of Sciences*, **104(12)**, 4814-4819.

Wennberg, P.O., R.C. Cohen, R.M. Stimpfle, J.P. Koplow, J.G. Anderson, R.J. Salawitch, D.W. Fahey, E.L. Woodbridge, E.R. Keim, R.S. Gao, C.R. Webster, R.D. May, D.W. Toohey, L.M. Avallone, M.H. Proffitt, M. Loewenstein, J.R. Podolske, K.R. Chan, and S.C. Wofsy, 1994: Removal of stratospheric O_3 by radicals: *in situ* measurements of OH, H_2O, NO, NO_2, ClO, and BrO. *Science*, **266(5184)**, 398-404.

WMO (World Meteorological Organization), 2003: *Scientific Assessment of Ozone Depletion: 2002.* Global Ozone Research and Monitoring Project report no. 47. World Meteorological Organization, Geneva, Switzerland, 498 pp.

WMO (World Meteorological Organization), 2007: *Scientific Assessment of Ozone Depletion: 2006.* Global Ozone Research and Monitoring Project report no. 50. World Meteorological Organization, Geneva, Switzerland, 572 pp. <http://www.esrl.noaa.gov/csd/assessments/2006/>

Zeng, G. and J.A. Pyle, 2003: Changes in tropospheric ozone between 2000 and 2100 modeled in a chemistry-climate model. *Geophysical Research Letters*, **30(7)**, 1392, doi:10.1029/2002GL016708.

CHAPTER 5 REFERENCES

AFEAS (Alternative Fluorocarbons Environmental Acceptability Study), 2007: Data tables at <http://www.afeas.org/>

Daniel, J.S., S. Solomon, and D.L. Albritton, 1995: On the evaluation of halocarbon radiative forcing and global warming potentials. *Journal of Geophysical Research*, **100(D1)**, 1271-1285.

Eyring, V., N. Butchart, D.W. Waugh, H. Akiyoshi, J. Austin, S. Bekki, G.E. Bodeker, B.A. Boville, C. Brühl, M.P. Chipperfield, E. Cordero, M. Dameris, M. Deushi, V.E. Fioletov, S.M. Frith, R.R. Garcia, A. Gettelman, M.A. Giorgetta, V. Grewe, L. Jourdain, D.E. Kinnison, E. Mancini, E. Manzini, M. Marchand, D.R. Marsh, T. Nagashima, P.A. Newman, J.E. Nielsen, S. Pawson, G. Pitari, D.A. Plummer, E. Rozanov, M. Schraner, T.G. Shepherd, K. Shibata, R.S. Stolarski, H. Struthers, W. Tian, and M. Yoshiki, 2006: Assessment of temperature, trace species and ozone in chemistry-climate model simulations of the recent past. *Journal of Geophysical Research*, **111**, D22308, doi:10.1029/2006JD007327.

Eyring, V., D.W. Waugh, G.E. Bodeker, E. Cordero, H. Akiyoshi, J. Austin, S.R. Beagley, B.A. Boville, P. Braesicke, C. Brühl, N. Butchart, M.P. Chipperfield, M. Dameris, R. Deckert, M. Deushi, S.M. Frith, R.R. Garcia, A. Gettelman, M.A. Giorgetta, D.E. Kinnison, E. Mancini, E. Manzini, D.R. Marsh, S. Matthes, T. Nagashima, P.A. Newman, J.E. Nielsen, S. Pawson, G. Pitari, D.A. Plummer, E. Rozanov, M. Schraner, J.F. Scinocc, K. Semeniuk, T.G. Shepherd, K. Shibata, B. Steil, R.S. Stolarski, W. Tian, and M. Yoshiki, 2007: Multimodel projections of stratospheric ozone in the 21st century. *Journal of Geophysical Research*, **112**, D16303, doi:10.1029/2006JD008332.

Hadjinicolaou, P., J.A. Pyle, and N.R.P. Harris, 2005: The recent turnaround in stratospheric ozone over northern middle latitudes: a dynamical modeling perspective. *Geophysical Research Letters*, **32**, L12821, doi:10.1029/2005GL022476.

IPCC (Intergovernmental Panel on Climate Change), 1999: *Aviation and the Global Atmosphere*. Special report of Working Group I and Working Group III of IPCC. [Penner, J.E., D.H. Lister, D.J. Griggs, D.J. Dokken, and M. McFarland (eds.)]. Cambridge University Press, Cambridge, UK, 373 pp.

IPCC (Intergovernmental Panel on Climate Change), 2001: *Climate Change 2001: The Scientific Basis*. Contribution of Working Group I to the Third Assessment Report of the Intergovernmental Panel on Climate Change. [Houghton, J.T., Y. Ding, D.J. Griggs, M. Noguer, P.J. van der Linden, X. Dai, K. Maskell, and C.A. Johnson (eds.)]. Cambridge University Press, Cambridge, UK, and New York, 881 pp.

IPCC (Intergovernmental Panel on Climate Change), 2007: *Climate Change 2007: The Physical Science Basis*. Contribution of Working Group I to the Fourth Assessment Report of the Intergovernmental Panel on Climate Change. [Solomon, S., D. Qin, M. Manning, Z. Chen, M. Marquis, K.B. Averyt, M. Tignor, and H.L. Miller (eds.)]. Cambridge University Press, UK, and New York, 996 pp. <http://www.ipcc.ch>

IPCC/TEAP (Intergovernmental Panel on Climate Change/Technology and Economic Assessment Panel), 2005: *IPCC/TEAP Special Report on Safeguarding the Ozone Layer and the Global Climate System: Issues Related to Hydrofluorocarbons and Perfluorocarbons*. [Metz, B., L. Kuijpers, S. Solomon, S.O. Andersen, O. Davidson, J. Pons, D. de Jager, T. Kestin, M. Manning, and L. Meyer (eds.)]. Cambridge University Press, New York, 478 pp.

Joshi, M., K. Shine, M. Ponater, N. Stuber, R. Sausen, and L. Li, 2003: A comparison of climate response to different radiative forcings in three general circulation models: towards an improved metric of climate change. *Climate Dynamics*, **20(7-8)**, 843-854.

Nakićenović, N. and R. Swart (eds.), 2000: *Special Report on Emissions Scenarios*. A special report of Working Group III of the Intergovernmental Panel on Climate Change. Cambridge University Press, Cambridge, UK, and New York, 599 pp.

Newman, P.A., E.R. Nash, S.R. Kawa, S.A. Montzka, and S.M. Schauffler, 2006: When will the Antarctic ozone hole recover? *Geophysical Research Letters*, **33**, L12814, doi:10.1029/2005GL025232.

UNEP (United Nations Environment Programme), 2007: Ozone Secretariat ozone depleting substances production and consumption data tables on the web at <http://ozone.unep.org/Data_Reporting/>

UNEP/TEAP (United Nations Environment Programme/Technology and Economic Assessment Panel), 2007: *Report of the Task Force Response on HCFC Issues and Emissions Reduction Benefits Arising from Earlier HCFC Phase-Out and Other Practical Measures*. United Nations Environment Programme, Ozone Secretariat, Nairobi, Kenya, 132 pp. <http://ozone.unep.org/Assessment_Panels/TEAP/Reports/TEAP_Reports/index.shtml>

Weatherhead, E.C., G.C. Reinsel, G.C. Tiao, C.H. Jackman, L. Bishop, S.M.H. Frith, J. DeLuisi, T. Keller, S. Oltmans, E. Fleming, D. Wuebbles, J. Kerr, A. Miller, J. Herman, R. McPeters, R. Nagatani, and J. Frederick, 2000: Detecting the recovery of total column ozone. *Journal of Geophysical Research*, **105(D17)**, 22201-22210.

WMO (World Meteorological Organization), 1995: *Scientific Assessment of Ozone Depletion: 1994*. Global Ozone Research and Monitoring Project report no. 37. World Meteorological Organization, Geneva, Switzerland.

WMO (World Meteorological Organization), 1999: *Scientific Assessment of Ozone Depletion: 1998*. Global Ozone Research and Monitoring Project report no. 44. World Meteorological Organization, Geneva, Switzerland, 486 pp.

WMO (World Meteorological Organization), 2003: *Scientific Assessment of Ozone Depletion: 2002*. Global Ozone Research and Monitoring Project report no. 47. World Meteorological Organization, Geneva, Switzerland, 498 pp.

WMO (World Meteorological Organization), 2007: *Scientific Assessment of Ozone Depletion: 2006*. Global Ozone Research and Monitoring Project report no. 50. World Meteorological Organization, Geneva, Switzerland, 572 pp. <http://www.esrl.noaa.gov/csd/assessments/2006/>

Yang, E.S., D.M. Cunnold, R.J. Salawitch, M.P. McCormick, J. Russell III, J.M. Zawodny, S. Oltmans, and M.J. Newchurch, 2006: Attribution of recovery in lower-stratospheric ozone. *Journal of Geophysical Research*, **111**, D17309, doi:10.1029/2005JD006371.

WMO (World Meteorological Organization), 2007: *Scientific Assessment of Ozone Depletion: 2006*. Global Ozone Research and Monitoring Project report no. 50. World Meteorological Organization, Geneva, Switzerland, 572 pp. <http://www.esrl.noaa.gov/csd/assessments/2006/>

CHAPTER 6 REFERENCES

IPCC (Intergovernmental Panel on Climate Change), 2007: *Climate Change 2007: The Physical Science Basis*. Contribution of Working Group I to the Fourth Assessment Report of the Intergovernmental Panel on Climate Change. [Solomon, S., D. Qin, M. Manning, Z. Chen, M. Marquis, K.B. Averyt, M. Tignor, and H.L. Miller (eds.)]. Cambridge University Press, UK, and New York, 996 pp. <http://www.ipcc.ch>

Nakićenović, N. and R. Swarz (eds.), 2000: *Special Report on Emissions Scenarios*. A special report of Working Group III of the Intergovernmental Panel on Climate Change. Cambridge University Press, Cambridge, UK, and New York, 599 pp.

WMO (World Meteorological Organization), 2003: *Scientific Assessment of Ozone Depletion: 2002*. Global Ozone Research and Monitoring Project report no. 47. World Meteorological Organization, Geneva, Switzerland, 498 pp.

PHOTOGRAPHY CREDITS

Cover/Title Page/Table of Contents
Image for Chapter 1, page 23, (Ozone cartoon), Used with permission from D. W. Fahey and the World Meteorological Organization, Geneva, Switzerland, 2007.

Image for Chapter 2, page 29, (Visible light spectrum), Used with permission from the Scientific Assessment of Ozone Depletion, 2006.

Image for Chapter 3, page 79, (Ozone hole), Paul Newman, NASA's Goddard Space Flight Center.

Image for Chapter 4, page 111, (Polar stratospheric clouds), David J. Hofmann, NOAA.

Chapter 2
Page 72, (Polar stratospheric clouds), Lamont Poole, NASA's Langley Research Center.

Contact Information

Global Change Research Information Office
c/o Climate Change Science Program Office
1717 Pennsylvania Avenue, NW
Suite 250
Washington, DC 20006
202-223-6262 (voice)
202-223-3065 (fax)

The Climate Change Science Program
incorporates the U.S. Global Change Research
Program and the Climate Change Research
Initiative.

To obtain a copy of this document, place
an order at the Global Change Research
Information Office (GCRIO) web site:
http://www.gcrio.org/orders

Climate Change Science Program and the Subcommittee on Global Change Research

William Brennan, Chair
Department of Commerce
National Oceanic and Atmospheric Administration
Acting Director, Climate Change Science Program

Jack Kaye, Vice Chair
National Aeronautics and Space Administration

Allen Dearry
Department of Health and Human Services

Jerry Elwood
Department of Energy

Mary Glackin
National Oceanic and Atmospheric Administration

Patricia Gruber
Department of Defense

William Hohenstein
Department of Agriculture

Linda Lawson
Department of Transportation

Mark Myers
U.S. Geological Survey

Timothy Killeen
National Science Foundation

Patrick Neale
Smithsonian Institution

Jacqueline Schafer
U.S. Agency for International Development

Joel Scheraga
Environmental Protection Agency

Harlan Watson
Department of State

EXECUTIVE OFFICE AND OTHER LIAISONS

Stephen Eule
Department of Energy
Director, Climate Change Technology Program

Katharine Gebbie
National Institute of Standards & Technology

Stuart Levenbach
Office of Management and Budget

Margaret McCalla
Office of the Federal Coordinator for
Meteorology

Rob Rainey
Council on Environmental Quality

Daniel Walker
Office of Science and Technology Policy